Pharmaceutical Chemistry of Antihypertensive Agents

Authors

György Szász, D.Sc.
Professor and Chairman
Institute of Pharmaceutical Chemistry
Semmelweis Medical University
Budapest, Hungary

Zsuzsanna Budvári-Bárány, M. Pharm.
Senior Lecturer
Institute of Pharmaceutical Chemistry
Semmelweis Medical University
Budapest, Hungary

CRC Press
Boca Raton Ann Arbor Boston

Library of Congress Cataloging-in-Publication Data

Szász, György.
Pharmaceutical chemistry of antihypertensive agents/authors, Görgy Szász, Zsuzsanna Budvári-Bárány.
p. cm.
Includes bibliographical references.
Includes index.
Contents: v. 1. Antihypertensive agents.
ISBN 0-8493-4724-6
1. Cardiovascular agents. 2. Hypotensive agents. 3. Pharmaceutical chemistry. I. Budvári-Bárány, Zsuzsanna. II. Title.
[DNLM: 1. Antihypertensive Agents. 2. Cardiovascular Agents. 3. Chemistry, Pharmaceutical. QV 150 S996p]
RS431.C25S93 1990
615′.71—dc20
DLC 90-2630
for Library of Congress CIP

This book represents information obtained from authentic and highly regarded sources. Reprinted material is quoted with permission, and sources are indicated. A wide variety of references are listed. Every reasonable effort has been made to give reliable data and information, but the author and the publisher cannot assume responsibility for the validity of all materials or for the consequences of their use.

Direct all inquiries to CRC Press, Inc., 2000 Corporate Blvd., N.W., Boca Raton, Florida 33431.

International Standard Book Number 0-8493-4724-6

Library of Congress Card Number 90-2630
Printed in the United States

PREFACE

Even a superficial glance at books published under such titles as ''Pharmaceutical Chemistry'' or ''Medicinal Chemistry'' is enough to demonstrate the basic differences that can be found in the editorial concepts and contents of these books. Extreme cases are frequently encountered, e.g., the appearance of a ''Pharmaceutical Chemistry'' with an editorial concept predominated by either analytical or organic synthetic chemistry. Besides books of this nature, there exists an even more extreme type of ''Pharmaceutical (Medicinal) Chemistry'' in which chemistry seems to be pushed entirely into the background, and the pharmacologically, pharmacokinetically, and therapeutically related aspects play the dominant role.

In view of these fundamental differences in editorial conceptions of pharmaceutical chemistry, it seems reasonable to raise the question as to what the term ''pharmaceutical chemistry'' should be intended to cover. The authors of the present book answer this by giving two equivalences: Pharmaceutical chemistry = Chemistry of pharmaceutical products = Chemistry for pharmacists. In both cases, pharmaceutical chemistry has to provide chemical knowledge to the experts (pharmacists, chemists, biologists, physicians) engaged in the different spheres of pharmacy (synthesis, pharmaceutical technology, quality control of drugs, biopharmacy, etc.), and hence pharmaceutical chemistry is to be regarded as a discipline with complex interdisciplinary features. This complexity is mirrored in this book. Besides the antihypertensive agents par excellence, the book discusses the group of diuretics used either in combination with antihypertensives, or alone in hypertensive therapy.

The close relationship elucidated between the actions of important antihypertensive agents and their interactions with the hormones and enzymes of the renin-angiotensin-aldosterone system have led to the inclusion of a chapter summarizing the pharmaceutical chemistry of these bioregulator substances. On the other hand, certain groups of agents having additional therapeutic indications other than antihypertension had to be excluded from this volume, and will receive a detailed treatment in a later volume in this series.

The chemical orientation of the book is to be observed in the articles on the individual compounds, but the insertion of a short introductory chapter on the physiological and pharmacological basis of blood pressure regulation also seemed reasonable.

An appreciable amount of attention is paid to the topics of structure-activity relationship (SAR) and analysis. Through the parts on SAR, the reader can get a historical view of the discovery of the compounds and the trends in further development in that particular field of drug research. This overview tends to generate new ides for SAR research too.

The analytical sections of the chapters treat pharmaceutical analysis in a broad sense: besides the quality control of drugs, their detection and assay in biological media are also included. In this complex treatment of the subject, the metabolites of the drug compounds are listed, and their analytical detection is also discussed. Similarly, not only the parameters of drug stability are included, which are important in pharmaceutical technology and biopharmacy, but also analytical methods indicating the stability.

The list of references includes all the basic and significant papers published up to 1987—88.

Budapest, 1989

György Szász
Zsuzsanna Budvári-Bárány

THE AUTHORS

György Szász, M. Pharm., D.Sc., is Chairman of the Institute and Professor of Pharmaceutical Chemistry at the Semmelweis Medical University of Budapest.

Dr. Szász received his pharmaceutical training at the Semmelweis Medical University between 1947 and 1952. In 1953 he joined the staff of the Department of Pharmaceutical Chemistry. He was appointed Assistant Professor in 1969, and was Dean of the Pharmaceutical Faculty at the Semmelweis Medical University between 1972 and 1982. In 1974 he assumed his present position.

Dr. Szász defended his thesis in the field of pharmaceutical analysis of amines and alkaloids, and obtained his C.Sc. (1968) and D.Sc. (1980) degrees. He was Chairman of the Chemical Committee of the Editorial Board of the Hungarian Pharmacopoeia VII. He is now a member of the Committee of the Hungarian Academy of Sciences on Analytical Chemistry.

The current major research interests of Dr. Szász include the structure-property-activity relationship among nitrogen-bridged compounds.

Dr. Szász has published over 160 papers. He is the Editor of the *Textbook of Pharmaceutical Chemistry* (Vol. 1 and 2, 4th ed., Medicina, Budapest, 1990).

Zsuzsanna Budvári-Bárány, M. Pharm., is Lecturer in Pharmaceutical Chemistry at the Semmelweis Medical University, Budapest.

Dr. Budvári-Bárány obtained her training at the Medical University of Budapest, receiving the B.Sc. degree in 1958, and the degree of University Doctorate in 1967. In 1972, she gained the title Specialist in Analytical Chemistry. She is a member of the Hungarian Pharmaceutical Society.

Dr. Budvári-Bárány joined the staff of the Institute of Pharmaceutical Chemistry in 1958, dealing with the teaching of special chapters of pharmaceutical chemistry. She received the Excellent Teaching Award from the Ministry of Health, Hungary, in 1983.

Dr. Budvári-Bárány is the author of 30 papers, mainly in the field of drug analysis by chromatographic methods. She was a member of the Editorial Board of the Hungarian Pharmacopoeia VII, and one of the authors of its general chapter "Chromatography". She is co-author and vice-editor of the *Textbook of Pharmaceutical Chemistry* (Medicina, Budapest, 1990).

TABLE OF CONTENTS

Section I: Introduction

Section II: Antihypertensive Agents

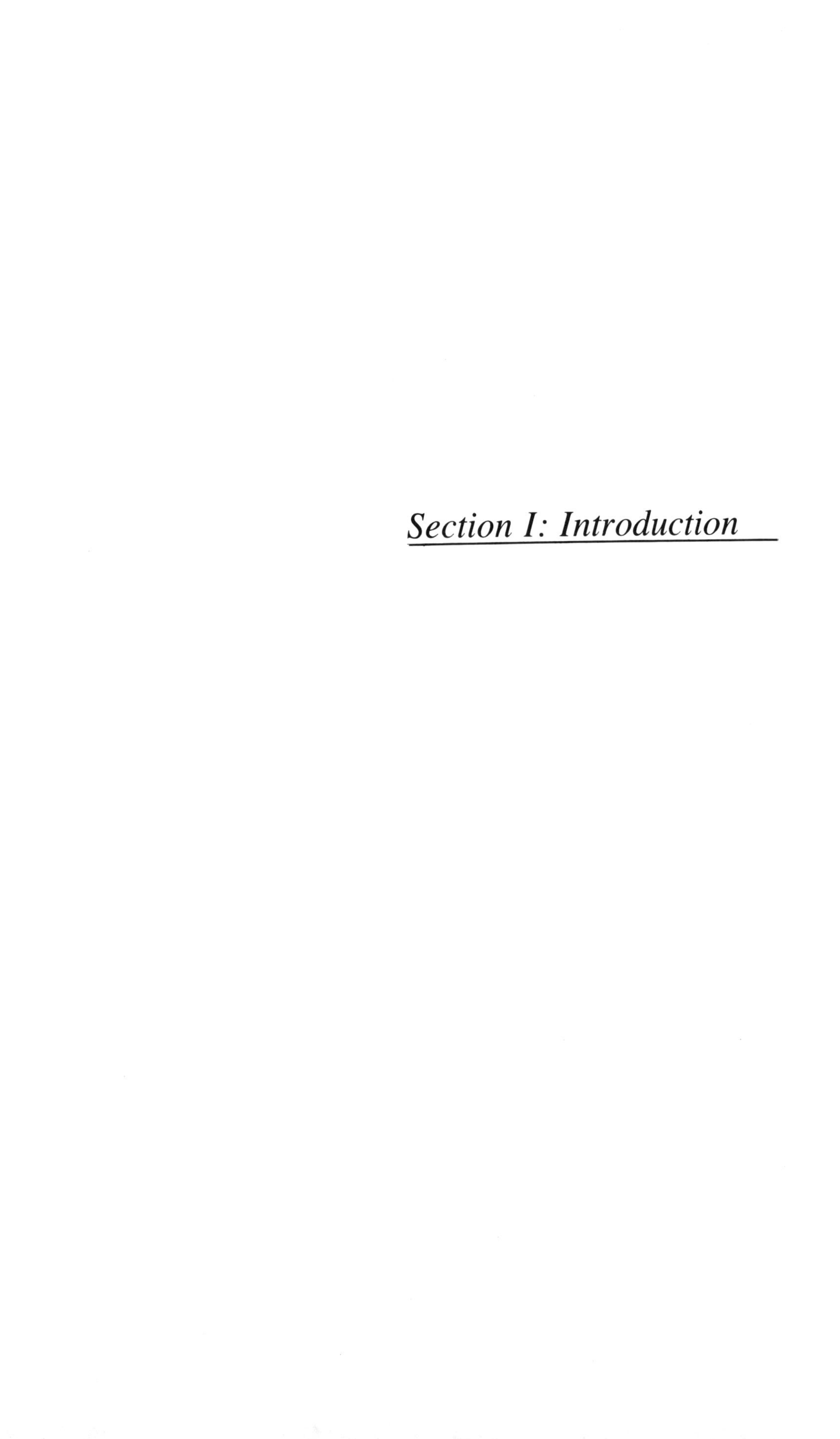

Section I: Introduction

Chapter 1

BLOOD CIRCULATION AND BLOOD PRESSURE

The blood pressure is a numerical reflection of the blood flow, a basic process in the human organism; it is expressed in mmHg. Its average physiological value is determined jointly by the concerted functions and parameters of the heart, which generates the flow of blood, and the associated network of blood vessels. A change in arterial pressure may be a physiological response to a different loading of the organism. The relationship

$$\text{blood pressure} = \text{output} \times \text{peripheral resistance}$$

may be regarded as the ''basic equation'' of blood pressure. The two variants in the equation arise from the concerted effects of many factors (Figure 1). Mathematically, the mean arterial pressure may be calculated as the sum of the diastolic pressure and one third of the difference between the systolic and diastolic pressures. For example, in the case of a blood pressure of 110/80 mmHg, the arterial pressure is 90 mmHg. Although it is rather difficult to make a sharp distinction between normal (physiological) blood pressure and hypertension, certain authors consider that the blood pressure is excessive for an adult when the diastolic pressure permanently and reproducibly exceeds[1] or attains[2] 90 mmHg.

Experience has demonstrated the importance of the conditions of measurement[1]* in order to gain correct blood pressure values: the conditions relating to the patient (posture, dietary circumstances, smoking, resting, etc.), and the conditions relating to the equipment and technique (number of readings, rate of inflation and deflation of the bladder, etc.).

Statistical data from different geographical regions clearly indicate the significant correlation of morbidity and mortality vs. the blood pressure and the existence of hypertension.[1,3]

In certain countries (U.S., Australia, Finland, etc.), the increasing effectiveness of the campaign against smoking, together with suggestions concerning a reasonable fat diet, have led to a significant reduction in the number of deaths caused by coronary insufficiency. Notwithstanding, hypertension remains the major risk factor for coronary and cerebral disease and death. As a result, intensive research into the pathological basis of hypertension and into new therapeutic methods has continued in the eighties. The same is true for the relevant drug research, which will be surveyed in some detail in this book.

Hypertension (elevated arterial pressure) occurs as a result of a dysfunction of the circulatory control mechanism, i.e., the harmonious interplay of local, humoral, and neural factors. The dysfunction is manifested by an increased blood pressure in the vessels.

The pathogenesis of hypertension generally involves a slow and gradually developing process. The initiating causes of hypertension remain hidden for a time. The high blood pressure is manifested when one or more factors of the regulatory system undergo a functional defect,[4] and the system can no longer compensate the blood pressure elevating effects of certain main factors, such as the decrease in cardiac output or the increase in peripheral resistance. The factors directly initiating hypertension are interpreted in terms of different hypotheses by different authors. However, the general view is that hypertension as a pathological state is introduced and accompanied by characteristic hemodynamic abnormalities. Additionally characteristic deviations may be found in the composition of the blood plasma.

The initial steps in the manifestation of a hypertensive state include increases in the cardiac output and plasma volume, as a consequence of which the peripheral resistance also increases. This rise in resistance may be a result of autoregulation, when the plasma volume increase initiates the contraction of blood vessels, with a shift in the earlier equilibria. This

* For References see end of Chapter 2.

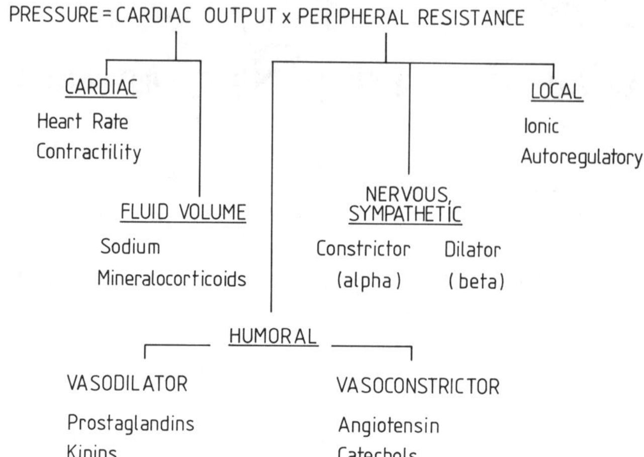

FIGURE 1. The basic equation and influencing factors of blood pressure (From Kaplan, N. M., in *Heart Disease*, 2nd ed., W. B. Saunders, Philadelphia, 1984. With permission.)

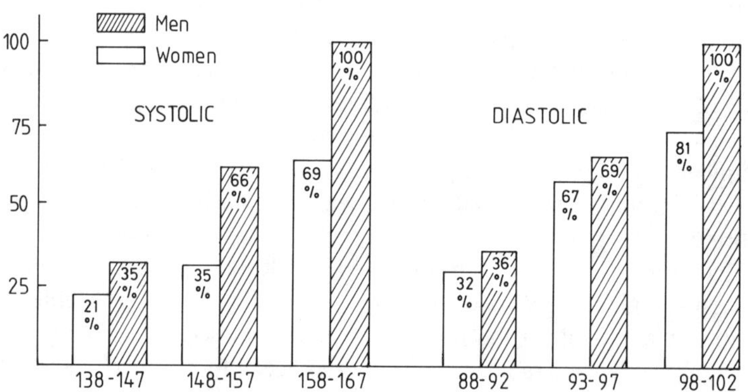

FIGURE 2. Excess mortality over 20 years (Society of Actuaries, 1980). (From Kaplan, N. M., in *Heart Disease*, 2nd ed., W. B. Saunders, Philadelphia, 1984. With permission.)

TABLE 1
Types and Etiologic Factors of Hypertension[a]

I. Primary (essential)
   - Neurogenic (sympathetic nervous activation, etc.)
   - Renal (defect of pressure-natriuresis process, excess of intracellar calcium, natriuretic hormone deficiency, etc.)
   - Hormonal (defect of renin-angiotensin-aldosterone axis, vasodepressor hormone deficiency, etc.)
   - Genetic (enhanced pressor sensitivity to norepinephrine, angiotensin, etc.)

II. Secondary
   - Renal (artery obstruction, parenchymal disease, renin secreting tumor, etc.)
   - Adrenal (primary aldosteronism, Cushing's syndrome, pheochromocytoma, etc.)
   - Neurogenic (excessive stress, encephalitis, etc.)
   - Miscellaneous causes (oral contraceptives, pregnancy, thyrotoxicosis, arteriosclerosis, etc.)

[a] Adapted from Reference 2.

TABLE 2
Chemical Composition of Human Blood, and Deviations in Heart and Kidney Diseases

Component		Value non-SI	Value SI	Note
Inorganics				
Ammonia		60—100 μg/100 ml	35.2—58.7 μmol/l	Increased in portal hypertension, lower in renal disease
Calcium (ionic)		4.5—5.2 mg/100 ml	1.08—1.30 mmol/l	Decreased in renal insufficiency
Potassium		137—184 mg/l	3.5—4.7 mmol/l	Higher in severe states of renal or/and cardiac insufficiency
Sodium			136—142 mmol/l	
Iron	man	100—150 μg/100 ml	17.9—26.9 μmol/l	
	woman	80—140 μg/100 ml	14.3—35.1 μmol/l	
Copper	man	70—140 μg/100 ml	11.0—22.0 μmol/l	
	woman	85—155 μg/100 ml	13.3—24.3 μmol/l	
Magnesium			0.75—1.25 mmol/l	
Lithium			0.5—1.5 mmol/l	
Chloride			97—105 mmol/l	Increased in essential hypertension
Phosphorus (inorganic)		3.0—4.5 mg/100 ml	0.97—1.45 mmol/l	
Zinc		50—150 μg/100 ml	7.7—23.0 μmol/l	
Hydrogencarbonate			21—28 mmol/l	
Organics				
Cholesterol		151—250 mg/100 ml	3.90—6.5 mmol/l	Increased in nephrosis, and possibly in hypertension
Biliary acids			3—30 mg/l	
Triglyceride		50—190 mg/100 ml	0.5—0.9 g/l	
Phospholipids		150—380 mg/100 ml	1.94—4.9 mmol/l	
Fatty acids			0.5—15.0 mmol/l	
Total fat content			5—8 g/l	
Protein			parts in protein total:	
Protein, total			60—78 g/l	Albumin level very markedly decreased, γ-globulin content low, and concentrations of α- and β- protein elevated in nephrosis
Albumin		52—65%	0.52—0.65	
α_1-globulin		2.5—5%	0.025—0.05	
α_2-globulin		7—13%	0.07—0.13	
β-globulin		8—14%	0.08—0.14	
γ-globulin		12—22%	0.12—0.22	
Insulin		8—24 μIU/ml	63—190 pmol/l	
Ceruloplasmin		23—50 mg/100 ml	1.52—3.30 μmol/l	
Fibrinogen		0.17—0.35 g/100 ml	4.9—10.2 μmol/l	
Hamoglobin	man	15.0 ± 2 g/100 ml	2.33 ± 0.3 mmol/l	
	woman	13.0 ± 2 g/100 ml	2.02 ± 0.3 mmol/l	
Immunoglobulin	IgGG		6—16 g/l	
	IgA		1.1—4.8 g/l	
	IgM		0.7—2 g/l	
Carbohydrate Blood sugar		60—80 mg/100 ml	3.3—4.4 mmol/l	
Metabolites				
Bilirubin, total		1.7—20.5 μmol/l	0.1—1.2 mg/100 ml	
Uric acid		2.7 mg/100 ml	118—413 μmol/l	Increased in severe cardiac insufficiency
Creatine		0.77—1.29 mg/100 ml	59—98 μmol/l	
Creatinine		0.6—1.2 mg/100 ml	53—106 μmol/l	Increased when rate of glomerular filtration reduced

TABLE 2 (continued)
Chemical Composition of Human Blood, and Deviations in Heart and Kidney Diseases

Component		Value non-SI	Value SI	Note
Urea		6—27 mg/100 ml	2.1—9.6 mmol/l	Increased when rate of glomerular filtration reduced
Pyruvic acid		0.3—0.9 mg/100 ml	34—103 μmol/l	Increased in cardiomyopathy, arterial thrombosis
Lactic acid		4.7—15.1 mg/100 ml	0.55—1.70 mmol/l	
Enzymes				
Aldolase-S		2.22—5.92 mIU/dl	0.37—0.99 nKat/l	Increased in heart infarct
CPK	man	10—100 mIU/ml (37°C)	0.17—1.67 μKat/l	Increased in heart infarct
	woman	10—60 mIU/ml (37°C)		Increased in heart infarct
Transaminase GOT		3.8—15.8 IU/l	63.5—264 nKat/l	Highly elevated in heart infarct
Transaminase GPT		0.48—17.3 IU/l	8-289 nKat/l	Mildly elevated in heart infarct

TABLE 3
Physiological (Normal) Hemodynamic Characteristics of Blood Circulation

Plasma volume		2900—3100 ml
Heart index (volume/min/m^2)	adult	3.0—3.2 l
	child	4.3—4.4 l
	elderly	2.5—2.7 l
Pulse volume		80—90 ml
Heart volume after diastole		800—1000 ml
Volume of blood in heart after diastole		500—600 ml
Mechanical heart output at rest		1.5—1.7 watt
Blood streaming rate	aortic	28—30 cm/s
	capillary	0.04—0.05 cm/s
	in vena cava	10—15 cm/s

Total peripheral resistance: 1.136 kPa/s/ml^{-1}

functional change may elicit certain structural changes, such as the thickening of the blood vessel walls. The latter change is partly reflected in a rise in peripheral resistance.[4]

In the hemodynamic changes mentioned above, habitual and environmental factors have been shown to play a role, and sodium has a primary influence in the changes in both factors.[1] The plasma volume increase that may be observed in the introductory stage of hypertension is a direct consequence of the increased sodium retention and the reduced secretion. The genetic nature of the defective sodium transport mechanism, and the reduced permeability of all membranes, are more or less experimentally proved facts.[5] For the development of essential hypertension, both endogeneous (genetic) and exogenous (environmental) factors are needed. A person predisposed by an inherited defect may not develop hypertension if the environmental factors are avoided.[1] A high sodium intake is to be regarded as the most important environmental factor. It is generally accepted that a sodium intake of less than 50 mmol (1 g) per day results in a moderately sodium-poor diet. Another environmental factor adding to the genetic defect in the sodium transport mechanism is stress,

TABLE 4
The Role of Sodium Excess and of the Sympathetic Nervous System in Primary Hypertension[a]

Evidence for the role of sodium excess in primary (essential) hypertension	Role of sympathetic nervous activation in primary (essential) hypertension
Elevation of sodium intake increased the number or individuals developing hypertension	By the release of catecholamines, acute hypertension can be induced in animals
Subjects consuming less than 50 mmol/day had little or no hypertension	In some hypertensives, a high plasma catechol was found, which correlated with the blood pressure
Prolonged restriction of sodium loading (60-90 mmol/day) lowers blood pressure in most people	Stress produces hypertension, and the extent of the latter corresponds to the extent of the stress
An increased concentration of sodium was found in the vascular tissue and blood cells of most hypertensives	Antiadrenergic agents reduce the blood pressure

[a] Adapted from Reference 1.

which is responsible for an increased degree of activation of the sympathetic nervous system. In this respect, very convincing argumentation may be found in the review by Kaplan (see Table 4).

The biochemical chain of blood pressure regulation is initiated in the plasma when renin cleaves a protein to yield angiotensin I. On the action of the angiotensin-converting enzyme (ACE), this decapeptide is transformed to the octapeptide hormone, angiotensin II, which occupies a key position in the establishment of the blood pressure, ensuring the physiological rhythm and extent of constriction of the blood vessels. Another physiological function of angiotensin II is to increase water and sodium retention. Accordingly, a defect in the sodium transport mechanism is usually associated with a disorder of the renin-angiotensin system,[6] but not always with a high renin level.[7] This latter finding is not surprising if it is considered that even a normal renin level is indicated as a high one by the feedback mechanism of a hypertensive state. In some hypertensive subjects, this feedback mechanism is defective.

ACE is an enzyme with a dual role: it also takes part in the activation of the vasodilator bradykinin. Thus, the blood pressure lowering achieved with ''ACE inhibitors'' does not necessarily reflect an inhibition of ACE; it may be a consequence of a bradykinin abundance.

In the development and preservation of primary hypertension, factors other than those already mentioned are also said to be of importance: the levels of prostaglandin[8] and calcium.[9] Further data are needed on the blood pressure-elevating effects of opioid peptides.[10]

Chapter 2

A BRIEF CHEMICAL SURVEY OF SOME ENZYMES AND HORMONES INVOLVED IN BLOOD PRESSURE REGULATION

I. RENIN

Although renin-like activity has been detected in several regions of the human and animal organism, the kidney is the most important organ as far as the occurrence and production of renin are concerned. Undoubtedly, the most significant additional finding is the identification of renin activity in the walls of certain main arteries and veins.[11] This must be a result of local renin production, as indicated by the fact that this finding could be reproduced when the individual in question was nephrectomized.[12] This observation is of interest in connection with the report[13] that pathological renin levels were not found in the blood of hypertensives. Renin activity has been observed in the brain, uterus, placenta, and lung of different animal species; the submandibular gland of mouse contains an especially high amount of renin.[14,15]

The crude extract gained from the kidney is generally used for the isolation and purification of renin. For a long period, its purification was performed by classical extraction-fractionation.[16]

This was followed from the sixties by the application of chromatographic and electrophoretic methods on a preparative scale. New perspectives were opened up by the use of gel chromatography,[17] which was succeeded by the different variants of affinity chromatography; pepstatin has been found suitable as an affinity ligand.[18,19] New synthetic specific affinity ligands have recently been developed.[20]

As regards the molecular mass of renin, rather controversial data have been published. Waldhause et al.[21] applied gel chromatography and determined a molecular mass of 42,000 for human renin. Four different forms of human renin were separated from an aqueous kidney extract[22] chromatographed on DEAE-cellulose (pH = 5.5). The M_r of each of the four isomers was found to be 39,500, but significant differences were observed in their protease activities. These molecular mass values are similar to the value of 40,000 obtained by ultracentrifugation.[23] The isoelectric point (pH = 5.25) was determined previously.[17]

The molecular mass of the monomeric enzyme was estimated by Misono et al.[24] to be 36,000. In spite of the fact that the active sites of renin and acidic proteases (e.g., pepsin) display similarities, renin does not exhibit general protease activity.[25] On the other hand, the sequences of the amino acids in renin and acidic proteases are closely similar. Renin has been found to consist of 336 amino acid residues: one heavy chain with 288 residues, and one light chain with 48 residues.[26] The two chains are connected by a disulfide bridge. A particularly fascinating parallelism is observed between the topography of the most important active sites of renin produced from the submandibular gland of mouse and human and that of certain acidic proteases (porcine pepsin, penicillo-pepsin, bovine chymosin, etc.). While renin and acid proteases do share several common features, the substrate specificity and neutral pH optimum distinguish renin from proteases (the latter cleave the protein chain with optimum activity in definitely acidic medium). Like the digestive acidic proteases in general, renin is an aspartyl protease, but its activation may follow a different mechanism. The process seems to involve the role of the paired basic amino acids (Lys, Arg) adjacent to the N-terminal of the active renin structure; this basic centre cannot be found in acidic proteases.

The amino acid sequences and substrate specificities of human kidney renin and mouse submandibular renin exhibit both close similarities and distinctive features. Only 68% of

TABLE 1
Sequence Homologies (%) between Asp Proteases[a]

	Mouse renin	Penicillo-pepsin	Porcine pepsin	Rhizopus pepsin	Endothia-pepsin
Human renin	68	25	41	31	28
Mouse renin		29	43	29	28
Penicillo-pepsin			34	43	62
Porcine pepsin				43	39
Rhizopus pepsin					39

[a] From Akahane, K. et al., *Hypertension*, 7, 3, 1985. With permission.

the sequence homology is shared by the two enzymes.[27,28] While human renin displays cross activity, mouse renin does not hydrolyse the human substrate. Nevertheless, a comparison of five proteases demonstrated the greatest degree of similarity between human and mouse renin (Table 1).

Besides comparing sequences, Akahane et al.[29] constructed the three-dimensional structure of human renin. As reference standard they used penicillo-pepsin because of the great similarities found between the three-dimensional arrangements of the hydrophobic and hydrophilic amino acid side-chains in penicillo-pepsin and human renin (similar length and folding of the polypeptide chains, etc.). The constructed model proved to be consistent with several properties of human renin described previously. The interesting structural feature suggested[29] that many of the ionic acids in human renin form ion pairs with the nearby residues of opposite charge.[30] As a difference, human renin contains 31 such basic amino acid residues, in contrast with only eight in penicillo-pepsin. The ion pair occurring most often in the human renin side-chains appears to form between the Lys and Glu moieties.

The substrate specificity of human renin can be interpreted via the difference found between the active sites in renin and other proteases. Two catalytically important sites in human renin are the aspartyl carboxyl groups in Asp 38 (strand C_1) and Asp 226 (strand C_1); these can come into close proximity to each other, and are therefore able to function as bidentate ligands. An outstanding distinguishing feature between other proteases and renin is the replacement of Thr 216 in penicillo-pepsin by Ala 229 in human renin. The former is able to form a hydrogen bond with Asp 213 in the catalytic site, whereas the latter fails to do so. With the exception of Val 36, in close proximity to the catalytic Asp 38, the amino acid sequence is identical in human renin and the other aspartyl proteases. This Val 36 is replaced either by Asp (Rhisopus chinensis pepsin, Endothia parasitica pepsin) or by Ile (mouse renin, porcine pepsin). Since Ala 317 in the renins (human and mouse) corresponds to Asp 300 in the pepsins, it was suggested[30] that Asp is not essential for the catalytic function of aspartyl proteases.

A. DETERMINATION OF RENIN CONTENT

Although the renin plasma level did not prove to be an adequate or direct parameter of the hypertensive state, a knowledge of the actual renin concentration is still important for clinical-pharmacological research. The plasma renin activity is most often determined via the angiotensin released by renin from angiotensinogen. The earlier used methods were based on measurement of the changes in blood pressure.[31-34] They have subsequently been replaced by different variants of radioimmunoassay (RIA).

The RIA method described by Goto et al.[35] permits the determination of both active and total renin content. (''Inactive renin'', containing a single polypeptide chain, may be activated by proteases; its apparent M_r = 50,000; see Reference 36.) Plasma is incubated with an excess of sheep angiotensinogen (pH = 6.5, 37°C, 4^h) in the presence of ethylenediaminetetraacetic acid (EDTA) and phenylmethanesulfonyl fluoride. These latter substances serve to inhibit the angiotensinases. For the determination of total renin, the inactive renins are activated with trypsin. The RIA of angiotensin-I (equivalent with the renin content) is carried out with a kit after deproteinization of the reaction mixture. In the total renin determination described by Yamagata et al.,[37] the activization is achieved by acidic treatment. The activity of the active renin is determined by RIA of the angiotensin-I produced without acid treatment. The activity of the inactive renin is then calculated by subtraction from the angiotensin content found following incubation after acidic treatment.

Ikeda et al.[38,39] proposed a ''solid-phase'' RIA in which the antibody is adsorbed on polystyrene spheres. The inhibition of dipeptidyl carboxypeptidase and angiotensinase during the enzymic reaction is achieved with a combination of toluene-α-sulphonyl fluoride and EDTA. For the inhibition of angiotensin-I formation during the RIA, pepstatin-A is used. The incubation may be carried out at room temperature. Labeled ^{125}I angiotensin is used as reagent.

A comparison of enzyme immunoassay (EIA) and RIA for plasma renin activity was made by Hubl et al.[40] The antiserum was raised in rabbits by angiotensin-bovine serum albumin conjugate. The detection limits for the EIA and the RIA were 4 and 2.5 pg angiotensin-I, respectively. A pepstatin-immunometric assay for activie renin in human plasma was recently reported by Higaki et al.[41]

The investigation and optimization of the conditions of RIA have been the subject of several works. The circumstances of sample collection and processing for assay of plasma renin activity were studied by Roulstone et al.[42] As anticoagulants, heparin lithium, heparin sodium, and EDTA were compared. With a longer delay at ambient temperature, samples treated with EDTA show a progressive increase in angiotensin-I blank.[43] The same was not observed when heparin was used as anticoagulant. Heparin is therefore preferable, but the final plasma concentration must be kept at < 40 units ml^{-1}. Higher heparin concentrations inhibit angiotensin-I formation. According to Matsunaga et al.,[44] plasma can be stored at $-20°C$ for at least 4 weeks without any significant changes in active renin. The effectiveness of inhibitors has also been the repeated subject of experiments.[39,41,44] Quinolin-8-ol was found to be a good angiotensinase inhibitor; during incubation for 1 h it preserved angiotensin-I almost without loss in activity.[45]

Simple calculator programs have been developed for the RIA of plasma renin activity and plasma angiotensin-II.[46] The optimum pH for the RIA of plasma renin activity was found to be 5.70 to 6.00[47] when toluene-α-sulfonyl fluoride, captopril, and quinolin-8-ol hemisulfate were applied as inhibitors, with incubation for 1 h.

Stirati et al.[48] established that the storage of plasma samples between ambient temperature and $-20°C$ does not influence the plasma renin activity. Processing at room temperature should therefore be preferred on the grounds of practicality.

For the determination of semipurified rat renal renin, a spectrofluorimetric method has been developed, which gave almost identical results with those of RIA.[49,50] Ethanolic 1.4-diamino-2,3-dichloroanthraquinone was used as fluorogenic reagent, and the fluorescence was measured at 340 nm (excitation at 284 nm) after an incubation period of 3 h. The calibration plot was linear in the range 0.05 to 0.75 μM angiotensin-I.

The spectrophotometric method is chemically interesting, but it seems to lack the selectivity.[51] The method is based on the addition of a synthetic peptide substrate with a β-naphthylamine terminal to the plasma sample. Plasma renin cleaves the Leu-Leu bonds of the peptide chain. The action of the added aminopeptidase-M leads to the formation of β-naphthylamine in a quantity equivalent to the renin, and this is measured spectrophotometrically.[51] An interesting plasma renin assay was suggested by Skeggs[52] who applied a tetradecapeptide labeled with [^{14}C]Val^3 as renin substrate. From the activity of angiotensin-I, formed in an equivalent amount, the renin activity can be deduced. The inactive and active renin contents may be determined by the method of Nielsen and Poulsen.[53] After separation of the inactive and active renin by high-performance liquid chromatography (HPLC) in a relatively simple system (Protein PAK 300 SW column, detection at 280 nm by UV spectroscopy), the inactive renin is eluted as a discrete peak before the active renin. Its molecular mass has been found to be 38,000. By incubation with trypsin at 4°C, renin is reactivated and is assayed by a known antibody trapping method.

II. ANGIOTENSINOGEN

Angiotensinogen represents the protein substrate of the human and several animal organisms; on the action of renin, it is converted to angiotensin-I. This is a decapeptide found in the α_2-globulin fraction of the plasma,[54] but it is generated by the liver.[55]

Campbell[56] published a review with 94 references on the production of angiotensin by peripheral tissue and on the role of the circulating renin-angiotensin system in this process.

In the extracts gained by countercurrent distribution or chromatography, three glycoproteins have been identified with a molecular mass of 58,000.[57] They are identical in amino acid sequence, but differ significantly in the carbohydrate moiety. The peptide chain is comprised of 17 different amino acids, Leu occuring most frequently (67 mol^{-1}).

A surprising observation in the early stage of research was that angiotensinogen could be cleaved not only by renin, but also by trypsin. It was discovered that a decapeptide can be split from angiotensinogen by simple alkaline hydrolysis. This finding initiated the hypothesis of the ester binding of a tetradecapeptide through a Ser unit to the bulk of the substrate:[58]

```
Tyr–C–NH–CH–CH2O–C–CH–...
    ||            |  |
    O             O  NH–...
```

The binding between the tetradecapeptide unit and the bulk of the angiotensinogen molecule.

The N-terminal part of the tetradecapeptide has been identified as Asp.

A. ANGIOTENSIN-I AND II

The biologically most active member of the renin-angiotensin system is undoubtedly angiotensin-II, which is formed from angiotensinogen via angiotensin-I:

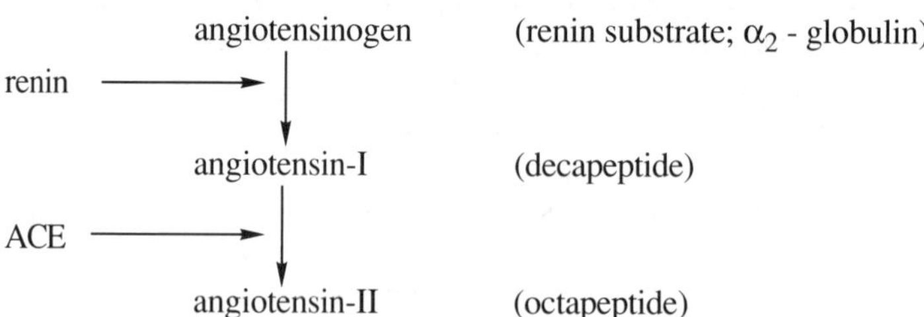

The pressor effect of angiotensin-II exceeds that of norepinephrine. Due to the small molecular masses of angiotensin-I and angiotensin-II, determination of the amino acid sequences was not too difficult:

Angiotensin-I

Asp–Arg–Val–Tyr–Ile–His–Pro–Phe–His–Leu
(Val)

Angiotensin-II

A new development in the field of structure analysis was the increasing prominence of the nuclear magnetic resonance (NMR) methods, which proved their applicability in elucidation of the sequences of the angiotensins.[59-62] The traditional methods, based on the chemical or chromatographic determination of the amino acid residues split off by stepwise hydrolysis of the protein chain (e.g., Reference 63), are rarely used nowadays. The evident advantage of the NMR methods is their nondestructiveness, whereby the danger of amino acid degradation, which must be faced in work with the traditional hydrolytic methods, can be avoided. NMR became a method of choice for the structure analysis of angiotensin-I.[64] Among the early investigations, Glickson et al.[59] published assigned resonances of the 220 MHz proton magnetic resonance (PMR) spectra of angiotensin-II in D_2O and H_2O, based on structural analogies and the pH-dependence of the chemical shifts. A synthetic congener of angiotensin-II, angiotensinamide (Asn^1-Arg^2-Val^3-Tyr^4-Val^5-His^6-Pro^7-Phe^8), served as model substance; it elicits similar biological responses to those of angiotensin, and differs chemically from the latter only in replacement of the N-terminal Asp residue of angiotensin-II by Asn. Measurements were made near the solvent freezing point, which allowed the good observation of all CH resonances. The restricted rotation about the His^6-Pro^7 peptide bond gives rise to *cis* and *trans* isomers. The observed chemical shift of 4.36 ppm of the Pro α-CH peak of angiotensin-II suggests an anomalous but still *cis* orientation of the peptide bond. However, a carbon-13 NMR study was made a few years later by Deslauries et al.[65] and the chemical shifts demonstrated the *trans* conformation of the Pro residue about the His-Pro bond and the fact that the His-imidazole ring has a strong preference for the N^T-H tautomer. The ^{13}C spin-lattice relaxation times show that the α-carbons of the peptide backbone are restricted in their motion. Overall, it was suggested[65] that Ile^5-angiotensin-II has a restricted backbone conformation.

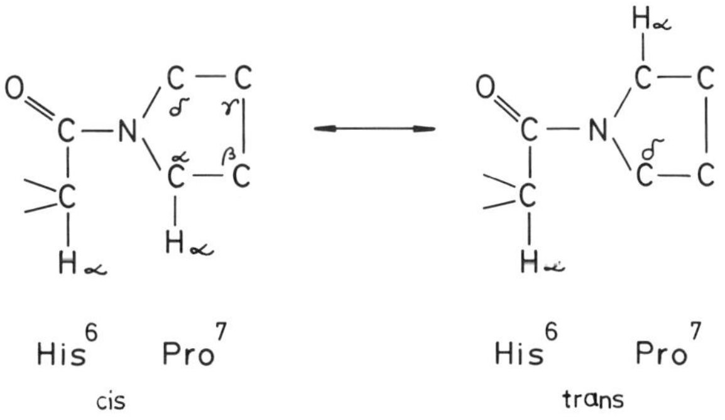

His^6 Pro^7
cis

His^6 Pro^7
trans

A subsequent paper from the same research center[66] convincingly proved that conformational changes and interactions between side-chain groups are detectable; these became comprehensible through the determination of ^{13}C relaxation times. The Monte Carlo method was applied to four different molecular models of angiotensin-II.[67,68] Analysis of the results

TABLE 2
Mean End-to-End and Tyr⁴-Phe⁸ Side-Chain Centroid Distances in the Clusters[a]

Cluster	End-to-end distance (Å)	Tyr⁴-Phe⁸ side-chain centroid distance (Å)	Cluster appearance %
1	9.8	14.9	5
2	21.3	15.8	25
3	19.6	6.9	15
4	17.8	12.9	25
5	23.4	12.6	30
In the whole sample	20.3	12.8	

[a] From Marchionini, C. et al., *Res. Commun. Chem. Path. Pharm.*, 112, 339, 1983. With Permission.

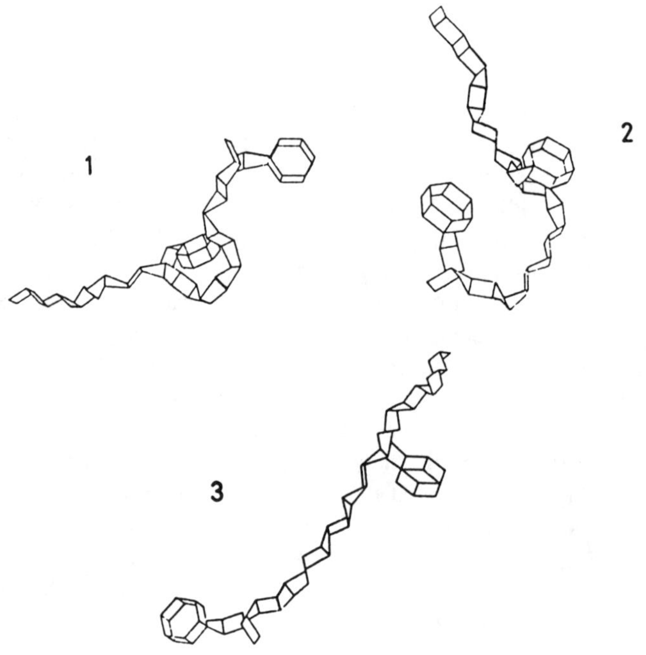

FIGURE 1. Typical conformations in some clusters. The peptide backbone is represented ribbon-like. (Adapted from Reference 69.)

showed that an extended conformation is more likely to be present in acidic solution, whereas folded conformations due to interactions of charged terminal groups are favored at neutral pH. The Monte Carlo method has also been used by the same research team in a very interesting approach relating to clustering of conformations of angiotensin-II that may exist in acidic medium.[69] As characteristics of the five cluster groups, the ''end-to-end'' distance and the average distance between the Tyr⁴ and Phe⁸ side chains are listed. The results (Table 2), in accordance with the previous ones, are indicative of the predominance of extended forms.

The quantitatively most important group, cluster 3, involving 30% of the sample, exhibits a completely extended conformation with distant aromatic rings. Calculations based on dihedral angles reveal the most probable conformations in clusters 1 and 2.[69] A projection of the conformations in the cluster may be seen in Figure 1.[69] Fujiwara et al.[70] worked with

TABLE 3
Dissociation Constants of Angiotensinamide

Amino acid residue	Group	Dissociation constant PMR[59]	Alkalimetry[71]
Phe^8	$-COOH$	3.07 ± 0.03	3.84
His^6	$-NH_{imidazole}$	6.26 ± 0.04	6.30
Asn^1	$-H_2N$	6.98 ± 0.04	6.92
Tyr^4	$-OH_{phenolic}$	10.2 ± 0.2	10.4

conformational energy minimization and found ten local minima. This also evidenced the bend structure in the backbone of angiotensin-II.

The chemical shift corresponding to the Arg guanidino protons could be assigned by following the pH-dependence. The four-proton resonance at pH 6.53 can be assumed to be generated by the four equivalent hydrogens of the protonated Arg guanidino group, and it is similarly reasonable to identify the resonance at 6.64 ppm as a consequence of the guanidino resonance structure:

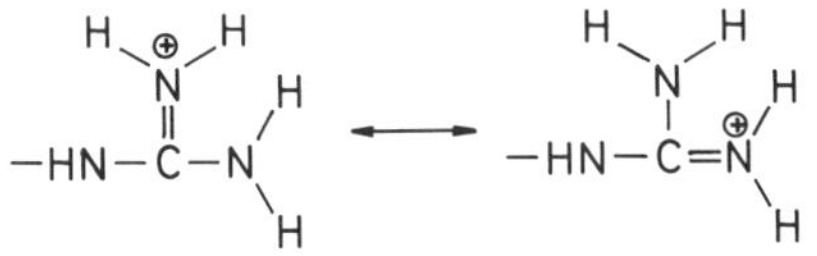

$$-HN-C(=N^{\oplus}H_2)-NH_2 \longleftrightarrow -HN-C(-NH_2)=N^{\oplus}H_2$$

Meaningful information relating to proton dissociation from the peptide chain of angiotensin-II could be obtained via PMR and ^{13}C-NMR spectroscopy. The constants found agree reasonably well with those determined by alkalimetric titration (Table 3).

In the region of phenol titration, the PMR spectrum indicated conformational changes by the displacement of resonances originating from the His CH, Pro CH and Val CH_3 hydrogens.[59] Mention should be made of the work of Zalitis et al.,[72] in which RIA and radioreceptor analytical methods allowed an insight into the conformational properties of angiotensin-II during its binding to protein.

The treatment of angiotensin-II with acetic anhydride showed[73] that acetylation and desacetylation of the Tyr OH and His side-chain in angiotensin-II take place with higher rates than in the free amino acids; this appears to indicate an increased nucleophilicity and the existence of an intrinsic desacetylating mechanism. This observation is in accordance with the assumption that the Asp COOH group of angiotensin-II functions as a charge control unit, which allows the hypothesis that angiotensin-II interacts with the membrane receptors according to the general mechanism for ser proteases.

The δ-H shift of the α-C atoms in the ^{13}C-NMR spectrum proved to be a useful parameter for a description of the aliphatic side-chain properties, especially when it was correlated with biological activity.[74]

Besides NMR, circular dichroism (CD) also gave important information about the conformational relations in angiotensin-II.[75] The application of CD determination is of additional importance, for the results of conformation analysis performed by more than one method may be correlated with several physical and chemical properties. The effects of the individual amino acids on the conformation could be concluded.[75] It was found that the alkyl side-chains have a definite influence on the steric positions of the aromatic nuclei: the steric orientation of the Tyr^4 ring is influenced relatively weakly by Val^3 but more strongly by Ile^5. This is reflected by the siginificant change in the CD spectrum when Ile^5 is replaced by Ala. Analysis of the CD spectra has confirmed the previous conclusion of the key position of Pro^7 in the conformation of angiotensin-II.

Investigations on complex formation between angiotensin-II and metals are informative as regards not only elucidation of the angiotensin-II structure, but also the biological activity of angiotensin-II.[76-78]

The interaction between angiotensin-II and Ca^{2+} (Mg^{2+}) ions was postulated several years ago when it was observed that changes in the concentration of Ca^{2+} or angiotensin-II give rise to changes in the concentration of free angiotensin-II or Ca^{2+}. The removal of extracellular calcium (or the administration of verapamil) inhibited the development of the contraction initiated by angiotensin-II.[79] The same response was elicited by another Ca-antagonist, diltiazem.[80] These experiments were carried out with angiotensin-II and its [Asn^1, Val^5] analogue. The very interesting but not fully understood observation was made that in certain cases Ca^{2+} (Mg^{2+}) could potentiate only the pressor effect of the angiotensin-II analogue [Asn^1, Val^5], and not that of angiotensin-II.

Degani and Lenkinski[81] provided proof about the ion-pairing (complexing) property of both compounds. The through-membrane (phosphatidylcholine) transport of Mn^{2+} ions took place at the same rate regardless of whether it was stimulated by the presence of angiotensin-II or of [Asn^1, Val^5] angiotensin-II. It should be mentioned here that angiotensin-III (Arg-Val-Tyr-Ile-His-Pro-Phe) had no influence on Mn^{2+} transport. In accordance with previous experience,[80] this fact reflects the participation of a terminal carboxy group (Asp^1) in the complex formation by angiotensin-II. On the basis of the membrane transport rate vs. Mn^{2+} concentration relationships, the compositions of the Mn^{2+}-angiotensin-II and Mn^{2+}-[Asp^1,Val^5] angiotensin-II complexes were found equimolecular (molecular ratio 1:1). The same composition was established for the Ca^{2+}-angiotensin-II complex.[83]

Since the membrane-penetration ability of Mn^{2+} and certain other trace elements has been proved,[81] it was reasonable to assume that the formation of angiotensin-II-transition metal complexes plays a role in the through-membrane penetration of the hormone. The complex formation between angiotensin-II and Mn^{2+} ions was studied by Degani and Lenkinski,[81] and by Basosi et al.[82] by means of NMR relaxation and ESR relaxation methods, respectively. The changes in the electron spin resonance (ESR) specrum proved the existence of outer and inner sphere complexes.[82-85] The NMR relaxation data suggest that Mn^{2+} interacts with the His^6 of angiotensin-II, but this interaction should be strongly pH-dependent and takes place in a narrow pH range. In conclusion, these investigations did not succeed in proving the rate control role of the Mn^{2+}-His^6 reaction on the membrane transport of angiotensin-II. Nevertheless, the majority of the findings indicate a cation-potentiated feature of the binding of angiotensin-II and of [Asn^1,Val^5] angiotensin-II.[81] It is not surprising, therefore, that the presence of EDTA or other Ca^{2+} chelators has a considerable influence on the effect of Ca^{2+}(Mg^{2+}, Mn^{2+}).

As regards the cause of the pH-dependence of the membrane transport of Mn^{2+} stimulated by angiotensin-II, Degani and Lenkinski[81] presume a conformational change related to the protonation of the amino group. The point of inflexion of the rate constant vs. pH curve was found at pH 7.7 (Figure 2), while the pK of the terminal amino group in angiotensin-II is 7.6. This means that 44% of this amino group is protonated. From the experimental results acquired so far, it can be concluded that angiotensin-II has the ability to convey Ca^{2+}, Mn^{2+} and presumably other cations too across the cell membranes. The most probable background of this transport is the 1:1 complex formation.

1. Determination of Angiotensin-II Content

Quantitation of the angiotensin content in plasma is a substantial task in laboratories dealing with etiology, diagnostics, drug research, etc. Basic advances in this field were achieved through (1) the increasing selectivity of immuno- and radioimmuno-analytical methods, and (2) the developments in peptide and, especially, angiotensin chromatography. Consequently, the most important methods may be divided into these two types of analysis.

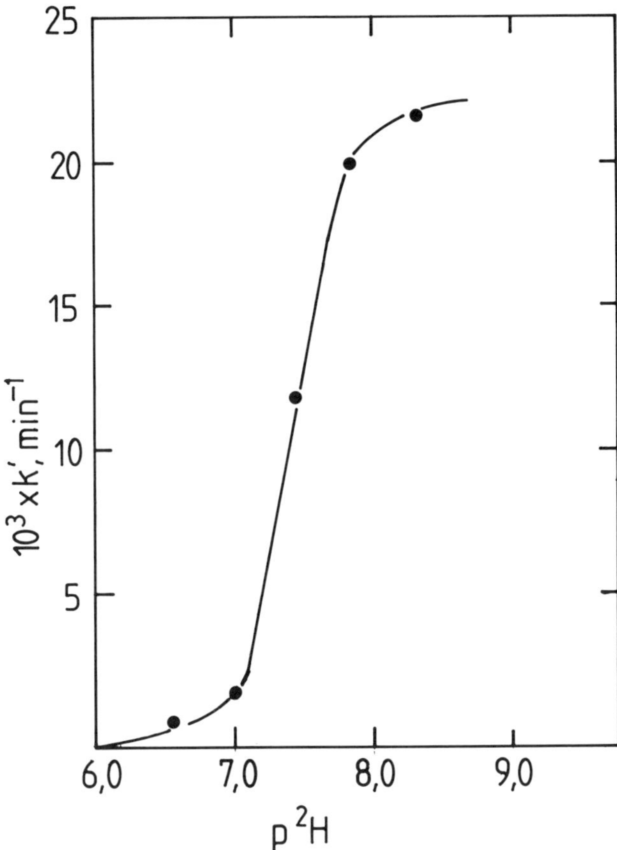

FIGURE 2. pH dependence of the rate of Mn(II) transport. The solutions contained 7.45×10^{-2} *M* EPC vesicles in the presence of 3.96 m*M* $MnCl_2$ and 6.95 m*M* A-II at 55°C. EPC = pure phosphatidylcholine. (From Degani, H. et al., *Biochemistry*, 19, 3430, 1980. With permission.)

The submicro amounts of angiotensin-II in circulating blood (10 to 100 pg ml^{-1}) do not allow its determination with satisfactory accuracy by the biological method (blood pressure bioassay) or early methods of immunoanalysis. A rapid advance in this field was provided by the introduction of RIA.[86]

The labeling of angiotensin itself is a fairly easy task; due to the high Tyr content, its molecule may readily be labeled with an iodine isotope (^{125}I or ^{131}I). The most widely used Greenwood method[87] is claimed to achieve an incorporation of approximately 90% (1 mCi ^{125}I) after 10 s, but some reserve is to be noted in the literature. Thus, it has been mentioned that the labeling procedure[87] has a side reaction involving the formation diiodoangiotensin,[88] a compound with a greatly reduced immunological activity.[89] To reduce this danger of iodination as much as possible, certain authors[88] suggest the application of a higher pH (~9.0) than usual (~7.5), while others[90] suggest the preparation of a product with relatively low specific activity. For the purification of labeled angiotensin-II (removal of diiodoangiotensin and nonlabeled angiotensin), chromatographic methods[91] are most suitable. A special problem discussed in the literature is the storage of labeled angiotensin. A partial loss of activity may occur due to the adsorption of angiotensin on the walls of the glass or plastic containers. Attempts have been made to overcome this phenomenon by such precautions as the use of an alkaline buffer, tensides (sodium laurylsulfate[92]), siliconized containers, etc.

A precondition of the precise and sensitive determination of angiotensin-II is the preliminary separation and isolation of the peptide. After the methods used in the early period (countercurrent distribution,[93] paper chromatography (PC),[93-95] thin layer chromatography (TLC),[96] ion exchange column chromatography (IEC)[97]), HPLC gained predominance in this field.[98-100] This method is of importance even in the RIA of relatively pure angiotensin-II because of the cross reactions between the antibody used and the structurally related impurities of angiotensin-II (angiotensin-I, Asp[1]-angiotensin-I, angiotensin-III, angiotensin C-terminal hexapeptide, etc.). The detection of such impurities provides information in connection with their different biological activities.

The effective applicability of HPLC is excellently reflected by the separation of seven immunologically related angiotensins.[101] The plasma sample was prepared for chromatography by C_{18}-cartridge extraction (mini-column 9 cm in length, methanol-formic acid 4 : 1 mixture as solvent, adjusted to pH 7.4). The reversed-phase high-performance liquid chromatography (RPHPLC) column was C_2-silica, and 60 to 70% m*M* ammonium formate in methanol was applied as mobile phase (gradient plus isocratic elution). The isolated angiotensin-II was quantitated by RIA (label: ^{125}I). The antiserum was raised in rabbits. Essentially the same task, at the level of sensitivity of RIA, could be solved by using HPLC[102] on a cation-exchange resin stationary phase bearing a carboxyl group. The first (isocratic) development was made with an ammonium formate-acetonitrile (90 : 10) developing solvent (the pH was adjusted to 4.2). The second (gradient elutional) development was performed qualitatively with the same eluent (final composition: acetonitrile-ammonium formate 80:20, adjusted to pH 4.2) At this pH, all the investigated peptides are protonated and will be bound via salt formation on the stationary phase. By means of this method, the following substances could be separated:

angiotensin-I	H–Asp–Arg–Val–Tyr–Ile–His–Pro–Phe–His–Leu–OH
des-Asp^1-A-I	H–Arg–⌇ ⌇–Leu–OH
angiotensin-II	H–Asp–Arg–Val–Tyr–Ile–His–Pro–Phe–OH
angiotensin-III	H–Arg–⌇ ⌇–Phe–OH
C-terminal hexa-) peptide	H–Val–Tyr–Ile–His–Pro–Phe–OH
C-terminal penta-) peptide	H–Tyr–⌇ ⌇–Phe–OH
C-terminal tetra-) peptide	H–Ile–⌇ ⌇–Phe–OH

Angiotensin-II and its metabolites are separable by the HPLC method of Nussberger et al.[103] Plasma samples were extracted through phenylsilica cartridges and, after separation by HPLC, angiotensin was determined by RIA. The limit of detection in plasma was 0.2 p*M*.

Though different angiotensin-II analogues could be separated by reversed-phase HPLC, the separation of β-Asp[1]-angiotensin-II and angiotensin-II did not succeed. This analytical task was solved by Kumagaye et al.,[104] using a cation-exchange HPLC method. A TSK gel CM-2SW column was employed, with 0.2 *M* phosphate buffer (pH 6 to 7) containing 10% acetonitrile and 0.5 *M* sodium chloride as mobile phase.

Accurate characterization of the plasma angiotensin content is a difficult task, and demands very careful work. Hoshino et al.[105] described an HPLC method for the separation, and RIA for the determination of angiotensins in plasma. They give the following mean blood levels:

Angiotensin-I	72—86 pg ml^{-1}
Angiotensin-II	10—12 pg ml^{-1}
Angiotensin-III	2—7 pg ml^{-1}

For the extraction of angiotensin-II from plasma before RIA, two methods (cation-exchange resin, C_{18} cartridge) were used and compared by Morton and Webb.[106]

Hermann et al.[107] used an RPHPLC-RIA combined method, which was suitable for the determination of angiotensin-I in fentomol amounts. The same was true for angiotensin-II determination by the RPHPLC methods of Nussberger and Brunner.[108,109] Normal persons on a free diet averaged 4.2 fmol angiotensin-II ml^{-1} serum. The administration of furosemide increased this value to 22 fmol ml^{-1}. In hypertensive patients receiving long-term treatment with enalapril, the angiotensin-II level fell from 2.7 to 0.9 fmol ml^{-1}. The method is sensitive and specific, and demonstrates that angiotensin-II virtually disappears during ACE inhibition. For the assay, 2 ml of serum is needed. Angiotensin-II was extracted reversibily onto bonded silica, isolated by isocratic RPHPLC, and finally quantified by RIA according to the procedure of Nussberger et al.[109] Chappell et al.[110] modified the cleaning-up process of the latter method and described a procedure for the HPLC separation and RIA of several angiotensin peptides.

The separation of angiotensins-I, II, and III could also be performed by gel electrophoresis.[111] Polyacrylamide gel was used as stationary phase, and the gel electrophoresis was carried out at pH 6.41 in buffer systems of positive polarity. The separation of the three peptides was based not on the molecular dimension differences (M_r = 1295, 1045, and 930, respectively), but on the net charges. The HPLC separation of angiotensins-I, II, and III was also performed in the work of Joergen and Knud.[112] A comparison of HPLC retentions proved the identity of human and mouse angiotensins, and rat angiotensin-I was identified as the Ile[5] derivative. A basically new method was developed by Ishida et al[113] for the determination of angiotensins-I, II, and III on the basis of the Tyr contents of the peptides. A pre-column was used on which the Tyr **(1)** was formylated in alkaline medium, and the aldehyde **(2)** formed was transformed with diaminodimethoxybenzene **(3)** to a fluorescent derivative **(4)** which was chromatographed on a LiChrosorb RP18 column with acetonitrile-potassium chloride-hydrochloric acid (pH 2.2) + Na-1-hexanesulfonate as mobile phase:

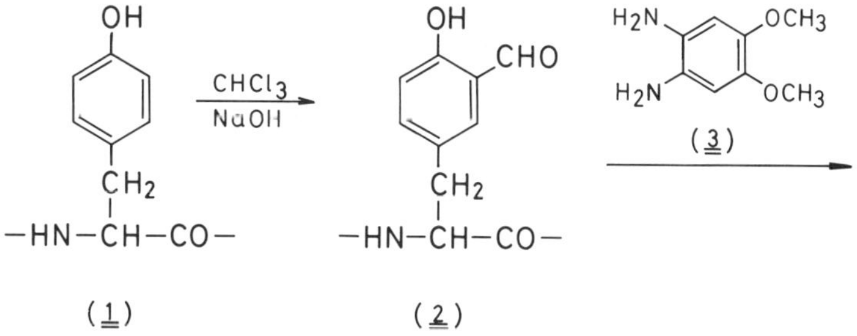

$CHCl_3$ / NaOH

(1) (2) (3)

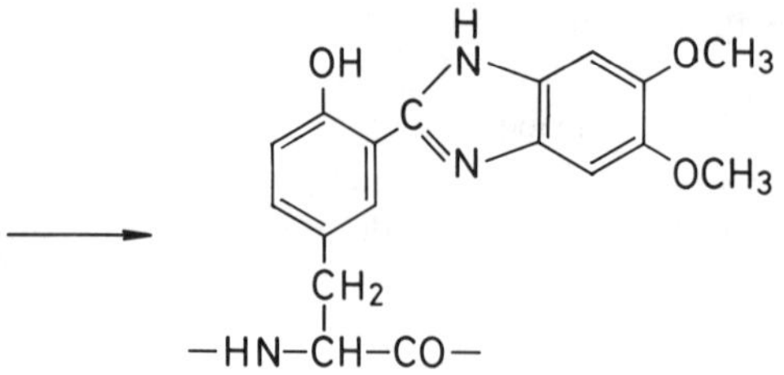

(4)

The reaction products, benzimidazole derivatives in protonated form, move and are separated as fluorescent ion pairs. Linearity is obeyed in the range 0.02 to 60 nmol. Pre-column derivatization was also used by Sakamoto et al.[114] for the determination of ACE and the angiotensins in blood. In its reaction with benzoin, angiotensin-II forms an intensely fluorescent derivative. The detection limit for angiotensin-II is 0.66 pmol. Important features of the method are that it is also suitable for the separation of angiotensin analogues, while it requires a plasma volume of only 5 μl. The separation of [Ile^5] angiotensins-I and II and related peptides has been solved by HPLC on a C_{18} column using a pre-column with the same packing.[115] An acetonitrile-sodium perchlorate-phosphoric acid mixture was applied as mobile phase, with monitoring of the eluate at 200 nm. Amounts down to 10 pmol in 100 μl could be determined.

A fairly frequently investigated and debated problem of angiotensin analysis is the stability of angiotensin in the presence of EDTA, which is the most commonly used anticoagulant for samples for renin-activity assay. Closely related problems that arise concerning the reproducibility of renin activity and angiotensin content determinations are (1) the stability of the renin sample (see p. 9) and (2) the stability of angiotensin during the storage and processing of the samples. The finding of Roulstone et al.[42,116] in favour of the applicability of EDTA for the solution of both problems is not in full accord with the conclusion of Stockigt and Hewett,[43] who preferred Li-heparinate for renin activity conservation. (vs.).

III. ANGIOTENSIN-I CONVERTING ENZYME (KININASE-II)

Angiotensin-I converting enzyme (ACE) has an important role within the blood pressure regulatory system of the vertebrate organism. Besides its function in the cleavage of angiotensin-I to the blood pressure-elevating angiotensin-II (p.10), ACE also cleaves the blood pressure-lowering nonapeptide bradykinin to inactive lower peptides. The synonymous name kininase originates from the latter function.

ACE was recognized in horse plasma and named by Skeggs et al. (1954 to 1956), and its peptidyl-dipeptidase property was also pointed out.[117,118] Subsequently, ACE was isolated from human plasma too.[119] It was later reported that many other organs contain relatively high concentrations of ACE, particularly the epithelial cells of the kidney,[120] placenta,[121] and choroid plexus.[122] The highest concentration of ACE is found in the kidney,[120] but the enzyme isolated from this organ manifested a higher molecular mass than that purified with trypsin.[123] The difference (~ 10,000) may be the mass of the polypeptide unit removed by the enzymic treatment. Of several peptides tested by Skidgel,[124,125] bradykinin has the highest affinity for ACE. The difference between the affinity parameters of bradykinin and angiotensin-I is one order of magnitude in favor of bradykinin.

ACE is a glycoprotein with a relatively high, but not exactly quantitated, carbohydrate content,[126,127] and a fairly high proportion of aromatic amino acids (especially Try). This has been identified as N-terminal, and Ala as C-terminal amino acid.[126] The divergent values obtained for the molecular mass can be attributed to the analytical difficulties caused by the high carbohydrate content. As the most plausible value for M_1, 1,29,000 may be accepted.[126]

ACE is a metalloenzyme, containing one Zn^{2+} ion per molecule. The range 4.6 to 4.8 has been reported as the isoelectric point.[128] In accordance with this, chelators such as EDTA, 8-hydroxyquinoline and mercapto derivatives can inactivate ACE.[129]

One of the active sites of ACE is a protonated (positively charged) Arg unit, which presumably interacts with the C-terminal carboxyl group of the substrates,[124,130,131] irrespective of the blocked or free state of this group.[124]

A. DETERMINATION OF ACE CONTENT

In a large group of methods of ACE determination, hippuryl-His-Leu is used as substrate:

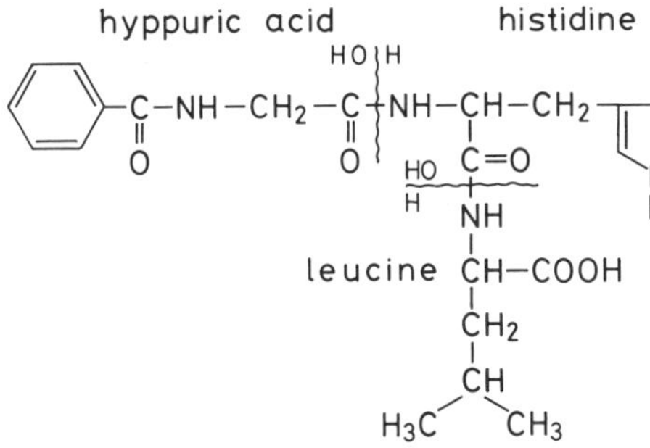

```
          hyppuric acid                 histidine
                        HO|H
(C6H5)-C-NH-CH2-C-|-NH-CH-CH2-(imidazole, N, NH)
       ||       ||        |
       O        O   HO   C=O
                    H  --|--
                         NH
              leucine    CH-COOH
                         |
                         CH2
                         |
                         CH
                        /  \
                     H3C    CH3
```

In general, the methods are based on the measurement of the hippuric acid released in the ACE-substrate reaction. A relatively simple way is to determine hippuric acid by HPLC. Horiuchi et al.[132] used HPLC for the determination of ACE in human and animal blood, and different tissues. To the blood or tissue extract (prepared with a detergent), the hippuryl-His-Leu (H-H-L) substrate was added (solvent: phosphate buffer, pH 8.3) and, after incubation at 37°C for 30 min, the reaction was terminated with metaphosphoric acid. Hippuric acid in the supernatant solution was determined by HPLC (column: C_{18}, mobile phase: methanolic aqueous phosphate buffer, pH 3), detection at 228 nm.

Methods were developed by Maurich et al.[133-135] for the estimation of human serum and urine ACE via the ion pair HPLC determination of hippuric acid. After incubation (at 37°C, pH 8.3) for 30 min and subsequent termination of the reaction, the hippuric acid released from the substrate (H-H-L) was determined by HPLC (column: C_{18}, mobile phase: water-methanol, with tetrabutylammonium phosphate as ion pairing agent).

Pre and Bladier[136] described the measurement of ACE in serum or plasma by a spectrophotometric method based on the reaction of the hippuric acid released from the H-H-L substrate with cyanuric chloride. After incubation for 15 min, the reaction was stopped by the addition of EDTA. A similar method was developed by Hayakari et al.[137] After 5 + 30 min of incubation (substrate: H-H-L), the reaction was terminated by acidification; 2,4,6-trichloro-1,3,5-triazine was used as a chromogenic reagent. The absorption was measured at 382 nm.

Hippuric acid has been determined by using 4-dimethylaminobenzaldehyde for the direct conversion of hippuric acid into acetic anhydride and pyridine.[138] The sensitivity was increased when incubation was conducted at pH 7 instead of pH 8 and H-H-L substrate containing [^{14}C] hippuric acid was used.[139] The [^{14}C] hippurate released from the labeled H-H-L substrate was extracted into ethyl acetate, and the radioactivity was measured. A similar solution was applied previously by O'Brien et al.[140] who compared their radiometric results with those of spectrophotometry based on the measurement of inactive hippuric acid following the addition of cyanuric chloride reagent. A good correlation was found between the results of the methods. Ultraviolet spectrophotometry of hippuric acid was performed and discussed by Stoner.[141] The ϵ value (at 228 nm) of hippuric acid was 10,300. The recovery of hippuric acid was 89 to 99%, depending on the amount of ethyl acetate used for the extraction. The sensitivity of ACE determination was increased significantly by the method of Santos et al.[142] The serum was incubated with the H-H-L substrate at pH 8.3 (37°C, 15 min), and the reaction was terminated with sodium hydroxide. The His-Leu released was measured fluorimetrically at 495 nm 10 min after the addition of phthalaldehyde reagent.

Siems et al.[143] also used the H-H-L substrate and the phthalaldehyde reagent for transformation of the His-Leu released to a fluorescent product. This permitted the very sensitive and widely applicable (serum, tissues, cell suspensions, biological fluids) measurement of ACE. An HPLC evaluation of urinary and blood ACE was introduced by Pitotti et al.[144] The sample was incubated with the H-H-L substrate, and the reaction product was applied to the HPLC system (column: LiChrosorb RP18; mobile phase: 40% aqueous methanol containing tetrabutylammonium cation and K_2HPO_4).

Critical reviews have been published on spectrophotometric and spectrofluorometric ACE assays.[145,146]

Another substrate kit quite often used for ACE assay contains 4-hydroxyhippuryl-His-Leu as substrate. The 4-hydroxyhippuric acid released is hydrolyzed by the method of Hurst and Lovell-Smith[147] to 4-hydroxybenzoic acid, which is oxydatively coupled with 4-aminoantipyrine to produce a quinoneimine dye in a reaction characteristic of phenol-type compounds.[148] The absorption is measured at 505 nm. A high degree of correlation was found when the results were compared with those obtained with the method based on the hippuric acid-cyanuric chloride reaction. The sensitivity of the kit was found to be about a quarter of that of the latter method, but the precision of the kit was higher. A similar comparative study was made by Boomsma and Schalekamp,[149] in which the method based on quinoneimine formation was evaluated with reference to a modification of the method of Cushman and Cheung.[150] Haemoglobin, lipids, bilirubin, and prednisone did not interfere, but uric acid at a concentration of 600 μM caused negative error. The assay is rapid: 20 samples could be analyzed in 1 h, and no pre-extraction was necessary. Among other substrates, Neels et al.[151] applied hippuryl Gly-Gly. After incubation for 30 min, the liberated Gly-Gly was reacted with 2,4,6-trinitrobenzene-sulphonic acid in alkaline medium (pH 9.3). The yellow product was determined at 420 nm. The reaction was quantitative only after 1 h. The results correlate well with those of an HPLC method involving the determination of hippuric acid. The method was modified appreciably by Mayr[152] in an effort to optimize the experimental conditions. Another substrate, 3-(2-furylacryloyl)-L-Phe-Gly-Gly was used by Holmquist[153] for ACE determination. Some of the parameters of the latter method were modified by Ronca-Testoni.[154] During enzymatic hydrolysis of the substrate, a bathochromic effect can be observed. The reaction was monitored at 340 to 345 nm. The same procedure has been studied and applied for the determination of serum ACE content by Neels et al.[155]

Microcentrifugal analysis and a program from a cassette tape were used. The concentration vs. photoabsorption relationship allows the determination of ACE in the range 0 to 250 IU l^{-1}. The use of a mixture of the substrate 3-(2-furylacryloyl)-L-Phe-Gly-Gly and N-3-(2-furylacryloyl)-L-Phe to establish theoretical net values in ACE determination in order to obtain standard results was studied by Buttery and Chamberlain.[156,157] A continuous-monitoring spectrophotometric method utilizing furylacryloyl-Phe-Gly-Gly as substrate was developed by Maguire et al.[158] The detection limit was 12.2 IU ml^{-1}.

Kinetic fluorometric methods suitable for use in a microcentrifugal analyzer were reported almost simultaneously by Maguire et al.[159] and Neels et al.[160] The procedures are based on the ACE-catalyzed hydrolysis of the intramolecularly quenched fluorogenic substrate 2-aminobenzoyl-Gly-4-nitro-L-Phe-L-Pro. The fluorescence of the hydrolysis product

2-aminobenzoyl-Gly is measured with the use of a microcentrifugal analyzer. The detection limit was found[160] to be 0.4 IU l^{-1}. A fluorometric assay based on hydrolysis of the substrate dansyl-Phe-Ala has been described.[161] The fluorescent products were extracted into chloroform, where the fluorescence was measured at 510 nm. As little as 1 liter of serum could be analyzed.

A very sensitive inhibitor binding assay for ACE was developed,[162] which was based on the specific binding of the ^{125}I-labeled inhibitor (the 4-hydroxybenzamidine derivative of a peptide) to the active center of ACE. The bound inhibitor was separated, and the unbound one remaining in the solution was counted. The sensitivity was 200 IU l^{-1}, corresponding to 5 ng of the enzyme.

A sensitive assay for serum ACE assay by HPLC with fluorimetric detection has been described.[163] The method is based on the transformation of angiotensin-I to angiotensin-II via the action of ACE. The latter was derivatized before the HPLC development. The detection limit was 0.66 pmol of angiotensin-II.

Automated methods based on different principles have recently been proposed for the sensitive determination of ACE.[164-167]

IV. ATRIAL NATRIURETIC FACTOR

The real history of atrial natriuretic factor (ANF), a hormone, started when Kisch published his paper "Electron microscopy of the atrium of the heart."[168] The electron microscopic observation of special storage granules in the atrium of guinea pig was followed by an analogous observation in the atrial cardiocytes of mammalians and humans.[169-171] These findings led to the discovery of ANF by de Bold.[172] The site of storage of ANF was closely related to the granules, as demonstrated by co-sedimentation and cytochemical experiments.[172,173,175] The postulates relating to the biochemical properties and chemical (structural) features of the content of the granules were confirmed by the results obtained during the next few years, when crude atrial muscle extracts induced very significant natriuretic (diuretic) and hypotensive effects in experimental subjects.[174] One year later, de Bold[175] verified that the intensity of the natriuretic activity roughly corresponds to the number of atrial granules, and the protein character of the ANF content and the amino acid sequences of different ANFs were established. Further investigations revealed that a pre-propeptide with 151 amino acid residues is present in the human atrium; from this, the storage form of ANF containing 126 amino acid residues is formed[176] and stored in the atrial cardiocytes.[177] A relatively short-chained peptide with 28 amino acid residues was claimed to be the main circulating form of ANF.[178-180] The final episodes in the history of the discovery were the purification processes developed by de Bold et al.[178] which resulted in the isolation of pure peptides suitable for amino acid sequence analysis.

Before this period, several pioneering efforts were made to isolate the active components of cardiac tissues. The valuable results of this time include the experimental findings of a team from Charing Cross Hospital in London. Two natriuretic substances were isolated by Clarkson et al.[181] from salt-loaded normal human urine through the use of Sephadex G50 filtration. Their results demonstrated[181] that, when normal human urine was extracted by the two methods they described, it could be shown to contain two natriuretic substances. The one obtained with Sephadex G50 had a molecular mass of over 30,000, while the one obtained with Sephadex G25 was a peptide with a molecular mass of less than 3000. The natriuretic action of this latter substance was manifested immediately. The greatest value of this paper[181] was its revelation of the mistakes in the earlier separation experiments. The main conclusion (two natriuretic substances were detectable in the urine of both salt-loaded and salt-depleted subjects) could serve as the starting point for further purification-isolation experiments. It is worth quoting the final text of the paper (from 1976!): "Whether there

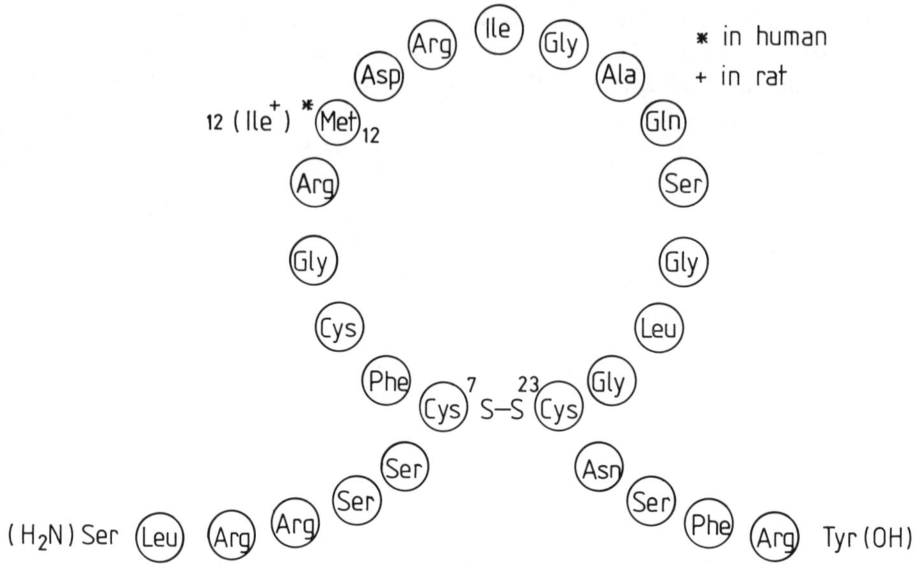

FIGURE 3. Amino acid sequence of human atrial natriuretic factor (αANF).

is any relationship between the long and short-acting natriuretic materials is not known. It is possible that the long-acting material is a precursor to the short-acting material, which would explain why the natriuretic activity of the long-acting material is delayed after its injection into the animal.'' This was undoubtedly a postulate of historical value, including the prediction of the existence of two opposing hormones controlling the sodium metabolism and excretion rate: ''aldosterone on the one hand, and natriuretic hormone on the other''.

In contrast to the very intensive research focused on the isolation, sequence analysis, and structure elucidation of atrial cardioactive substances, clear conclusions of the results in this field cannot be given at present. There are quite significant differences in the atrial contents of the human and the various animal atrial granules, and there is also a need for standardization of the nomenclature relating to ANF peptides.[182] In spite of the fact that the active forms of the natriuretic and diuretic atrial substances have a verified hormone character, the relevant Committee[182] favoured the term ''atrial natriuretic factor'', keeping the name given by de Bold et al. at the time of its discovery. The two main types of ANF should be designated by the prefixes ''human'' and ''rat'', and not by the designations ''Met'' and ''Ile'', which mirror the difference in amino acid composition. Thus, the standard names of the active (circulating) forms are ''human atrial natriuretic factor- (99-126)'' and ''rat atrial natriuretic factor-(99-126)'', or in abbreviated forms hANF (99-126) and rANF (99-126), respectively. The storage form of ANF (the biosynthetic precursor of the circulating form) should be named ''human pro-ANF'' (number of amino acid residues: 126). The precursor of ANF (1-126), which is the primary translational product from messenger RNA, has the trivial name ''Human prepro-ANF''.

The circulating form of ANF is a peptide with 28 amino acid residues, found in both humans and rats. The only difference between the compositions is the identity of amino acid 12 (Figure 3). de Bold and Flynn[183] applied reversed-phase HPLC to investigate atrial extracts and assigned four definite peaks in the chromatograms. In a subsequent paper[178] they reported the isolation of a peptide with diuretic-natriuretic activity, containing 28 amino acid residues. During this work a BioGel P10 column was used for the rough fractionation of the atrial extract, and preparative reversed-phase HPLC was performed (column: Vydac ODS, mobile phase: 20 to 40% acetonitrile in 0.1% aqueous tetrahydrofuran). Repetition of the chro-

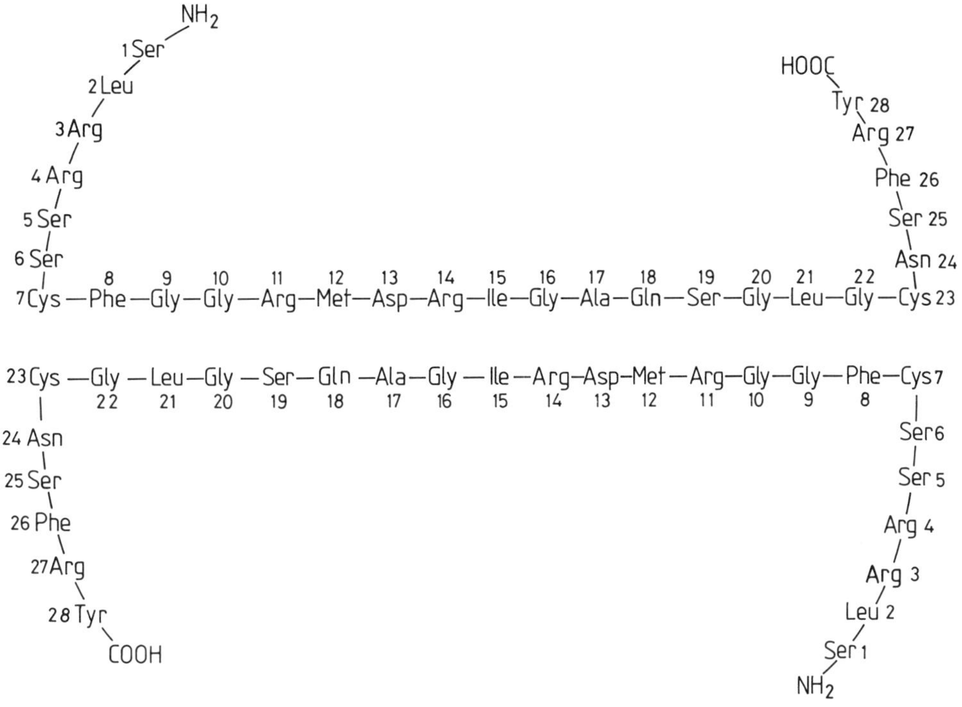

NH2
1 Ser
2 Leu
3 Arg
4 Arg
5 Ser
6 Ser
7 Cys—Phe—Gly—Gly—Arg—Met—Asp—Arg—Ile—Gly—Ala—Gln—Ser—Gly—Leu—Gly—Cys 23
8 9 10 11 12 13 14 15 16 17 18 19 20 21 22
Asn 24
Ser 25
Phe 26
Arg 27
Tyr 28
HOOC

23 Cys—Gly—Leu—Gly—Ser—Gln—Ala—Gly—Ile—Arg—Asp—Met—Arg—Gly—Gly—Phe—Cys 7
22 21 20 19 18 17 16 15 14 13 12 11 10 9 8
24 Asn
25 Ser
26 Phe
27 Arg
28 Tyr
COOH
Ser 6
Ser 5
Arg 4
Arg 3
Leu 2
Ser 1
NH2

FIGURE 4. Primary structure of β-human ANF (α-human dimer).

matographic procedure three times resulted in one definite, distinct peak. The substance gained from this peak displayed very intense diuretic-natriuretic activity and had the sequence seen in Figure 3.

Kangawa and Matsuo[184] succeeded in isolating human cardiac polypeptide via the extraction of human atrial tissue. The postmortem samples, stored for less than 10 h, were homogenized, repeatedly purified, and finally subjected to preparative reversed-phase HPLC. The substance with the smallest (~3000) molecular mass was extracted and submitted to amino acid sequence analysis. The primary structure was found to be almost identical with that of ANF isolated by Flynn et al.[178] from rat atria. The difference in amino acid 12 can be seen in Figure 3. This substance, identical with human ANF(99-126), proved to be a very effective inhibitor of Na-reabsorption and exerted much stronger effects than those of furosemide on the urine volume and Na^+, Cl^--excretion. The two other substances found[184] in the crude atrial extract had greater molecular masses, and were identical with "cardionatrins".[178] The experimental results indicate that the human atria contain the β form of ANF (β-ANF), with 56 amino acid residues (Figure 4). This substance has an M_r of 6000, twice that of α-ANF. Although α- and β-ANF separated from each other very sharply, on a cation-exchange or reversed-phase column they had surprisingly identical amino acid compositions. This fact suggested the idea that β-ANF might be a dimeric form of α-ANF.[185] The further idea that β-ANF could be an antiparallel dimer of α-ANF originated from the results of comparative trypsinization of intact α- and β-ANF. Two sets of intermolecular crosslinks between Cys^7 and Cys^{23} could be isolated by digestion of β-ANF, besides Cys^7-Cys^7 and Cys^{23}-Cys^{23} crosslinking (Figure 5). It should be noted that the β-form of ANF is missing in the atria of rat and other mammals.

Miyata et al.[186] made very interesting findings via a combination of RIA and HPLC. They demonstrated that the plasma of healthy volunteers contains two immunoreactive peaks, identified as α- and β-ANF. This finding shows that human β-ANF is not only present in

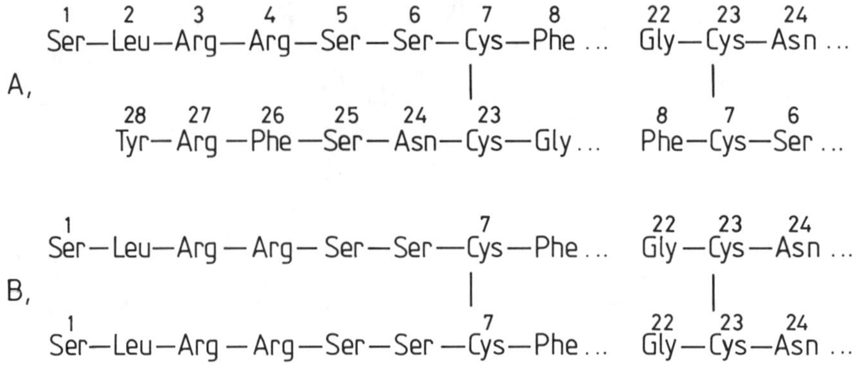

```
    1     2     3     4     5     6     7     8          22    23    24
    Ser—Leu—Arg—Arg—Ser—Ser—Cys—Phe ...    Gly—Cys—Asn ...
A,                                |                        |
         28    27    26    25    24    23               8     7     6
         Tyr—Arg—Phe—Ser—Asn—Cys—Gly...      Phe—Cys—Ser ...

    1                             7                22    23    24
    Ser—Leu—Arg—Arg—Ser—Ser—Cys—Phe ...    Gly—Cys—Asn ...
B,                                |                        |
    1                             7                22    23    24
    Ser—Leu—Arg—Arg—Ser—Ser—Cys—Phe ...    Gly—Cys—Asn ...
```

FIGURE 5. (A) Antiparallel dimeric structure for β-ANF. (B) Assumed structure for dimeric form of β-ANF.

the atria, but also circulates in the blood. In summary, it can be seen that in most of the plasma specimens (66%) human α-ANF was identified as a major immunoreactive component. The human α-ANF level was between 50 and 110 pg ml^{-1} in six cases, while one result was extremely high (480 pg ml^{-1}), and one moderately high (290 pg ml^{-1}). For both of these plasma samples, a second major peak appeared, which proved to be a macromolecular bound form of ANF. High plasma levels of α-ANF are characteristic for patients with renal disease (200 to 460 pg ml^{-1}); only in one of the eight samples was an ANF content near to 100 pg ml^{-1} measured. The situation is similar (with a somewhat higher deviation: from 85 up to 590 pg ml^{-1}) in patients with heart disease. In six of the 24 patients, the β (dimer) form of ANF was found in smaller amounts. Only the plasma of two patients contained the macromolecular form. These findings clearly show that no direct relationship can be demonstrated at present between the molecular form of ANF and the pathophysiology.[186] The situation is somewhat different in the case of the amount of α-ANF. Extremely high levels strongly suggest a pathological state.

An experimentally supported hypothesis for the biosynthesis of ANF was outlined recently by Bloch et al.[187] They postulated that rat serum (but not the plasma) contains a protease that specifically cleaves the 17-kilodalton (kDa) pro-ANF (126 amino acid residues). The products were analyzed by immunoprecipitation, and a 14-kDa amino-terminal peptide and a 3-kDa carboxy-terminal peptide mixture were identified. The latter are the circulating forms of ANF, and are present in a ratio of 3:1.

Although several data are available about the propeptide ANF containing 126 amino acid residues (Asn^1-Tyr^{126}), the exact nature of this ''intracellular factor'' was first established only recently.[188] For this purpose, secretory granules from rat atria were isolated, and the extracted ANF was purified and chemically investigated.[188] The final isolation of the granule content was performed by means of semipreparative HPLC on a C_{18} Bondapak column and gradient elution with acetronitrile. The found amino acid composition of the purified pro-ANF was very close to the theoretical one. These results[188] revealed that the ANF content of the rat atria granules corresponds to the propeptide ANF (Asn^1-Tyr^{126}), and a significant amount of certain short forms (α-ANF = ANF Ser^{99}-Tyr^{126}) is not stored in the granules. It seems plausible that the high molecular mass form (Asn^1-Tyr^{126}) is converted to the circulating form at the time of secretion from the granules into the blood stream. Inagami et al.[189] reported localization of the enzyme responsible for the generation of ANF (99-126) in the membrane fraction of the atrial homogenate, and identified it as Ser protease. The structural identification of Asn^1-Tyr^{126} (γ-ANF) was reported by Kangawa et al.[185] undoubtedly, the most important region of γ-ANF is the N-terminal-99-126, which is common with α-ANF and also includes the disulfide bridge between Cys^{105} and Cys^{120} (corresponding to the Cys[7]-Cys[23] link in α-ANF). The N-terminal sequence of γ-ANF was found to be almost identical with the reported N-terminal 30 residues of the peptide cardiodilatin isolated from porcine atrial tissue.[190]

Although endogenous human and other mammalian ANFs are known to differ in chain length and share the core sequence, most publications agree that there are three distinct peptides in human atrial extracts (α-, β- and γ-ANF), with a common precursor containing about 150 amino acid residues. Because of the utmost physiological and possibly therapeutical significance of ANF, its synthesis and structure-activity relationship study were of outstanding value. A solid-phase synthesis of ANF was proposed by Wade et al.[191] The preparation of α-ANF was undertaken by the two principal methods of solid-phase synthesis popular today, based on the work of Merrifield.[192] The synthesis was carried out fully automatically with a peptide synthesizer. The protecting groups were *t*-butoxycarbonyl (Boc) (cleaved with acid) and fluorenylmethoxycarbonyl (Fmoc) (cleaved with piperidine). After the cleavage (hydrofluoric or trifluoracetic acid), the peptides were purified by gel filtration, ion-exchange chromatography, and finally semi-preparative HPLC. They were then characterized by reversed-phase HPLC and also by adsorption TLC. The recovery of the purified peptides was approximately 20% in both cases. The theoretical amino acid analysis values and the biological activities (natriuretic responses) of the two synthetized peptide samples were in good agreement with those of the standard hANF purchased commercially. The recovery achieved with these methods surpassed those of the other published methods.[193-195] The Fmoc method was found[191] to be more advantageous because of the mild experimental conditions.

Within the field of chemical synthesis of ANF peptides, the work of Lyle et al.[196] proved successful. As the model target of their synthesis route, they chose the 26-residue peptide (rat ANF 8-33). The synthesis was carried out on a polystyrene resin support. The chloromethyl-functionalized resin was treated with the caesium or triethylammonium salts of the Boc-amino acid derivatives to bind the different peptide fragments to the resin surface. The amino acids with the most suitable side-chain protection (Arg-NO_2, Ser-benzyl, Asp-cyclohexyl ester, Cys-acetamidomethyl, Tyr-2,6-dichlorobenzyl ester) were used, and the coupling process (the binding of amino acids to the resin) was achieved with a standard Merrifield solid-phase method. The very careful purification process afforded a final product with a purity of over 97%. Promising results on biological activity in animal experiments and in clinical trials on human subjects were reported following the use of ANF products.[197,198] Schiller et al.[199] described the biologically more active ANF derivatives with chains shorter than that of α-ANF itself. Three analogues of ANF (105-126) were prepared, where Cys[105] was substituted with mercaptoacetic acid (I), mercaptopropionic acid (II) or mercaptobutyric acid (III) residues. These structural changes resulted in deletion of the N-terminal amino group, and the products could therefore be expected to be resistant to degradation by aminopeptidases. Comparative data relating to the potency of ANF(101-126), which proved equipotent with ANF(99-126) and I-III, revealed the high *in vivo* activities of II and III. This result very clearly indicates the role of degradation by aminopeptidase in the inactivation of ANFs.

As regards the urine volume and sodium excretion-increasing potency, derivatives II and III displayed effects two to four times stronger than those of ANF(101-126).

The conformational properties of atriopeptin-III (a 24-amino acid peptide, lacking the N-terminal Ser-Leu-Arg-Arg segment of α-ANF), were investigated by Fourier-transform infrared spectroscopy by Surewicz et al.[200] Their paper illustrates the great value of resolution-enhanced infrared spectroscopy in yielding more information on the conformation of vital peptide horomones. This technique disclosed that the predominant structure of both self-associated (the aggregation of the peptide was provoked by incubation at 30°C) and lipid-bound ANF-III is a β-sheet; the conformational differences between the two forms of the peptides were also detectable. The lipid-bound peptide (model membrane substance: dimyristoyl-phosphatidylglycerol) has a significantly more random structure than that of the self-associated form (21 vs. 11%). Both forms involve β-sheets, but the spectrum of the self-associated peptide was richer in signs of turns than the spectrum of the complexed peptide (24 vs. 10%). The demonstrated conformational flexibility of ANF-III enhances the planning experiments, highlighting the physiological relevance of physiologically preferred conformations. Together with the earlier used techniques (CD, NMR, etc.), infrared spectroscopy might be a good tool to provide new evidence concerning the hypothesis that the lipid membranes of the target cells prepare their "meeting" with the receptors by complexation of the peptide molecules, i.e., the lipid facilitates the correct ("physiologically preferred") folding of the peptide hormone.

A. DETERMINATION OF ANF CONTENT

The quantitative determination of ANF is based almost exclusively on RIA methods. An excellent review of circulating forms and RIA for ANFs was published by Gutkowska et al.[201,202]

As common features of the RIA methods for ANFs, it may be mentioned that antibody is generally raised in rabbits following injection with synthetic ANF-human-α-globulin. The interaction is promoted by incubation for 24 h. The unbound antibody is determined after a repeated 24 h incubation with ^{125}I-labeled ANF. In general, the limit of sensitivity is in the pg-region.[203-206]

A highly sensitive RIA method for ANF in human plasma and urine was forwarded by Marumo et al.[207] The limit of detection was 0.3 pg of α-hANF. The within- and between-assay coefficients of variation were determined for the calibration range up to 30 pg.

Nishiuchi et al.[203] described an RIA for ANF determination in normal subjects and patients, after extraction of the ANF from the plasma to a Sep-Pak C_{18} cartridge. The detection limit was 4 pg of ANF, with intra- and interassay coefficients of variation of 4.6 to 11.4%. A similarly performed RIA for ANF had a detection limit of 0.8 pg of ANF.[204] Gutkowska et al.[202] reported an RIA method for ANF which displayed linearity from 2.08 fmol upwards. The detection limit was 2 fmol. Iinuma et al.[208] described an RIA for ANF in heat-treated plasma. The detection limit was 0.4 pg. The effects of haemolysis and prolonged cold storage on the human plasma ANF content were studied by Lijnen et al.[209] by means of RIA. A radioreceptor assay for ANF was devised by Gutkowska et al.[210] using rat glomerular membrane as receptor source and ^{125}I as labeling substance. The receptor-bound and free radioactivity were separated by filtration, and the radioactivity of the dry filter was counted. The detection limit was 2 fmol. A good correlation was found with the results of RIA. The results of radioreceptor assay (using bovine adrenal membrane) and of RIA were also compared by Sagnella et al.,[205] who found a good correlation between them.

V. BRADYKININ

The main physiological action of bradykinin is an intense dilator effect on capillary blood vessels. It functions as a counterpart to the physiological agents that constrict blood vessels, such as angiotensin-II, norepinephrine, etc. Bradykinin is a product of the cleavage of high molecular mass kininogen by plasma kallikrein. Besides this process, Lys-bradykinin has been reported[221] to be produced by the cleavage of low molecular mass kininogen by tissue kallikrein. Lys-bradykinin is rapidly converted by a plasma aminopeptidase to bradykinin.[212] Bradykinin is metabolized and inactivated by ACE, which is also named kininase.

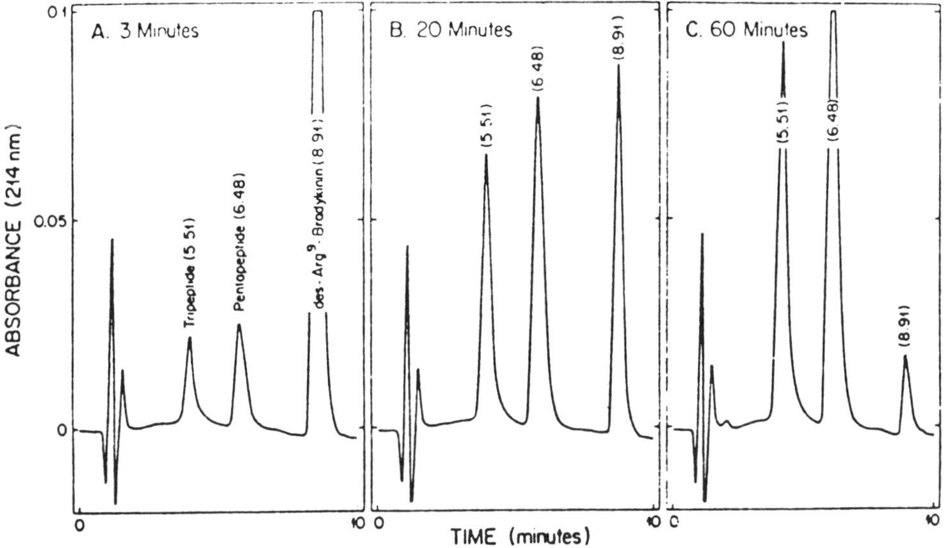

FIGURE 6. Elution pattern of the degradation of des-Arg[9] bradykinin by ACE showing time points obtained after (A) 3, (B) 20, and (C) 60 min, respectively. The major peaks are identified in A, and retention times, in minutes, are shown in parentheses. (From Sheikh, I. A. et al., *Biochem. Pharmacol.*, 35, p.1951, 1986. With permission.)

Bradykinin is a nonapeptide with the following amino acid sequence:

Kininogen

↓ ← Kallikrein

H–Arg(1)–(Pro)$_2$–Gly–Phe–Ser–Pro–Phe–Arg(9)–OH

bradykinin

↓ ← ACE

inactive degradation products

ACE (kininase) cleaves bradykinin, beginning from the C-terminal and removing a dipeptide unit. The heptapeptide is degraded further, with the formation of biologically inactive, smaller peptides. Sheikh et al.[213] studied the digestion of bradykinin and some other metabolites with ACE, and identified the degradation products of kinins in human plasma and serum. As initial bradykinin degradation product, they detected des-Arg[9]-bradykinin, which was shown to exert activity upon the blood vessels.[214] By monitoring the compounds appearing (cleavage products) by HPLC, Sheikh et al.[213] were able to discover interesting details of the influence of ACE on the bradykinin degradation process. When bradykinin was incubated, the normal dipeptidase activity of ACE was recognized primarily through identification of the dipeptides Phe^8-Arg^9 and Ser^6-Pro^7 plus the corresponding hexapeptide and pentapeptide. The end products of degradation by ACE were the pentapeptide and the two dipeptides. Monitoring of the degradation process in time gives valuable information on the bradykinin-ACE interaction. Therefore, the chromatograms are shown here without any further comment (Figure 6).

When the first cleavage product, des-Arg^9-bradykinin as a model substance, was further digested with ACE,[213] a tripeptide product (Ser-Pro-Phe) appeared in addition to the pen-

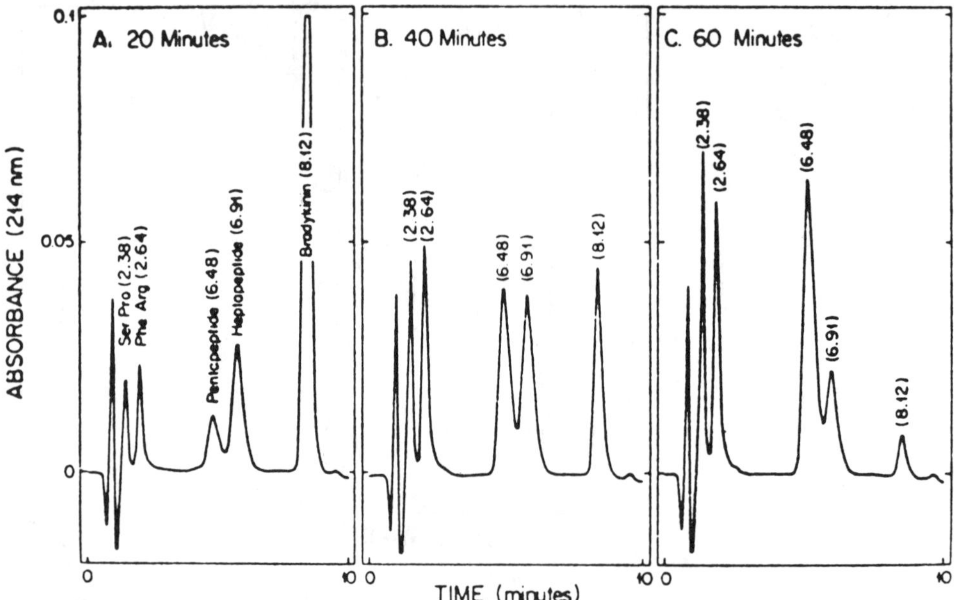

FIGURE 7. Elution pattern of bradykinin degradation by ACE on a reverse-phase HPLC column. Assessment of time points at 20, 40, and 60 min are shown in A, B, and C respectively. In A the identity of each peak is indicated, and its retention time, in minutes, is given in parentheses; only the latter is indicated in B and C. (From Sheikh, I. A. et al., *Biochem. Pharmacol.*, 35, p.1951, 1986. With permission)

tapeptide Arg-Pro-Pro-Gly-Phe. This finding confirms the tripeptidase activity of ACE reported previously:[215]

$$\text{H–Arg}^1\text{–Pro}^2\text{–Pro}^3\text{–Gly}^4\text{–Phe}^5\text{–Ser}^6\text{–Pro}^7\text{–Phe}^8\text{–OH}$$

des–Arg^9–bradykinin

ACE

H–Arg–Pro–Pro–Gly–Phe–OH H–Ser–Pro–Phe–OH

The corresponding chromatogram relating to degradation vs. time may give an insight into the mechanism of the bradykinin-ACE interaction (Figure 7).

Oshima et al.[216] have also published kinetic data on the cleavage of des-Arg^9-bradykinin by ACE. The Michaelis constants (K_m values) of ACE for des-Arg^9-bradykinin and bradykinin were found to be 240 μM and 4.4 μM, respectively; and the maximum velocities (V_{max}) 3.24 and 0.34 μmol min^{-1}, mg protein^{-1}, respectively. The half-lives of the hydrolyses by ACE were consistent with these data and showed a hydrolysis rate sequence des-Arg^9-bradykinin > bradykinin > angiotensin-II (t values in minutes: 4.3, 20.1, and 354, respectively). It was concluded, contradictory to previous findings, that ACE can hydrolyze the Pro residue in the penultimate position. This is very interesting, and may be of value from other aspects of the blood pressure regulation mechanism, but it needs further proof.

A. DETERMINATION OF BRADYKININ CONTENT

The most accurate method for bradykinin determination at present seems to be the combination of HPLC and RIA techniques. An example of this was published on the measurement of urinary kinins.[217] After purification of the urine on an Amberlite CG-50 pre-column, the kinins were eluted with 12 *M* formic acid. Thereafter, the kinins were separated on a Hypersil opthalmic Doppler sonogram (ODS) column with 17% acetonitrile in triethylamine-formate buffer as mobile phase. The separated bradykinin and Lys-bradykinin were determined by RIA with a sensitivity of 0.1 fmol for each. RIA is the method of choice for bradykinin (kinin) determination in blood. Down to 0.5 pg of kinin could be determined by the method of Shimamoto et al.[218] after the extraction of acidified blood with butanol and water. The double antibody (rabbit + ovine) solid-phase RIA of Nielsen et al.[219] involves the extraction of bradykinin from blood with acetone and purification on a QAE-Sephadex A-25 column. The recovery of added [^{125}I-Tyr^8]bradykinin was 27.8%. The sensitivity was 3 pg ml^{-1}. An RIA method with the same aim as above gave linearity for bradykinin in the range 0.022 to 87 μM.[220] The RIA method of Van Leeuwen et al.[221] includes the preliminary purification of blood extracts to prevent cross reactions of bradykinin with endogenous kininogen.

For the separation of bradykinin and other kinins, alternative procedures (affinity chromatography, ion-exchange chromatography) have been described.

An RIA for [des-Arg^9]bradykinin was developed by Odya et al.[222] with a low limit of detection (67 pg ml^{-1}), but with a certain antiserum cross reaction by bradykinin and lysilbradykinin.

HPLC may serve as a direct tool for the separation and determination of bradykinin in biological fluids. A method which is suitable for the measurement of bradykinin in human saliva[223] includes twofold chromatography in two different systems. With UV detection at 214 nm, a detection limit of 2 ng was claimed. The calibration graph was linear in the range 0 to 50 ng. Saliva bradykinin was determined fluorimetrically after HPLC separation, with post-column derivatization with phthalaldehyde.[224] The calibration graph was linear from 0.1 to 0.8 ng ml^{-1}.

A fluorimetric determination (λ_{exc} 290 nm, λ_{emiss} 475 nm) has been developed for the assay of bradykinin (kininase activity) in rat tissue.[225] Before fluorimetry, fluorescamine was added. The calibration graph was linear for 0.25 to 20 nmol of bradykinin. Bradykinin, Lys-bradykinin, and Met-Lys-bradykinin have been separated as their fluorescamine derivatives by HPLC on an ODS column with gradient elution (methanol-ammonium formate, pH 4.4). The detection limit was 2.5 ng. In the HPLC of the native kinins with UV detection, the limit of sensitivity was 1 ng.[226] Bradykinin and its fragments (Arg-Pro-Pro-Gly-Phe, Ser-Pro-Phe-Arg) could be measured fluorimetrically via the formation of dansyl derivatives.[227] This permitted the detection of 0.25 to 2 nmol after separation of the dansylated products (peptides) on a silica gel layer. Ninhydrin was used as fluorogenic reagent when bradykinin and two of its degradation products (des-Arg, and des-Phe-des-Arg analogs) were separated by HPLC.[228] The calibration graphs were linear in the range 0.05 to 0.1 to 5.0 nmol. The detection limits were 25, 50, and 25 pmol, respectively. The determination of bradykinin, kallidin (Lys-bradykinin), and Met-Lys-bradykinin by HPLC on a column of Nucleosil 5 C_8 has been solved. The detection limit was 20 pmol when UV detection was applied at 215 nm.[229]

EIA methods with a detection limit similar to that of RIA ($\gtrsim$30 pg) have been developed for the determination of bradykinin. As labeling enzymes, galactosidase[230] and peroxidase[231] were used. The enzyme activity associated with the antigen-antibody precipitate was measured with the use of a fluorogenic substrate.

VI. KALLIKREINS

Kallikreins are a subset of the Ser protease family of esteropeptidases; they liberate kinin from kininogen. They occur in the pancreas, salivary glands, and kidney,[232] and are also present in the plasma.[233,234]

The amino acid sequence of glandular kallikrein from the guinea pig prostate was investigated by Dunbar et al.[235] The extent of sequence identity with the kallikrein family of Ser proteases was found to be greater than 60%. This substance is comprised of 239 amino acids in a single polypeptide chain with carbohydrate attachments on Asn[78] and Asn.[169] The characteristic kallikrein features include ten half-cystine-residues and a C-terminal Pro. The molecular mass was determined as 36,000, and the protein moiety (i.e., deglycosylated protein) exhibited a molecular mass of 26,300.

A. DETERMINATION OF KALLIKREINS

The main methods for kallikrein determination may be grouped as follows: The RIA of glandular kallikrein in human plasma was described recently by Kazutaka et al.[236] Before the antibody-antiserum reaction, kallikrein was purified on an immunoaffinity column. In healthy patients, the glandular kallikrein content of the plasma was 1.36 ± 0.39 ng ml^{-1}. In patients with acute pancreatitis, its plasma concentration was 8.02 ± 6.15 ng ml^{-1}. The range of linearity was between 2.5 and 100 ng. RIA methods have been developed for the measurement of kallikrein in different biological fluids,[237] human urine,[238] rat urine,[239] and ewe blood plasma.[240] With the use of purified rat urinary kallikrein to raise antibodies in rabbits, a specific RIA was developed for the direct measurement of active plus inactive renal kallikrein.[241] The method is based on the observation that the antibody recognizes both forms of kallikrein but does not cross-react with trypsin, esterase-A, or human urine. The same paper[241] describes an HPLC method for the separation of active and inactive kallikrein, which thus permits assay of the inactive form.

Human urinary inactive kallikrein (pre-kallikrein) was determined by an indirect immunological method by Shimamoto et al.[242] The EIA of human urinary kallikrein was based on the measurement of liberated peroxidase activity[243] or luciferin via its luminescence.[244] An EIA of rat glandular kallikrein was based on incubation of the kallikrein-antibody complex with a chromogenic substrate and spectrophotometric measurement of the supernatant solution at 405 nm.[245] A similar principle, but with 4-nitrophenyl-β-D-galactoside as chromogenic substrate, was utilized to develop an EIA for the measurement of rat glandular kallikrein in tissues of diabetic and hypotensive animals.[246] The calibration graph was linear in the range 40 to 1000 ng of kallikrein.

The complex of human urinary kallikrein and antibody was adsorbed on polystyrene beads before addition of the chromogen, Pro-Phe-Arg-4-methylcoumarinyl-7-amide.[247] The same chromogen substrate was used for coupling in the determination of rat urinary kallikrein separated by HPLC.[248] Peptide- nitroanilides and thioesters have been applied as chromogenic substrates for porcine pancreatic kallikrein,[249] and also nitroanilides and anilides for kallikrein determination in plasma and blood products.[250-252] The esterase activity of kallikrein (substilisin) was determined by HPLC.[253] Benzoyl-L-Arg ethyl ester was added as substrate, which could be separated clearly from the benzoyl-L-Arg product in a simple chromatographic system (C_{18} Corasil pre-column, μBondapak C_{18} column, phosphate buffer, pH 2.8 acetonitrile mobile phase, UV detection at 254 nm). The calibration graphs were linear from 0.1 to 6.0 nmol. The detection limit was 1 ng ml^{-1} kallikrein.

In connection with an EIA of kallikreins,[254] kallikrein-A and B have been separated by column chromatography (DEAE-Sephadex-A-50 column, ammonium acetate as mobile phase) and purified by HPLC (column: TSK gel G 2000 SW, mobile phase: phosphate buffer, sodium chloride, sodium deoxycholate, detection at 280 nm).

REFERENCES

1. **Kaplan, N. M.,** Systemic hypertension: mechanisms and diagnosis, in *Heart Disease: A Textbook of Cardiovascular Medicine,* 2nd ed., Baumwald, E., Ed., W. B. Saunders, Philadelphia, 1984.
2. **De Quattro, V., Campese, V., and Antonaccio, M. J.,** Hypertension, ethyology, pathology and control, in *Cardiovascular Pharmacology,* Antonaccio, M. J., Ed., Raven Press, New York, 1977.
3. **Carragh, J. H., Ed.,** *Hypertension Manual,* Yorke Medical Books, New York, 1973.
4. **Folkow, B.,** Physiological aspects of primary hypertension, *Physiol. Rev.,* 628, 347, 1982.
5. **Meyer, P.,** *Br. Med. J.,* 282, 1114, 1981.
6. **Brock, T. A., Lewis, L. J., and Smith, J. B.,** *Proc. Natl. Acad. Sci. USA,* 79, 1438, 1982.
7. **Helmer, O. N.,** *Can. Med. Assoc. J.,* 90, 221, 1964.
8. **Romero, J. C. and Strong, S. G.,** *Circ. Res.,* 40, 35, 1977.
9. **Kestelsot, H. and Geboers, J.,** *Lancet,* 813, 1982.
10. **Lang, R. E., Bruckner, U. B., Kempf, B., Rascher, W., Strum, V., Unger, T., Speck, G., and Ganten, D.,** *Clin. Exper. Hyper. Theory and Practice,* 249, 1982.
11. **Gould, A. B., Skeggs, L. T., Jr., and Kahn, J. R.,** *J. Exp. Med.,* 119, 389, 1964.
12. **Ganten, D., Hayduk, K., Brecht, H. N., Boucher, R., and Genest, J.,** *Nature London,* 226, 551, 1970.
13. **Brown, J. J., Davies, D. L., Lever, A. F., and Robertson, J. I. F.,** *Can. Med. Assoc. J.,* 90, 201, 1964.
14. **Page, I. H. and Bumpus, M. F.,** Angiotensin, in *Handbuch experimentelle Pharmakologie,* Vol. XXXVII, Springer-Verlag, Berlin, 1974, 7.
15. **Alisono, K. S., Holladay, L. A., Murakami, K., Kuromizu, K., and Inagami, T.,** *Arch. Biochem. Biophys.,* 217, 574, 1982.
16. **Haas, E., Goldblatt, H., and Gibson, E. C.,** *Arch. Biochem. Biophys.,* 110, 584, 1965.
17. **Lucas, C. P., Fukutchi, S., Konn, J. W., Berlinger, F. G., Waldhausl, W. K., Cohen, E. L., and Rowner, M. R.,** *J. Lab. Clin. Med.,* 76, 689, 1970.
18. **Murakami, K. and Inagami, T.,** *Biochem. Biophys. Res. Commun.,* 62, 757, 1975.
19. **Yokosawa, N., Takahashi, N., Inagami, T., and Pge, D. L.,** *Biochem. Biophys., Acta* 569, 211, 1979.
20. **Szelke, M., Leckie, B., Hallet, A., Jones, D. M., Sueiras, J., Atrash, B., and Lever, A. F.,** *Nature London,* 299, 555, 1982.
21. **Waldhausl, W. K., Lucas C. T., Conn, J. W., Lutz, J. H., Cohen, E. L.,** *Biochim. Biophys. Acta,* 221, 536, 1970.
22. **Skeggs, L. T., Lontz, K. E., Kahn, J. R., Levine, M., and Dover, F. E.,** Multiple forms of human kidney renin, in *Hypertension,* Springer Verlag, Berlin, 1972.
23. **Rubin, I.,** *Scand. J. Clin. Lab. Invest.,* 29, 51, 1972.
24. **Inagami, T., Misono, K. S., Chang, J. J., and Takii, Y.,** *Biochem. Soc. Trans.,* 12, 951, 1984.
25. **Misono, K. S. and Inagami, T.,** *Biochemistry,* 19, 2616, 1980.
26. **Misono, K. S., Chang, J. J., and Inagami, T.,** *Proc. Natl. Acad. Sci. USA,* 79, 4858, 1982.
27. **Imai, T., Miyazaki, H., and Hirose, S.,** *Proc. Natl. Acad. Sci. USA,* 80, 7405, 1983.
28. **Panthier, J. J., Foote, S., Chambrieaud, B., Strosberg, A. D., Corvol, P., and Rougeon, F.,** *Nature London,* 298, 90, 1982.
29. **Akahane, K., Umeyama, H., Nakagawa, S., Moriguchi, I., Hirose, S., Iizuka, K., and Murakami, K.,** *Hypertension,* 7, 3, 1985.
30. **Morrish, B. J., Guss, M. J., Hunter, W. N., and Catanzaro, W. F.,** *Clin. Exp. Pharmacol. Physiol.,* 12, 299, 1985.
31. **Blundell, T., Sibanda, D. L., and Pearl, L.,** *Nature London,* 304, 273, 1983.
32. **Helmer, O. M. and Judson, W. E.,** *Circulation,* 27, 1050, 1963.
33. **Boucher, R., Weyrat, R., DeChamplain, J., and Genest, J.,** *Can. Med. Assoc. J.,* 90, 194, 1964.
34. **Dickens, T. T., Bumpus, F. M., Lloyd, A. M., Soneby, R. R., and Page, I. H.,** *Circ. Res.,* 17, 438, 1965.
35. **Goto, T., Imai, M., Hirose, S., and Murakami, K.,** *Clin. Chim. Acta,* 138, 87, 1984.
36. **Takii, Y. and Iragami, T.,** *Biochem. Biophys. Res. Commun.,* 104, 133, 1982.
37. **Yamagata, K., Maruyama, S., and Ishido, T.,** *Rinsho Kensa,* 25, 919, 1981.
38. **Inuma, K., Ikeda, I., Takai, M., Yanagawa, Y., Kurata, K., Ogihara, T., and Kumahara, Y.,** *Nippon Naibumpi Gakkai Zasshi,* 57, 1645, 1981.
39. **Ikeda, I., Inuma, K., Takai, M., Yanagawa, Y., Kurata, K., Ogihara, T., and Kumahara, Y.,** *J. Clin. Endocrinol. Metab.,* 54, 423, 1982.
40. **Hubl, W., Haussig, K., Hofmann, F., Buechner, M., Rohde, W., and Doerner, G.,** *Endokrinologie,* 77, 333, 1981.
41. **Higaki, J., Ogihara, T., Nishiura, M., Murakami, K., and Kumahara, Y.,** *Clin. Chim. Acta.,* 174, 345, 1988.

42. **Roulstone, I. E., Wathen, C. G., Sanger, B., and Muir, A. I.,** *Ann. Clin. Biochem.*, 20, 217, 1983.
43. **Stockigt, J. R. and Hewett, M. J.,** *Pathology,* 13, 603, 1983.
44. **Matsunaga, M., Suzuki, Y., Nakagawa, K., Wada, M., and Nishihata, S.,** *Clin. Chim. Acta,* 154, 213, 1986.
45. **Tikkanen, I., Fyhrquist, F., and Puntula-Rasanen, I.,** *Clin. Sci.*, 59, 381, 1980.
46. **Roulstone, J. F.,** *Lab. Prakt.*, 32, 83, 1983.
47. **Boer, P., Sleumer, J. H., and Spriensma, M.,** *Clin. Chem Winston-Salem N.C.*, 31, 149, 1985.
48. **Stirati, G., DeMartino, A., Mene, P., Pierucci, A., Simonetti, B. M., Feriozzi, S., Manzi, M., and Cinotti, G. A.,** *J. Clin. Chem. Clin. Biochem.*, 21, 529, 1983.
49. **Narvaez, J. A., Jiminez, E., Reyes, A., Garcia-Sanchez, F., and Morell, L.,** *Rev. Esp. Fisiol.*, 38, 143, 1982.
50. **Jimenez, E., Montiel, M., Garcia-Sanchez, F., Narvaez, J. A., Laserna, J., and Morell, M.,** *Anal. Lett.*, 14, 1669, 1981.
51. **Roth, M. and Reinharz, A.,** *Helu. Physiol. Pharmacol. Acta,* 24, 115, 1966.
52. **Skeggs, L. T., Lentz, K. E., Kahn, J. R., Doker, F. E., and Levine, M.,** *Circ. Res.*, 25, 451, 1969.
53. **Nielsen, A. H., and Poulsen, K.,** *J. Hypertens.*, 5, 25, 1987.
54. **Plentl, A. A., Page, I. H., and Davis, W. W.,** *J. Biol. Chem.*, 147, 143, 1943.
55. **Nasljetti, A. and Masson, G. M. C.,** *Circ. Res.*, 30-31 (Suppl. II), 187, 1972.
56. **Campbell, D. J.,** *J. Hypertens.*, 3, 199, 1985.
57. **Skeggs, L. T., Jr., Lentz, K. E., Hochstrasser, H., and Kahn, J. R.,** *J. Exp. Med.*, 118, 73, 1963.
58. **Lee, H. J. and Wilson, I. B.,** *Biochim. Biophys. Acta,* 243, 530, 1971.
59. **Glickson, J. D., Cunningham, W. D., and Marshall, G. R.,** *Biochemistry,* 12, 3684, 1973.
60. **Piriou, F., Lintner, K., Fermandian, S., Fromageot, P., Khosla, M. C., Smeby, R. R., and Bumpus, F. M.,** *Proc. Natl. Acad. Sci. U.S.A.*, 77, 82, 1980.
61. **Khosla, M. C., Stachowiak, K., Smeby, R. R., Bumpus, F. M., Piriou, F., Lintner, K., and Fermandian, S.,** *Proc. Natl. Acad. Sci. U.S.A.*, 78, 757, 1981.
62. **Lenkinski, R. E. and Stephens, R. L.,** *J. Inorg. Biochem.*, 15, 95,, 1981.
63. **Benson, J. R. and Hare, P. E.,** *Proc. Natl. Acad. Sci. U.S.A.*, 72, 619, 1975.
64. **Margolis, S. A. and Coxon, B.,** *Anal. Biochem.*, 141, 355, 1984.
65. **Deslauries, R., Paiva, A. C. M., Schaumburg, K., and Smith, I. C. P.,** *Biochemistry,* 14, 878, 1975.
66. **Deslauries, R., Ralston, E., and Somorjai, R. L.,** *J. Mol. Biol.*, 113, 697, 1977.
67. **Premilat, S. and Maigret, B.,** *J. Chem. Phys.*, 66, 3418, 1977.
68. **Premilat, S. and Maigret, B.,** *Biochem. Biophys. Res. Commun.*, 91, 534, 979.
69. **Marchionini, C., Maigret, B., and Premilat, S.,** *Biochem. Biophys. Res. Commun.*, 112, 339, 1983.
70. **Fujiwara, T., Iwai, T., Uyeda, M., and Tanimoto, O.,** *Mem. Fac. Eng. Osaka City Univ. 25,* 187, 1984; C.A., 103, 101, 1985.
71. **Paiva, T. B., Paiva, A. C. M., and Scheraga, H. A.,** *Biochemistry,* 2, 1327, 1963.
72. **Zalitis, G., Kublis, G., and Ancans, J.,** *Biokhimiya,* 50, 1083, 1985.
73. **Moore, G. J.,** *Int. J. Dept. Protein Res.*, 26, 469, 1985.
74. **Faucherre, J. L.,** QSAR Strategy Design of Bioactive Compounds, Proc. Eur. Symp. Quant. Struct, Activity Relat. 5th 1984 ed.: Seydel, J. K., VCH Publishers, Weinheim, 1985.
75. **Lintner, K., Fermandian, S., Fromageot, P., Khosla, M. C., Smeby, R. R., and Bumpus, F. M.,** *Biochemistry,* 16, 806, 1977.
76. **Schaechtelin, G., Walter, S., Salomon, H., Jelinek, J., Karen, P., and Cort, H.,** *Mol. Pharmacol.*, 10, 57, 1974.
77. **Schaechtelin, G., Surovec, D., and Walter, R.,** *Experientia,*, 31, 346, 1975.
78. **Blanc, E., Sraer, J., Sraer, J. D., Band, L., and Ardaillon, R.,** *Biochem. Pharmacol.*, 27, 517, 1978.
79. **Ackerley, J. A., Moore, A. F., and Peach, M. J.,** *Proc. Natl. Acad. Sci. U.S.A.*, 75, 5725, 1978.
80. **Sato, R., Sawada, S., and Tanaka, S.,** *Jpn. Circ. J.*, 42, 571, 1978.
81. **Degani, H. and Lenkinski, R. E.,** *Biochemistry,* 19, 3430, 1980.
82. **Basosi, R., Gaggelli, E., and Valensin, G.,** *J. Inorg. Biochem.*, 20, 263, 1984.
83. **Lenkinski, R. E., Glickson, J. D., and Walter, R.,** *Bioinorg. Chem.*, 8, 363, 1978.
84. **Basosi, R., Niccolai, N., Tiezzi, E., and Valensin, G.,** *J. Am. Chem. Soc.*, 100, 8047, 1978.
85. **Niccolai, N., Tiezzi, E., and Valensin, G.,** *Chem. Rev.*, 82, 359, 1982.
86. **Vallotton, M. B., Page, L. B., and Haber, E.,** *Nature London,* 215, 714, 1967.
87. **Greenwood, F. C., Hunter, W. M., and Glover, J. S.,** *Biochem. J.*, 89, 114, 1963.
88. **Gandolfi, C., Malvano, R., and Rosa, U.,** *Biochim. Biophys. Acta,* 251, 254, 1971.
89. **Lin, S. Y., Ellis, H., Weisblum, B., and Goodfriend, T. L.,** *Biochem. Pharmacol.*, 19, 651, 1970.
90. **Boyd, G. W. and Peart, W. S.,** *Lancet,* II, 129, 1968.
91. **Düsterdieck, G. and Me Elvee, G.,** in *Radioimmunassay Methods,* Churchill Livingstone, 1971, 24.
92. **Page, I. H. and Bumpus, F. M.,** *Physiol. Rev.*, 41, 331, 1961.

93. **Skeggs, L. T., Marsch, W. H., Kahn, J. R., and Sumway, N. P.,** *J. Exp. Med.,* 99, 275, 1954.
94. **Wiesenbaugh, P. E., Wills, N. E., and Hill, R. W.,** *Am. J. Physiol.,* 207, 759, 1964.
95. **Semple, P. F., Boyd, S. S., Dawes, P. M., and Morton, J. J.,** *Circ. Res.,* 39, 671, 1976.
96. **Lepri, L., Desideri, P. A., and Heimler, D.,** *J. Chromatogr.,* 211, 29, 1981.
97. **Boucher, R., Kurihara, H., Grise, C., and Genest, J.,** *Circ. Res.,* 33, 26, 1970.
98. **Dizdanglu, M., Krutzsch, H. C., and Simic, M. G.,** *J. Chromatogr.,* 211, 29, 1981.
99. **Tonnaer, J. A. D. M. and Verhoef, J.,** *J. Chromatogr.,* 183, 303, 1980.
100. **Hancock, W. S., Ed.,** *Angiotensin Vol. II. Handbook of HPLC for the Separation of Amino Acids, Peptide and Proteins,* CRC Press, Boca Raton, FL, 1989, 174.
101. **Doris, P. A.,** *J. Liq. Chromatogr.,* 8, 2017, 1985.
102. **Doris, P. A.,** *J. Chromatogr.,* 336, 392, 1984.
103. **Nussberger, J., Brunner, D. B., Waeber, B., and Brunner, H. R.,** *Hypertension,* 8, 476, 1986.
104. **Kumagaye, K. Y., Takai, M., Chino, N., Kimura, T., and Sakakibara, S.,** *J. Chromatogr.,* 327, 327, 1985.
105. **Hoshino, T., Sakarai, H., Kurimoto, F., and Sakuma, M.,** Japan Kokai Tokyo Koho J. P. 60, 190.863 28.sep.1985.
106. **Morton, J. J. and Webb, B. D. J.,** *Clin. Sci.,* 68, 483, 1985.
107. **Hermann, K., Lang, R. E., Unger, T., Bayer, C., and Ganten, D.,** *J. Chromatogr.,* 312, 273, 1984.
108. **Nussberger, R. P. and Brunner, D. B.,** *Int. J. Environ. Anal. Chem.,* 25, 257, 1986.
109. **Nussberger, R. P., Brunner, D. B., Vaeber, B., and Brunner, H. R.,** *Hypertension,* 7, 1, 1985.
110. **Chappell, M. C., Brosnihan, K. B., Welches, W. R., and Ferrario, C. M.,** *Peptides,* 8, 939, 1987.
111. **Anzan, C., Menard, J., Corvol, P., and Clorambach, A.,** *Sep. Purif. Methods,* 14, 97, 1985.
112. **Joergen, J. and Knud, P.,** *J. Hypertens.,* 3, 155, 1985.
113. **Ishida, J., Kai, M., and Okura, Y.,** *J. Chromatogr.,* 356, 171, 1986.
114. **Sakamoto, Y., Miyazaki, T ., Masaaki, K., and Okura, Y.,** *J. Chromatogr.,* 380, 313, 1986.
115. **Klickstein, L. B., and Wintroub, B. U.,** *Anal. Biochem.,* 120, 146, 1982.
116. **Roulstone, J. E., Sanger, B., and Wathen, C. G.,** *J. Clin. Chem. Clin. Biochem,* 21, 703, 1983.
117. **Skeggs, L. T., Kahn, J. R., Marsh, W. H.,** *J. Exp. Med.,* 99, 275, 1954.
118. **Skeggs, L. T., Marsh, W. H., Kahn, J. R., and Shumway, N. P.,** *J. Exp. Med.,* 103, 296, 1956.
119. **Erdös, E. G. and Yang, H. Y. T.,** *Life Sci.,* 6, 569, 1967.
120. **Hall, E. R., Kato, J., Erdös, E. G., Robinson, C. J. G., and Oshima, G.,** *Life Sci.,* 18, 1299, 1976.
121. **Johnson, A. R., Skidgel, R. A., Gafford, J. T., and Erdös, E. G.,** *Peptides,* 5, 789, 1984.
122. **Rix, E., Ganten, B., Schull, B., Unger, T., and Taugner, R.,** *Neurosci. Lett.,* 22, 125, 1981.
123. **Erdös, E. G. and Gafford, J. T.,** *Clin. Exp. Hypertension,* A5, 1251, 1983.
124. **Skidgel, R. A., Engelbrecht, S., Johnson, A. R., and Erdös, E. G.,** *Peptides,* 5, 769, 1984.
125. **Erdös, E. G. and Skidgel, R. A.,** *Biochem. Soc. Trans.,* 13, 42, 1985.
126. **Das, M. and Soffer, K. I.,** *J. Biol. Chem.,* 250, 6762, 1975.
127. **Oshima, G., Gesce. A., Erdös, E. G.,** *Biochim. Biophys. Acta,* 350, 26, 1974.
128. **Siems, W. E. and Halle, W.,** *Pharmazie,* 37, 539, 1982.
129. **Funae, Y., Komori, I., Sasaki, D., and Yamamoto, K.,** *Jpn. J. Pharmacol.,* 28, 925, 1978.
130. **Riordan, J. F.,** *Biochemistry,* 12, 3915, 1973.
131. **Bünning, P., Holmquist, B., and Riordan, J. F.,** *Biochem. Biophys. Res. Commun.,* 83, 1442, 1978.
132. **Horiuchi, M., Fujimura, K., Terashima, T., and Iso, T.,** *J. Chromatogr.,* 233, *Biomed. Appl.,* 22, 123, 1982.
133. **Maurich, V., Moneghini, M., Pitotti, A., and Vio, L.,** *J. Pharm. Biomed. Anal.,* 3, 51, 1985.
134. **Maurich, V., Pitotti, A., Vio, L., and Mamolo, M. G.,** *J. Pharm. Biomed. Anal.,* 3, 425, 1985.
135. **Pitotti, A., Maurich, V., Moneghini, M. and Vianello, S.,** *J. Pharm. Biomed. Anal.,* 4, 677, 1986.
136. **Pre, J. and Bladier, D.,** *Med. Sci., Libr. Compend.,* 11, 220, 1983.
137. **Hayakari, M., Geito, R., Furugosi, A., Hasimoto, Y., and Murakami, S.,** *Clin. Chim. Acta,* 144, 71, 1984.
138. **Filidovic, N., Borcic, N., and Igic, R.,** *Clin. Chim. Acta,* 128, 177, 1983.
139. **Ryder, K. W., Thompson, H., Smith D., Sample, M., Sample, R. B., and Oei, T. O.,** *Clin. Biochem.,* Ottawa, 17, 302, 1984.
140. **O'Brien, J. F., Forsman, R. W., and Rohrbach, M. S.,** *Clin. Chem. (Winston-Salem N.C.),* 29p., 1990, 1983.
141. **Stoner, L.,** *Clin. Chem. (Winston-Salem N.C.),* 30, 592, 1984.
142. **Santos, R. A. S., Krieger, E. M., and Greene, L. J.,** *Hypertension (Dallas),* 7, 244, 1985.
143. **Siems, W. E., Heder, G., and Komissarowa, N. W.,** *Z. Med. Laboratoriumsdiagn.,* 26, 232, 1985.
144. **Pitotti, A., Maurich, V., Moneghini, M., and Vianello, S.,** *J. Pharm. Biomed. Anal.,* 4, 677, 1986.
145. **Maguire, G. A.,** *Clin. Chem. (Winston-Salem N.C.),* 31, 1575, 1985.
146. **Shisheva, A., Ikonomov, O. and Sirakov, L.,** *Collect. Czech. Chem. Commun.,* 51, 2280, 1986.

147. **Hurst, C. L. and Lovell-Smith, C. J.,** *Clin. Chem. (Winston-Salem N.C.),* 30, 817, 1984.
149. **Boomsma, F. and Schalekamp, M. A.,** *J. Clin. Chem. Clin. Biochem.,* 21, 845, 1983.
148. **Svoboda, D. and Gasparic, J.,** *Mikrochim. Acta,* 384, 1971.
150. **Cushman, D. W. and Cheung, H. S.,** *Biochem. Pharmacol.,* 20, 1637, 1971.
151. **Neels, H. M., Van Sande, M. E., and Scharpe, S. E.,** *Clin. Chem. (Winston-Salem N.C.),* 29, 1399, 1983.
152. **Mayr, K.,** *Lab. Med.,* 10, 313, 1986.
153. **Holmquist, B., Buenning, P., and Riordan, J. F.,** *Anal. Biochem.,* 95, 540 1979.
154. **Ronca-Testoni, S.,** *Clin. Chem. (Winston-Salem N.C.),* 29, 1093, 1983.
155. **Neels, H. M., Scharpe, S. L., VanSande, M. E., and Fonteyne, G. A.,** *Clin. Cem. (Winston-Salem N.C.),* 30, 163, 1984.
156. **Buttery, J. E. and Chamberlain, B. R.,** *Clin. Chem. (Winston-Salem N.C.),* 31, 645, 1985.
157. **Buttery, J. E. and Chamberlain, B. R.,** *Clin. Chem. (Winston-Salem N.C.),* 32, 2121, 1986.
158. **Maguire, G. A. and Price, C. P.,** *Ann. Clin. Biochem.,* 22 204, 1985.
159. **Maguire, G. A. and Price, C. P.,** *Ann. Clin. Biochem.,* 21, 372, 1984.
160. **Neels, H. M., Scharpe, S. L., Fonteyne, G. A., Yaron, A., and Van Sande, M. E.,** *Clin. Chim. Acta,* 141, 281, 1984.
161. **Kapiloff, M. S., Strittmatter, S. M., and Fricker, E. D.,** *Anal. Biochem.,* 140, 293, 1984.
162. **Fyhrquist, F., Tikkanen, I., Gronhagen-Riska, C., Hortling, L., and Hitchens, M.,** *Clin. Chem. (Winston-Salem N.C.),* 30, 696, 1984.
163. **Sakamoto, Y., Miyazaki, T., Kai, M., and Ohkura, Y.,** *J. Chromatogr.,* 380, *Biomed. Appl.,* 53, 313, 1986.
164. **Beneteau, B., Baubin, B., Morgant, G., Giboudeau, J., and Baumann, F. C.,** *Clin. Chem. (Winston-Salem N.C.),* 32, 884, 1986.
165. **Roulston, J. E. and Allan, D.,** *Clin. Chim. Acta,* 168, 187, 1987.
166. **Johansen, K. B., Marstein, S., and Aas, P.,** *Scand. J. Clin. Lab. Invest.,* 47, 411, 1987.
167. **Wong, Y. W. and Kinniburgh, D. W.,** *Clin. Biochem.,* 20, 323, 1987.
168. **Kisch, B.,** *Exp. Med. Surg.,* 14, 99 1956.
169. **Jamieson, J. D. and Palade, G. E.,** *J. Cell Biol.,* 23, 151, 1964.
170. **Bencosme, S. A. and Berger, J. M.,** in *Methods and Achievements in Experimental Pathology,* Bajusz, E. and Jasmin, G., Eds. Vol. 5, S. Karger, Basel, 1971, 173.
171. **Forssmann, W. G., Hock, D., and Mutt, V.,** *Klin. Wochenschr.,* 64, (Suppl. VI), 4, 1986.
172. **deBold, A. J.,** *Proc. Soc. Exp. Biol. Med.,* 161, 508, 1979.
173. **Garcia, R., Cantin, M., Thibault, G., Ong., H., and Genest, J.,** *Experientia,* 38, 1071, 1982.
174. **Sonnenberg, H., Veress, A. T., Borenstein, A. B., and deBold, A. J.,** *Physiologist,* 23, 13a, 1980.
175. **deBold, A. J.,** *Proc. Soc. Exp. Biol. Med.,* 170, 133, 1982, *Can. J. Physiol. Pharmacol.,* 60, 324, 1982.
176. **Misono, K. S., Fukumi, H., Grammer, R. T., and Inagami, T.,** *Biochem. Biophys. Res. Commun.,* 119, 524, 1984.
177. **Kangawa, K., Fukuda, A., and Matsuo, H.,** *Nature,* 313, 397, 1985.
178. **Flynn, T. G., deBold, M. L., and deBold, A. J.,** *Biochem. Biophys. Res. Commun.,* 117, 859, 1983.
179. **Thibault, G., Lazure, T., Schiffrin, E. L., Gutkowska, J., Chartier, I., Garcia, R., Seidah, N. G., Chretien, M., Genest, J., and Cantin, M.,** *Biochem. Biophys. Res. Commun.,* 130, 981, 1985.
180. **Schwarz, D., Gellen, D. M., Manning, P. T., Siegel, W. R., and Fok, K. F., Salith, C. E., and Needleman, P.,** *Science,* 229, 397, 1985.
181. **Clarkson, E. M., Sheelagh, M. R., and deWardener, H. E.,** *Kidney Int.,* 10, 381, 1976.
182. A summary from the recommendations of a committee of the International Society of Hypertension, American Heart Association and World Health Organization, *New. Eng. J. Med.,* 316, 1278, 1987.
183. **de Bold, A. J. and Flynn, T. G.,** *Life Sci.,* 33, 297, 1983.
184. **Kangawa, K. and Matsuo, H.,** *Biochem. Biophys. Res. Commun.,* 118, 131, 1984.
185. **Kangawa, K., Fukuda, A., and Matsuo, H.,** *Nature,* 313, 397, 1985.
186. **Miyata, A., Toshimory, T., Hashiguchi, T., Kangawa, K., and Matsuo, H.,** *Biochem. Biophys. Res. Commun.,* 142, 461, 1987.
187. **Bloch, K. D., Zisfein, J. B., Margolies, M. N., Horney, C. H. J., Seidman, J. G., and Graham, R. M.,** *Am. J. Physiol.,* 252.E, 147, 1987.
188. **Thibault, G., Garcia, R., Gutkowska, J., Bilodeau, J., Lazure, C., Seidah, M. G., Chretien, M., Genest, J., and Cantin, M.,** *Biochem. J.,* 241, 265, 1987.
189. **Inagami, T., Misono, K. S., Fukumi, H., Maki, M., Tanaka, I., Takayanagi, R., Imada, T., Grammer, R. T., Naruse, M., Naruse, K., Pandey, K. N., Parmentier, N., Yasujima, M., and Abe, K.,** *Hypertension,* 10 (Suppl. 1), I-113, 1987.
190. **Forssmann, W. G., Hock, D., Lottspeich, F., Henschen, A., Kreye, W., Christmann, M., Reinecke, M., Metz, J., Carlquist, M., and Mutt, V.,** *Anat. Embryol.,* 168, 307, 1983.

191. **Wade, J. D., Fitzgerald, S. P., McDonald, M. R., McDougall, J. G., Tregear, G. W.,** *Biopolymers,* 25, S 21, 1986.
192. **Merrifield, R. D.,** *J. Am. Chem. Soc.,* 85, 2149, 1963.
193. **Minamitake, Y., Kubota, I., Hayashi, Y., Furuya, M., Kangawa, M., and Matsuo, H.,** in *Peptide Chemistry 1984,* Izumiya, N., Ed., Protein Research Foundation, Osaka, 1985, 229.
194. **Shimokura, N., Hosoi, S., Okamoto, K., Fujiwara, Y., Yoshida, M., and Kiso, Y.,** in *Peptide Chemistry 1984,* Izumiya, N., Ed. Protein Research Foundation, Osaka, 1985, 235.
195. **Chino, N., Nishiuchi, Y., Masui, Y., Noda, Y., Watanabe, T. X., Kimuea, T., and Sakakibara, S.,** in *Peptide Chemistry 1984,* Izumiya, N., Ed., Protein Research Foundation, Osaka, 1985, 241.
196. **Lyle, T. A., Brady, S. F., Ciccarone, T. M., Colton, C. H. D., Paleveda, W. J., Veber, D. F., and Nutt, R. F.,** *J. Org. Chem.,* 52, 3752, 1987.
197. **Keegan, M. E., Faison, E. P., and Bagkin, E. P.,** *Clin. Exp. Hypertens.,* Part A7., 869, 1985.
198. **Biollaz, J., Nussberger, J., Waeber, B., Brunner-Faber, F., Gomez, H. J., Otterbein, E. S., and Brunner, H. R.,** *Hypertension,* 7, 845, 1985.
199. **Schiller, P. W., Mazic, L. A., Nguyen, T. M. D., Godin, J., Garcia, R., deLean, A., and Cantin, M.,** *Biochem. Biophys. Res. Commun.,* 143, 499,, 1987.
200. **Surewicz, I. K., Mantsch, H. H., Stahl, G. L., and Epand, R. M.,** *Proc. Natl. Acad. Sci.,* U.S.A., 84, 7028, 1987.
201. **Gutkowska, J.,** *Nucl. Med. Biol.,* 14, 323, 1987.
202. **Gutkowska, J., Genest, J., Thibault, G., Garcia, R., Larochelle, P., Cusson, J. R., Kutchel, O., Hamet, P., deLean, A., and Cantin, M.,** *Endocrinol. Metab. Clin. North Am.,* 16, 183, 1987.
203. **Nishiuchi, T., Saito, H., Yamasaki, Y., and Saito, S.,** *Clin. Cim. Acta,* 159, 45, 1986.
204. **Rosmalen, F. M. A., Tan, A. C., Tan, H. S., and Benraad, T. J.,** *Clin. Chim. Acta,* 165, 331, 1987.
205. **Sagnella, G. A., Buckley, M. G., Markandu, N. D., and MacGregor, G. A.,** *Clin. Chim. Acta.,* 166, 37, 1987.
206. **Storm, T. L., Thamsborg, G., Sykulski, R., Keller, N., Thompsen, J., and Larsen, J.,** *Scand. J. Clin. Lab. Invest.,* 47, 745, 1987.
207. **Marumo, F., Sakamoto, H., Ando, K., Ishigami, T., and Kawakami, M.,** *Biochem. Biophys. Res. Commun.,* 137, 231, 1986.
208. **Iinuma, K., Ikeda, I., Ogihara, T., Hata, H., Shima, J., Kurata, K., and Kumahara, Y.,** *Clin. Chem. (Winston-Salem N.C.),* 33, 674, 1987.
209. **Lijnen, P., Huysecom, J., Fagard, R., Staessen, J., and Amery, A.,** *Clin. Chim. Acta,* 171, 333, 1988.
210. **Gutkowska, J., Carrier, F., St. Louis, J., Thibault, G., Cantin, M., and Genest, J.,** *Anal. Biochem.,* 168, 100, 1988.
211. **Geiger, R., Hofmann, W., Franke, M., and Bauer, X.,** in *Advances in Experimental Medicine and Biology,* Fritz, H., Back, N., Dietze, G., and Haberland, G. L., Eds., Plenum Press, New York, 1983, 275.
212. **Guimaraes, J. A., Borges, D. R., Prado, E. S., and Prado, J. L.,** *Biochem. Pharmacol.,* 22, 3157, 1973.
213. **Sheikh, I. A. and Kaplan, A. P.,** *Biochem. Pharmacol.,* 35, 1951, 1986.
214. **Marceau, F., Lussier, A., Regoli, D., and Giroud, I. P.,** *Gen. Pharmacol.,* 14, 209, 1983.
215. **Inokuchi, J. I. and Nagamatsu, A.,** *Biochim. Biophys. Acta,* 662, 300, 1981.
216. **Oshima, G., Hiraga, Y., Shirono, K., Ohishi, S., Sakakibara, S., and Kinoshita, T.,** *Experientia,* 41, 325, 1985.
217. **Fejes-Tóth, G., Náray-Fe jes-Tóth, A., and Froelich, G. C.,** *Clin. Chim. Acta,* 140, 21, 1984.
218. **Shimamoto, K., Ando, T., Tanaka, S., Nakakashi, Y., Nishitani, T., Hosoda, S., Ishida, H., and Iimura, O.,** *Endocrinol. Jpn.,* 29 487, 1982.
219. **Nielsen, M. D., Nielsen, F., Kappelgaard, A. M., and Giese, J.,** *Clin. Chim. Acta,* 125, 145, 1982.
220. **Van Rosevelt, R. F., Mulda, A., Philipse-Hoorweg, E. M., Van Ginkel, C. J. W., and Van Monrik, J. A.,** *Clin. Chim. Acta,* 126, 81, 1982.
221. **Van Leeuwen, B. H., Millar, J. A., Hammat, M. T., and Johnston, C. E.,** *Clin. Chim. Acta,* 127, 340, 1983.
222. **Odiya, C. E., Moreland, P., Stewart, J. M., Barabe, J., and Regoli, D. C.,** *Biochem. Pharmacol.,* 32, 337, 1983.
223. **Omori, H., Takahashi, Y., Sakakibara, T., Ota, S., and Nakashizuka, T.,** *J. Pharm. Biomed. Anal.,* 4, 529, 1986.
224. **Omori, H., Watanabe, N., Nakashizuka, T., and Yamazaki, S.,** II. High Resolut. Chromatogr. Chromatogr. Commun., 9, 306, 1986.
225. **Maita, E., Endo, Y., and Ogura, Y.,** *Anal. Biochem.,* 128, 36, 1983.
226. **Narayanan, T. K. and Greenbaum, L. M.,** *J. Chromatogr. Biomed. Appl.,* 31, 109, 1984.
227. **Kariya, K., Iwaki, H., Kanemaru, T., Tsuda, Y., and Okada, Y.,** *Anal. Biochem.,* 115, 46, 1981.

228. **Hiraga, Y., Shirono, K., Ohishi, S., Sakakibara, S., and Kinoshita, T.,** *Bunseki Kagaku,* 33, E279, 1984.
229. **Geiger, R., Hell, R., and Fritz, H.,** *Hoppe-Seyler's Z. Physiol. Chem.,* 363, 527, 1982.
230. **Keno, A., Ohishi, S., Kitagawa, T., and Katori, M.,** *Biochem. Pharmacol.,* 30, 1659, 1981.
231. **Stahr, K., Hubl, W., and Scheuch, D. W.,** *Z. Med. Laboratoriumsdiagn.,* 25, 346, 1984.
232. **Fiedler, F.,** *Enzymology of glandular kallikreins,* in Bradykinin, Kallidin and Kallikrein, Erdös, E. G., Ed., Springer Verlag, Berlin, 1979, 103.
233. **Shimamoto, K., Mayfield, R. K., Margolius, H. S., Chao, J., Stroud, W., and Kaplan, A. P.,** *J. Lab. Clin. Med.,* 103, 731, 1984.
234. **Bagshaw, A. F., Bhoola, K. D., Lemon, M. J. C., and Whicher, J. T.,** *J. Endocrinol.,* 101, 173, 1984.
235. **Dunbar, J. C. and Bradshaw, R. A.,** *Biochemistry,* 26, 3471, 1987.
236. **Kazutaka, N., Yoshikazu, I., and Tatsuo, K.,** *Clin. Chim. Acta,* 162, 341, 1987.
237. **Bagshaw, A. F., Bhoola, K. D., Lemon, M. J. C., and Whicher, J. T.,** *J. Endocrinol.,* 101, 173, 1984.
238. **Tanaka, S. and Shimamoto, K.,** *Sapporo Igaku Zasshi,* 50, 29, 1981.
239. **Oza, N. B.,** *J. Clin. Chem. Clin. Biochem.,* 19, 1033, 1981.
240. **Somler, B. and Stankov, B.,** *Dokl. Bolg. Akad. Nauk,* 38, 1219, 1985.
241. **Girolami, J. T., Bascands, J. L., Pecher, C., and Suc, J. M.,** *J. Immunoassay,* 8, 115, 1987.
242. **Shimamoto, K., Chao, J., and Margolius, H. S.,** *Tohoku J. Exp. Med.,* 137, 269, 1982.
243. **Franke, M., Rohrschneider, S., and Geiger, R.,** *J. Clin. Chem. Clin. Biochem.,* 20, 621, 1982.
244. **Miska, W. and Geiger, R.,** *Z. Anal. Chem.,* 324, 266, 1986.
245. **Johansen, L., Nustad, K., Berg, T., and Pierce, J. V.,** *J. Immunol. Methods,* 69, 253, 1984.
246. **Sakamoto, W., Yoshikawa, K., Nishikaze, O., Handa, H., Hirayama, A., and Uehara, S.,** *J. Clin. Chem. Clin. Biochem.,* 23, 521, 1985.
247. **Kagayama, H., Okumura, Y., Adachi, T., Ito, Y., Hirano, K., Sugiura, M., and Sawaki, S.,** *Chem. Pharm. Bull.,* 31, 2366, 1983.
248. **Funae, Y., Akiyama, H., Imaoka, S., Takaoka, M., and Morimoto, S.,** *J. Chromataogr.,* 264, 249, 1983.
249. **Powers, J. C., McRae, B. J., Tanaka, T., Cho, K., and Cook, R. R.,** *Biochem. J.,* 220, 569, 1984.
250. **Scheja, J. W. and Bruester, H. T.,** *Lab. Med.,* 7, 276, 1984.
251. **Scheja, J. W. and Bruester, H. T.,** *Lab. Med.,* 7, 344, 1984.
252. **Pietta, T., Mauri, P., and Pace, M.,** *J. Chromatogr.,* 441, 431, 1988.
253. **Shibata, T., Hasegawa, R., and Motchida, K.,** *Yakugaku Zasshi,* 103, 186, 1983.
254. **Rodgers, G. M., Donaldson, V. H., Shuman, M. A.,** *Thromb. Res.,* 27, 641, 1982.

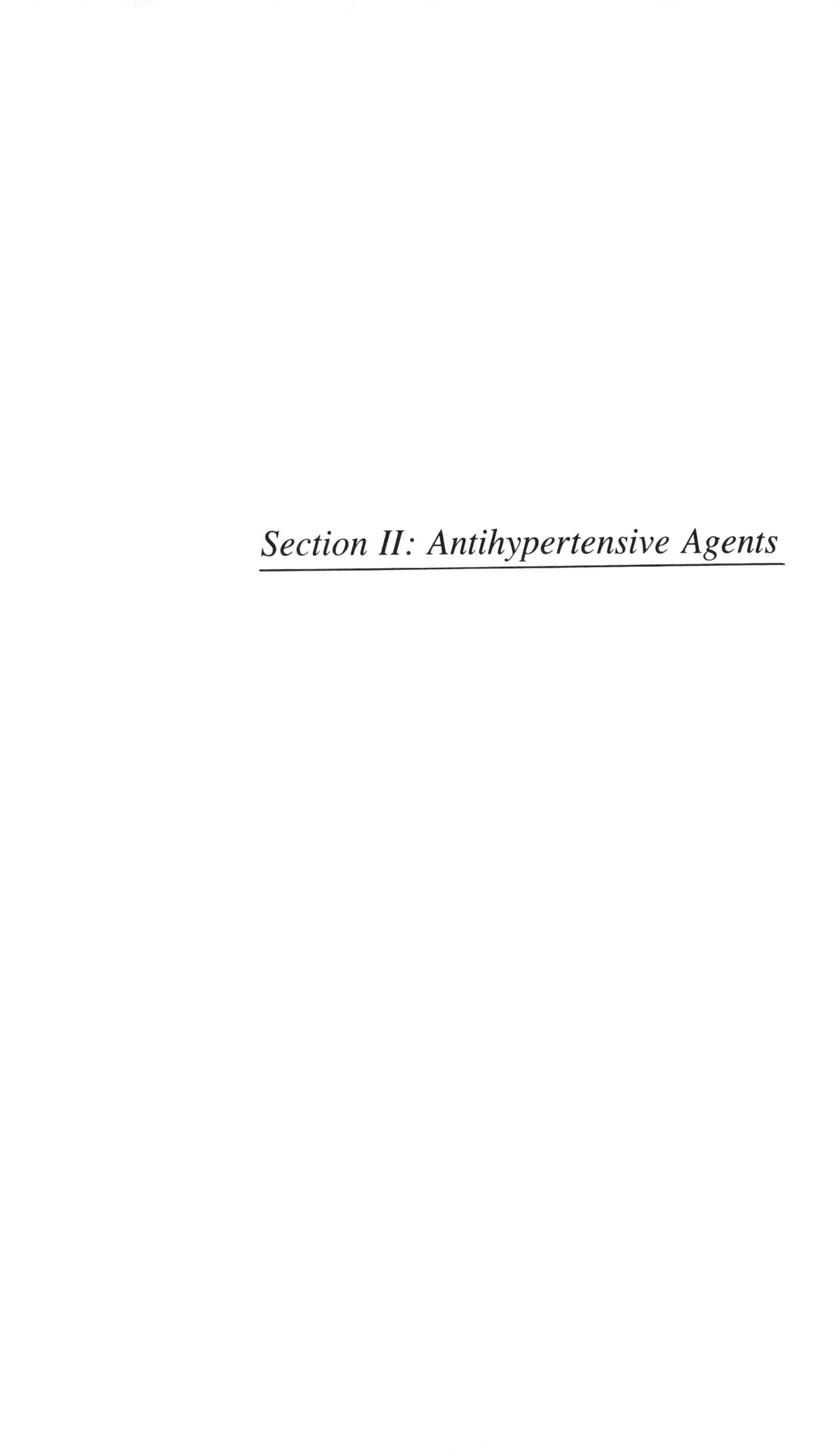

Section II: Antihypertensive Agents

Chapter 1

DIURETICS

I. INTRODUCTION

Diuretics are compounds with therapeutically useful effects on the mammalian nephral excretion balance; they can increase the net excretion of water and solutes. They have an important role in the treatment of edema and hypertension. This section deals only with diuretic agents involved in antihypertensive therapy. They can be grouped on the basis either of their chemical structures or of their site and mechanism of action. Here, the chemical treatment is predominant.

Diuretics became a cornerstone of hypertension therapy following the discovery of the breakthrough drug chlorothiazide in the late fifties. Since that time, diuretics have been the drugs most frequently utilized in cardiovascular practice. They were earlier used primarily in the treatment of heart failure, but their main role is now in the field of hypertension therapy. An annual worldwide consumption of diuretics to the value of about $ 1 billion has been estimated,[1] which means that diuretics rival the β-blockers in sales value but exceed them as regards the number of prescriptions.

The main question that arises in the use of diuretics against hypertension is whether they are suitable as monotherapeutics in mild or moderate cases of hypertension (diastolic blood pressure between 90 and 104 mm Hg), or whether they must be combined with, or even replaced by, some other antihypertensive drug.

The arguments against the use of diuretics as step-one drugs chiefly concentrate on one problem: the safety of diuretics. The main point in the propaganda against diuretics is the increase in the number of sudden deaths due to fatal coronary events and/or arrhythmic disturbances. In the biochemical, pathophysiological background, mention has been made of the hypopotassemic, hyperlipidemic, hyperglycemic, plasma uric acid content increasing, and other effects of antihypertensive diuretics. The rather sharp controversy on this problem is continuing; only a few of the numerous papers from this field are referred to here.[1-5]

For an effective discussion of the above problem, a knowledge of the sites and mechanisms of action of antihypertensive diuretics is needed. While numerous excellent physiological analyses of the latter problem are available,[6-8] a plausible molecular biological interpretation of the action of diuretics and a hypothesis suitable for generalization, remain problems to be solved. Figure 1, from the excellent paper by Kokko,[6] summarizes the natures and directions of the main transport processes and the sites of action of the most important diuretics.

The extreme complexity of the mechanism of antihypertensive action of diuretics suggests very different biochemical mechanisms of action, and this fact precludes the possibility of a generally valid structure-activity relationship. The extent to which the various factors contribute to the overall effect depend largely on the chemical nature of the diuretic used. The possibilities of drug design are therefore limited to a rather small circle of compounds within the various structural types. The main aims of the study of the structure-activity relationship are the design and synthesis of new antihypertensive diuretics with strong diuretic (natriuretic) action and a high potency, being effective even in the milligram dosing range, but at the same time which possess a very low profile of toxicity.

The thiadiazine-sulfonamide derivatives (thiazides) are the most widely used antihypertensive diuretics, and therefore receive a relatively detailed treatment. Nevertheless, the compilation in Table 1 also includes diuretics applied more rarely in hypertension therapy.

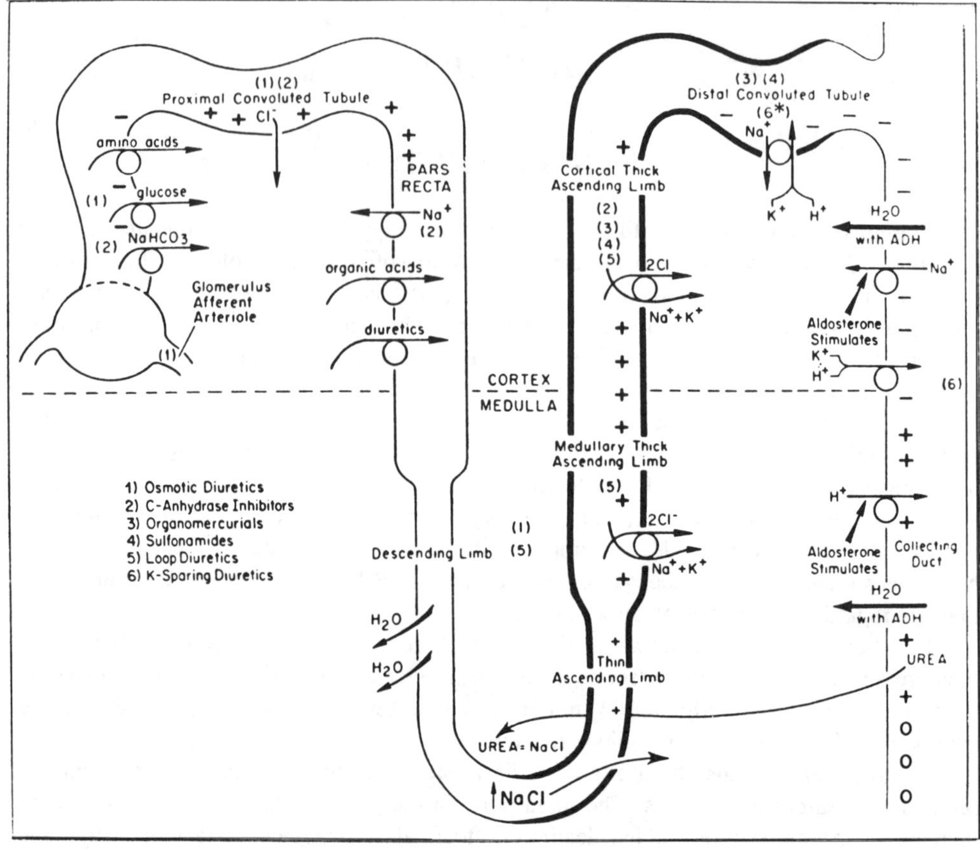

FIGURE 1. Schematic of major transport processes along the mammalian nephron and principal sites of action of the various groups of diuretics. The diuretics groups are referenced in the inset. (From Kokko, J.P., *Am. J. Med.*, 77 (Suppl.), 11, 1984. With permission.)

II. THIAZIDE DIURETICS

A. STRUCTURE

The skeleton of the thiazide diuretics is benzo-1,2,4-thiadiazine **(1)**. There is a sulfamoyl group at position C_7 and the S-1 atom is oxidized to a sulfone group **(2)**:

(1)
benzothiadiazine
skeleton

(2)

The presence of these two electron-attracting groups means that thiazide derivatives have a definitely acidic character. Other consequences of these structural features are the electron-attracting property of C-3 and the relatively low stability of the thiadiazine ring. Ring opening takes place even at room temperature in aqueous solution. This degradation

TABLE 1
Diuretics in Antihypertensive Therapeutic Use

Generic name	Chemical structure	Proprietary name	Therapeutic use	Doses
Carbonic anhydrase inhibitors				
Acetazolamide (USP)	$H_3C-OC-HN$, S, SO_2-NH_2, N—N	Diamox, Acetamox, Donmox, Fonurit, Lediamox, Sulfadiurine, Zolmox	Used in the treatment of glaucoma, for alkalinization of the urine in the management of periodic paralysis, even when associated with hypokalemia. Rarely used as a diuretic.	The range of effective dose: 250—500 mg once a day or every other day.
Methazolamide (USP)	CH_3CON, S, SO_2-NH_2, N—N, CH_3	Neptazane	Given in glaucoma to reduce intraocular pressure, and as an adjuvant in epilepsy. Can produce an electrolyte imbalance leading to acidosis.	50—100 mg 2 or 3 times daily.
Ethoxzolamide	C_2H_5O, S, SO_2-NH_2, N	Cardrase, Ethamide	Acts similar to methazolamide. Inhibitory potency is 2 times that of acetazolamide.	62.5—250 mg 2 to 4 times daily.
Dichlorphenamide (USP)	Cl, Cl, SO_2-NH_2, SO_2-NH_2	Daranide, Oratrol	Given in acute and chronic glaucoma to reduce intraocular pressure.	Usual dose: 100 to 200 mg daily.

TABLE 1 (continued)
Diuretics in Antihypertensive Therapeutic Use

Generic name	Chemical structure	Proprietary name	Therapeutic use	Doses
Benzothiazides				
Chlorothiazide (USP)		Aluren, Chlorosal, Chlorurit, Chlorothiazide, Diuril, Eudrin, Minzil, Niagarim, Saluren, Urinex	Induces sodium excretion, lowering the blood pressure. Also decreases the peripheral vascular resistance by direct action on the arteriolar smooth muscle. Used in mild to moderate hypertension, and to reduce the edemas.	500—1000 mg daily. Duration of action: 6—12 h.
Hydrochlorothiazide (USP)	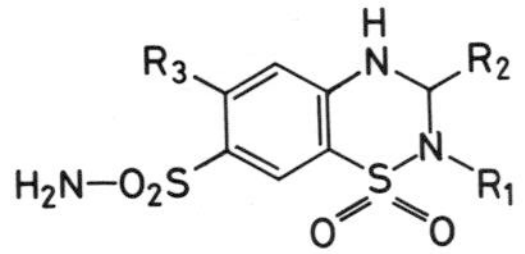 R_1: H; R_2: H; R_3: Cl	Aquarius, Chemiural, Chlorodiuril, Diclotride, Diurex, Esidrix, Hydril, Hypothiazid, Hydro-Diuril, Monuril, Nafril, Neo-Saluretic, Panurin, Tenzide, Oretic	Carbonic anhydrase inhibitory potency is milder, but the natriuretic activity is 5—10 times stronger than that of chlorothiazide.	25—200 mg daily/1 or 2 times. Duration of action: 12—18 h.
Benzthiazide (USP)	R_1: H; R_2: 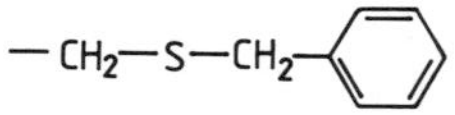 $-CH_2-S-CH_2-C_6H_5$; R_3: Cl	Aquapres, Benzthiazid, Dihydrex "Astra" Exosalt, Exua, Freeuril, Regulon, Urese	More active than chlorothiazide, but its electrolyte excretion pattern is similar to that of hydrochlorothiazide. Most serious side effects are hypokalemia and other electrolyte imbalances.	25—250 mg daily. Duration of action: 12—18 hr.

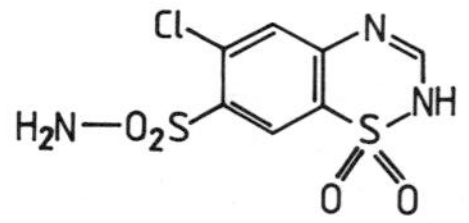

Bendroflumethiazide (USP)	H	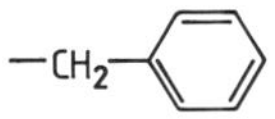 $-CH_2-C_6H_5$	CF_3	Naturetin, Aprinox, Bendroflux, Benuril, Diurex, Fluoro-Niagar, Plusuril, Salural, Sinesalin	One of the most potent diuretic and antihypertensive thiazide. Possible side effect: hypokalemia.	2.5—15 mg daily. Duration of action: more than 18 h.
Hydroflumethiazide (USP)	H	H CF_3		Saluron, Bristurin, Ademil, Diratyl, Diucardin, Fluoridril, Flutizide, Plusurin, Rodiuran	Qualitative and quantitative pattern equal to that of hydrochlorothiazide.	25—50 mg daily. Duration of action: 18—24 h.
Trichlormethiazide (USP)	H	$CHCl_2$	Cl	Metahydrin, Naqua, Datril, Hydropresse, Esmarin, Tridurex	About 10 times stronger diuretic than hydrochlorothiazide.	2—8 mg daily. Duration of action: more than 24 h.
Methyclothiazide (USP)	CH_3	CH_2Cl	Cl	Enduron, Duretic, Naturon, Aquatensen	Maximum action is developed quickly and is continuous, so a single daily dose is enough to obtain the therapeutic effect. Potentiates the action of other antihypertensives.	2.5—10 mg daily. Duration of action: more than 24 h.
Polythiazide (USP)	CH_3	$CH_2SCH_2CF_3$	Cl	Renese, Drenusil, Lotense, Nephril	In small dose is as effective as hydrochlorothiazide, but the principal effect is much rather the secretion of sodium and chloride than that of potassium and bicarbonate.	1—4 mg daily. Duration of action: 24—48 h.
Cyclothiazide (USP)	H	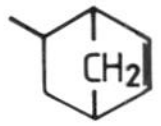 CH_2	Cl	Anhydron, Aquirel, Doburil, Renazide, Tensodiural	A potent, long acting diuretic.	1—6 mg daily. Duration of action: 18—24 h.

TABLE 1 (continued)
Diuretics in Antihypertensive Therapeutic Use

Generic name	Chemical structure	Proprietary name	Therapeutic use	Doses
Cyclopenthiazide	H —CH_2—(cyclopentyl) Cl	Navidrex	Indicated for the treatment of mild to moderate hypertension, in acute and chronic heart failure and edema associated with renal or liver disease. It is well tolerated, but can cause allergy, urticaria, nausea, vomiting, headache.	0.25—0.5 mg daily, increasing to a maximum of 1.5 mg.
Other sulphonamides Metholazone	Cl, H_2NO_2S, H, N, R_1, H, N, R_2, O R_1 CH_3; R_2 (2-methylphenyl) CH_3	Zaroxylin, Metenix, Zaroxolyn, Diulo	A diuretic for use in the treatment of mild and moderate hypertension, and also in cardiac, renal, and hepatic edema, ascites, or toxaemia of pregnancy	Usual dose 2.5—5 mg daily. Dosage range: 1—10 mg. Duration of action: 24 h.
Quinethazone (USP)	C_2H_5 H	Aquamox, Hydromox, Diurene, "Enteropica" Hydrox, Idrokin	Inhibits the Na and Cl reabsorption. In the management of hypertension it is used occasionally with other antihypertensive agents.	Usual dose: 50—100 mg once a day.

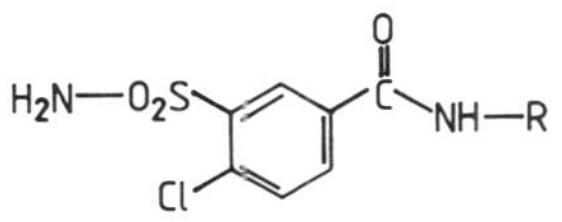

Alipamide	$H_2N—O_2S$, Cl, $C(=O)—NH—R$ R $N(CH_3)_2$			
Clopamid	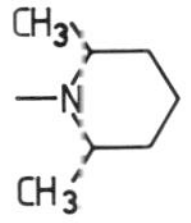 CH_3, —N, CH_3	Brinaldix, Adurix, Aquex "Sandoz"	Action is similar to that of high ceiling diuretics. Indications: hypertension, edema of cardiac, renal, or hepatic origin.	Usual dose: 5—10 mg daily, increasable up to 20 mg.
Indapamide	(See under "Loop diuretics")	(see under "Loop diuretics")		
Thiamizide	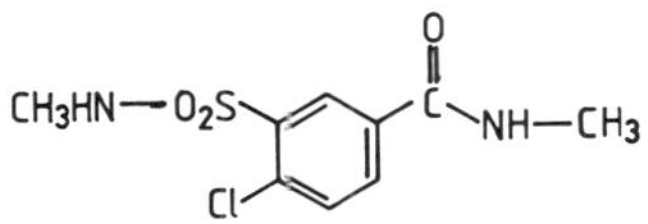 $CH_3HN—O_2S$, Cl, $C(=O)—NH—CH_3$	Diapamid, Vectren		
Xipamid	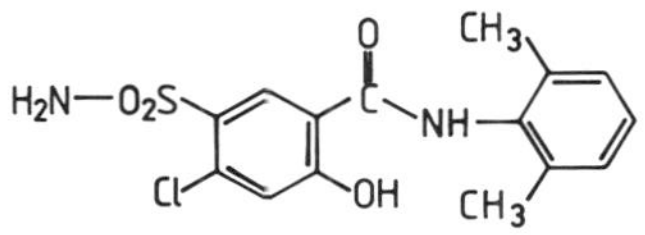 $H_2N—O_2S$, Cl, OH, $C(=O)—NH$, CH_3, CH_3	Aquaphor, Aquaphoril		

TABLE 1 (continued)
Diuretics in Antihypertensive Therapeutic Use

Generic name	Chemical structure	Proprietary name	Therapeutic use	Doses
Chlorthalidone (USP)	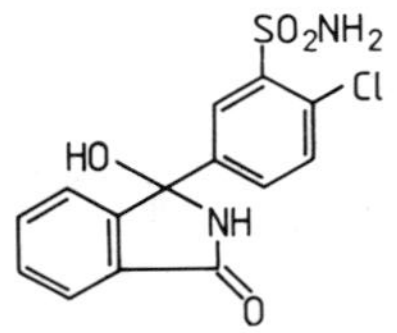 SO_2NH_2, Cl, HO, NH, O	Hygroton, Cloroftalidona, Famolin, Igrolina, Oradil, Saluretin, Urolin, Zambesil Thalitone	A potent, long acting diuretic and antihypertensive agent. Average therapeutic doses give primarily a saluretic effect with a minimal loss of potassium and bicarbonate. Electrolyte excretion pattern is similar to that of benzothiazides.	Usual dose: 100 mg once daily. Dose range: 50—100 mg/day. Duration of action: 24—72 h.
Loop/high ceiling/diuretics Furosemide (USP)	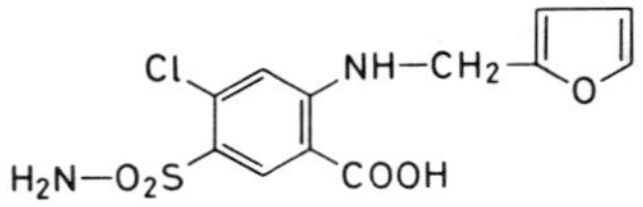 Cl, $NH-CH_2$, O, H_2N-O_2S, COOH	Lasix, Diuremid, Frusemide, Furix, Kinex, Protargen, Salix, Transit, Urosemid	Inhibits Na reabsorption through the renal tubules, including the loop of Henle, which may account for its high potency and effectiveness in case of glomerular filtration. Promotes potassium excretion. Has a potent blood pressure lowering property.	Usual dose: 20—80 mg once daily. Duration of action: 6—8 h.
Indapamide	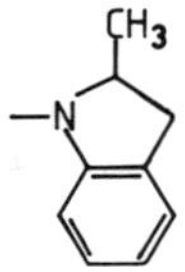 CH_3, —N	Natrilix, Metindamide, Fludex, Lozol	For treatment of essential hypertension. Can be combined with other antihypertensives too. Mechanism of action: reduction in peripheral arterial resistance and normalization of vascular hyperactivity. By overdosage electrolyte disturbances, hypertension, and muscular weakness may appear.	Usually a single dose daily of 2.5 mg is efficient.

Bumetanide	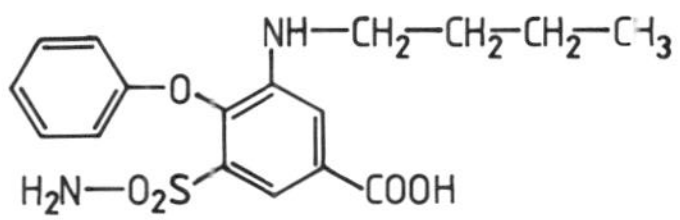	Fordiuran, Burinex	Activity is similar to that of Furosemide, but has to be administered in larger doses.	Usual dose: 1 mg daily (single dose)
Muzolimine	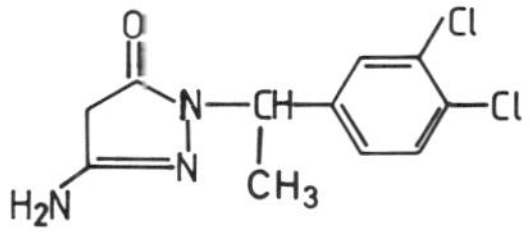	"Bai g 2821"	As effective as furosemide and ethacrynic acid. Active in the proximal tubulus and the middle part of loop of Henle. Quick and long acting.	Initial dose: 10 mg up to 80 mg/day.
Ethacrynic acid (USP)	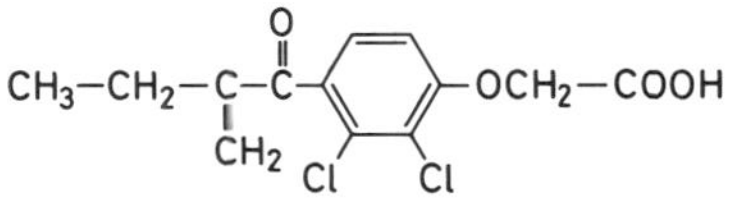	Edecrin, Uregyt	An extraordinary, active high ceiling diuretic. Acts very quickly, but the duration of action shorter than that of thiazides. Used in combination with K-sparing diuretics. Side effects: weakness/electrolyte loss/nausea, anorexia, and gastrointestinal disturbances, ototoxicity.	50—100 mg/day, occasionally 200 mg daily.
Potassium-sparing diuretics Triamterene (USP)	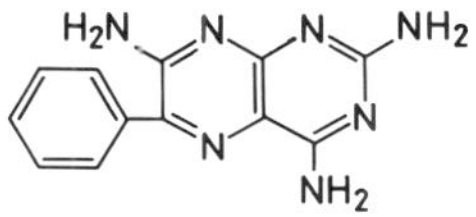	Dyrenium	As a potassium-sparing diuretic, increases the excretion of sodium. Co-administered with hydrochlorothiazide, optimum value of potassium has been measured in plasma.	Usual initial dose is 100 mg 2 times/day. The maximum dose is 300 mg.

TABLE 1 (continued)
Diuretics in Antihypertensive Therapeutic Use

Generic name	Chemical structure	Proprietary name	Therapeutic use	Doses
Amiloride Amiloride Hydrochloride (USP)	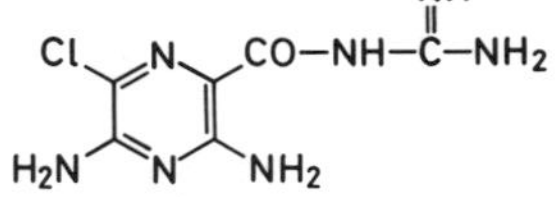	Moduretic, Midamor	Similar to triamterene, increases the excretion of sodium. Increases moderately the pH of the urine, which indicates exchange of H-ions. It is not an aldosterone antagonist, and does not inhibit the carbonic anhydrase.	Usual dose: 5—10 mg/day.
Aldosterone antagonist Spironolactone (USP)	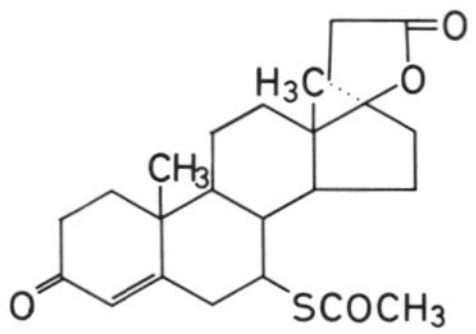	Aldactone, Aldonar, Espironolacton, Nefrolactone, Osiren, Spirolactone, Verospiron	A competitive antagonist of aldosterone, inhibiting sodium retention and not causing loss of potassium. Site of action: distal renal tubules.	25—100 mg 3 or 4 times daily. Duration of action: 8—12 h.

is enhanced in acidic or alkaline medium and/or by elevation of the temperature. This is important as regards the stability and expiration date of benzothiadiazine derivatives.

B. HISTORY AND SYNTHESIS

The diuretic effect of sulfonamide chemotherapeutics accompanied by acidosis was observed in the late thirties.[9] This action could be attributed to the inhibition of carbonic anhydrase.[10,11] Among the several hundred other compounds synthetized in the sulphonamide series,[13,14] acetazolamide was found to be a highly effective agent,[15] and it became the first clinically useful diuretic in the carbonic anhydrase inhibitor class. A rather large number of acetazolamide derivatives were prepared in the following years,[16-22] but few of them offered advantages (at least as diuretics) over the parent compound. Even so, their side-effects (e.g., strong bicarbonate diuresis) outweighed these advantages, as in the case of benzolamide **(3)**:

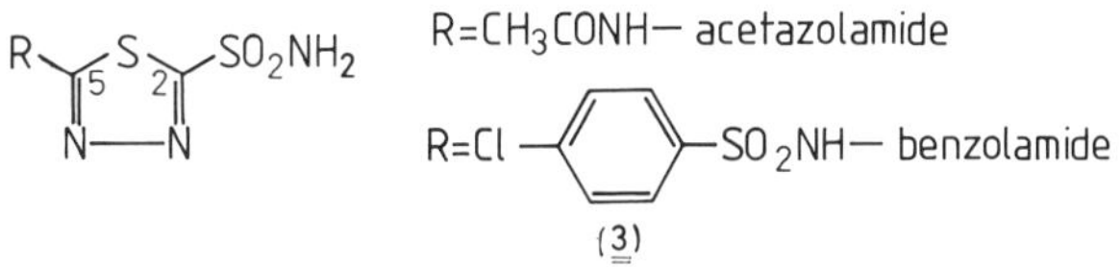

R=CH_3CONH— acetazolamide

R=Cl—C_6H_4—SO_2NH— benzolamide

(3)

Of a large number of prepared derivatives, acetazolamide and two other compounds were found to be suitable as therapeutics.

Etoxzolamide **(4)** is a benzothiazole derivative,[21,22] while methazolamide **(5)** can be regarded as a tautomeric N-methyl isomer of acetazolamide:[18]

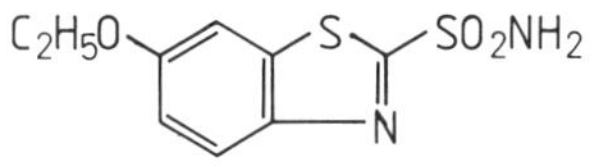

(4) 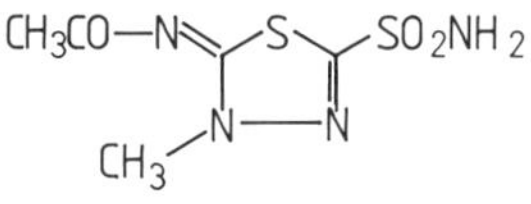 (5)

Although acetazolamide might be regarded as a safe, orally active diuretic, its side-effects have led to its use as an antihypertensive agent being almost negligible. Nevertheless, it played a significant role in the development of renal physiology and pharmacology.[23] Moreover, the *in vitro* carbonic anhydrase*-inhibiting activity remained one of the important parameters in diuretic effects for several years. At the same time, neglect of the *in vivo* saluretic effect was a source of some error in the drug research and evaluation of the activity.

The systematic testing of benzenesulfonamido derivatives revealed a very strong natriuretic effect of 4-carboxybenzenesulfonamide **(6)**, accompanied by a relatively weak carbonic anhydrase-inhibiting property.[24] Accordingly, attention in further research was directed to derivatives of benzenesulfonamide substituted with substituents similar to the carboxy group. Sprague[25] introduced a second sulfamoyl group into benzenesulfonamide: 1,4- and even more so 1,3-disulfonamidobenzene **(7,8)** displayed very strong carbonic anhydrase-inhibiting and diuretic effects.[26]

* Carbonic anhydrase (a more proper name is carbonic acid anhydrase) is a zinc-containing enzyme with a molecular mass of 30,000 (approximately 260 amino acid residues). It contains one Zn atom. The central Zn^{2+} is bonded to three imidazole (histidine) nitrogens and one water molecule or OH group. One of the important physiological roles of carbonic anhydrase is to catalyze the hydration of carbon dioxide in the following reaction:

$$CO_2 + H_2O \rightleftharpoons H_2CO_3 \rightleftharpoons HCO_3^- + H^+$$

Its acceleration of carbonic acid formation is very significant:
Rate constant k_1 (without carbonic anhydrase) $3{:}10^{-2}\ s^{-1}$
Rate constant k_2 (with carbonic anhydrase) $6{:}10^5\ s^{-1}$

TABLE 2
Diuretic Activity of Benzenedisulfonamide Derivatives

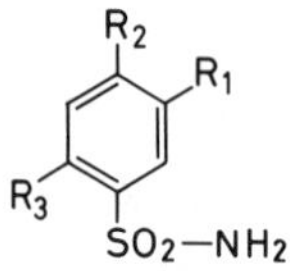

	R_1	R_2	R_3		Relative carbonic anhydrase-inhibitory activity
	H	NH_2	H		1
	H	COOH	H	**(6)**	3
	H	SO_2NH_2	H	**(7)**	40
	SO_2NH_2	H	H	**(8)**	
	SO_2NH_2	Cl	H	**(9)**	93
	SO_2NH_2	Cl	NH_2	**(10)**	3.7

The related investigations resulted in predicting the possibility of the preparation of saluretic diuretics with decreased carbonic anhydrase inhibition among the disulfonamides.[27] The breakthrough in the latter respect was the introduction of substituents into positions C-6 and C-3 of benzenedisulfonamide; the most successful derivatives were found to be the 6-chloro- and the 3-amino compounds **(9,10)**, as may be seen in Table 2.

During the following years, the study of the structure-activity relationship among the substituted disulfonamide derivatives continued,[28-33] but resulted in only a few compounds which are rarely used nowadays.

A cornerstone was reached in the development when the diuresis-enhancing effect of aminoacylation was explored. This proved even more interesting when the acylation-induced ring closure of **(10)** was observed. Acylation with formic acid resulted in 6-chloro-1,2,4-benzothiadiazine-7-sulphonamide 1,1-dioxide (chlorothiazide), a product which may exist in two tautomeric forms **(11/a,11/b)**. Further investigation revealed that form **11/b** predominates in ethanol solution, and form **11/a** in aqueous solution:

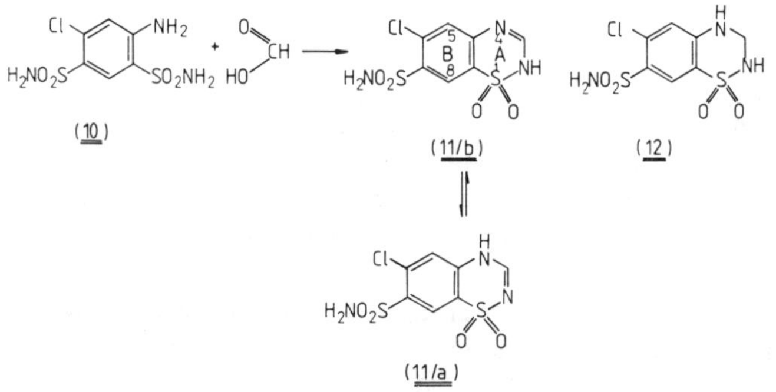

(10) (11/b) (12) (11/a)

The protons are exchanged by Na^+ and K^+, maintaining the reabsorption of these cations. The diuretic effect elicited by the inhibition of carbonic anhydrase is based on the decrease in Na^+ and K^+ reabsorption. The increasing $NaHCO_3$ concentration leads to an increase in urine volume. However, the urine becomes alkaline ($NaHCO_3$), accompanied by acidosis and potassemia as the primary side-effects.

As the first step in the enzyme-catalysed CO_2 hydration, the formation of a molecular complex (5/a) is assumed between carbonic dioxide and carbonic anhydrase.

The inhibitory effect of sulfonamides is based on the structural similarity of the sulfonamide group to carbonic dioxide and the competitive hindrance of the molecular complex formation:

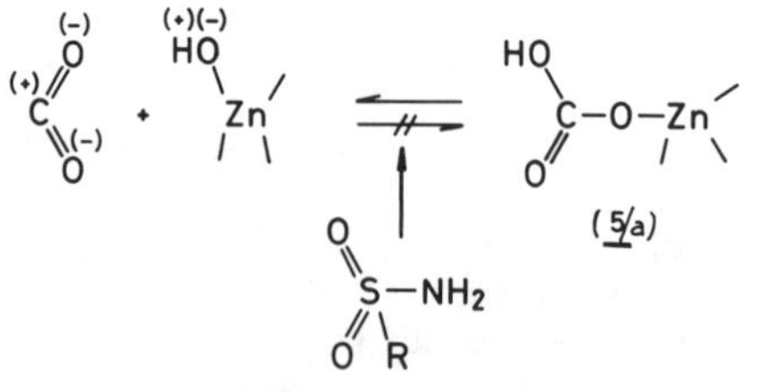

(5/a)

The idea of acylation was inspired by the experience that for aromatic or heteroaromatic sulfonamides bearing a primary amino group, the acylation of the latter resulted in a great increase in carbonic anhydrase-inhibitory activity.[13,24] This experience drew attention to the heterocycles formed in this acylation reaction.

These synthetic results and the subsequent structure-activity relationship research resulted in the preparation of hydrochlorothiazide **(12)** and some other effective diuretics. This development started a new chapter in the therapeutic application of diuretics.

C. STRUCTURE-ACTIVITY RELATIONSHIP

It is worth noting that a method for thiadiazine formation from sulfamoyl-aniline derivatives had been known previously,[35,36] but the early experiments did not lead to the development of new diuretics; this was due to the lack of knowledge about the sine qua non feature of the unsubstituted 7-sulfamoyl group. Although numerous laboratories carried out intensive research work to prepare diuretic sulfonamides, the pioneering role of the research team of Novello and Sprague (Merck, Sharp and Dohme, U.S.A.) should be emphasized.[37] Further, the research performed in the CIBA laboratories (Summit, NJ), which culminated within a year of Novello and Sprague's publication with the synthesis of hydrochlorothiazide, also merits special attention.[38] The condensation of formaldehyde with **(10)** resulted in a crystalline compound, hydrochlorothiazide **(12)**, instead of a polymeric product which might have been expected. The reaction took place almost spontaneously in anhydrous diethylene glycol dimethyl ether when a catalytic amount of hydrogen chloride was used. The reduction of chlorothiazide with sodium borohydride led to the same compound **(12)**, confirming the structural identity.

Hydrochlorothiazide was found to be 10 to 20 times as active as chlorothiazide and to have a lower toxicity.

At first sight, the great significance of the C-6 substituent seems surprising, since it can hardly have an active role in the interactions between the receptor and thiazides. Nevertheless, it is a fact that C-6-unsubstituted derivatives have no, or only a marginal, effect. Whereas substitution by Cl, Br or CF_3 on C-6 elicited definite and almost equivalent diuretic effects, the F and I derivatives showed much weaker activity. This suggests that certain C-6 substituents are essential for the diuretic activity, and may act via a certain steric effect and/or van der Waals forces; their action by electronic forces is hardly to be expected.

The above findings hold for the chlorothiazide and hydrochlorothiazide derivatives as well.[39] The qualitative picture about the role of the C-6 substituent seems to be confirmed by the quantum-chemical investigations of Orita et al.[40] The electronic state of thiazide and other diuretics was investigated in this research group during a period of several years.[41-43] Utilizing the geometric data on hydrochlorothiazide,[44] they obtained quantum-chemical parameters for hydrochlorothiazide by the CNDO/2 method. Further calculations revealed a close relationship between the diuretic activity and the van der Waals volume of the C-6 substitutent. The hydrophobic parameter of Hansch and the formal charge on position C-7 in hydrochlorothiazide seemed to contribute in part to the diuretic activity:[40]

$$\ln y = 0.2965 \cdot 10^{-3}\pi + 7.62535 \text{ V.W.} + 15.7681 \, F_{C\text{-}7} + 0.98501$$

$$r = 0.980 \; (p \; 0.01);$$

where y = diuretic activity;

π = hydrophobic parameter (Hansch);

V.W. = van der Waals volume (Bondi) of C-6 substituent;

$F_{C\text{-}7}$ = formal charge on C-7 substituent.

It was proposed by Orita et al.[40] that thiazide diuretics possess an electron-accepting property and that this should contribute to their diuretic activity.

The essential structural element of thiazides for significant diuretic activity is C-7. The best substituent at this position is the sulfamoyl group. Substitution (methyl, dimethyl, etc.) or a change to other, even stronger electronegative substituents (carboxy, carbamoyl, nitro, etc.) yielded compounds with markedly reduced or only negligible diuretic activity.

The best position for the free sulfamoyl group is C-7; the C-5 and C-6 sulfamoyl derivatives have much weaker activity.

The most interesting point in structure-activity relations is the olefinic bond between C-3 and C-4, which is the only structural difference between chlorothiazide and hydrochlorothiazide-type thiazides. Which of the physical-chemical properties is changed by hydrogenation of the C-3—C-4 bond, so as to cause an increase in diuretic and a decrease in carbonic anhydrase-inhibitory activity of more than tenfold? First of all, it is reasonable to consider the change in the acidic strength:

	pK_1		pK_2		% of dissociated form at pH 6	7.4	8
Chlorothiazide	6.5	$\sigma = +0.06$ n = 10	9.7	$\sigma = \pm 0.25$ n = 14	24.03	80	97.09
Hydrochlorothiazide	8.9	$\sigma = \pm 0.23$ n = 15	11.2	$\sigma = 0.30$ n = 19	0.126	3.07	11.19

Note: The determination of pK_1 and pK_2 was made by ultraviolet spectrophotometric method. λ: 311, 292, 270 nm. I = 0.2. (Private information by Takács-Novák, K., Dept. of Pharm. Chemistry, Semmelweis Medical University, Budapest, 1989).

A consequence of these pK values is a complete or significant dissociation (ionic state) of chlorothiazide at the almost neutral or mildly alkaline pH of the intestine, extracellularly and in the kidney. At the same time, at pH 7.4 hydrochlorothiazide is only 3% ionized. This sharp difference in ionizing ability may cause correspondingly sharp differences in such properties of the two compounds as partition, diffusion, reactivity (decisive in the absorption), penetration, receptor binding, etc. Another factor may be the possibility of two-point binding by the *meta*-NH groups in the case of hydrochlorothiazide.

The activity is enhanced not only by the substituents at C-6, but also by the substituents at C-3. Generally, the C-3-substituted thiazide derivatives are more active than the parent compounds in both the chlorothiazide and the hydrothiazide series. Clinically, the hydrothiazides are much more important. With regard to the ease of synthesis, large series of such compounds have been prepared[45-49] and tested for diuretic and for carbonic anhydrase-inhibitory potency. Numerous among these structural variants have been introduced into clinical practice (see Table 1).

One of the most generally used method for the preparation of the hydrothiazides is the condensation of a substituted orthanilamide **(13)** with the appropriate aldehyde or its acetal. Reaction conditions suitable for formation of the expected 3-substituted hydrothiazides vary,

TABLE 3
Comparative Data on Some Highly Potent Thiazides

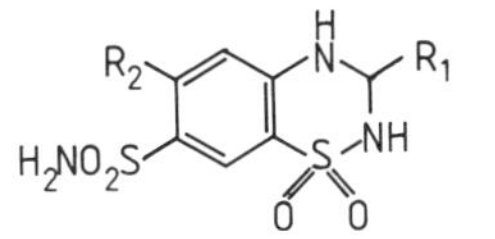

Relative activity in carbonic anhydrase inhibition[26]	Diuresis[26]	R_1	R_2	
0.6	10	-H	-Cl	hydrochlorothiazide
1.0	1000—2000	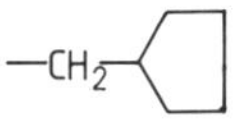 $-CH_2$-cyclopentyl	-Cl	cyclopenthiazide
0.2	100—200	$-CHCl_2$	-Cl	trichlormethiazide
0.04	100—200	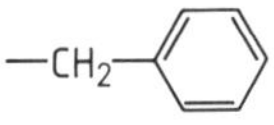 $-CH_2$-phenyl	$-CF_3$	bendroflumethiazide

depending on the nature of the aldehyde. In general, the following conditions were found to be the most appropriate:

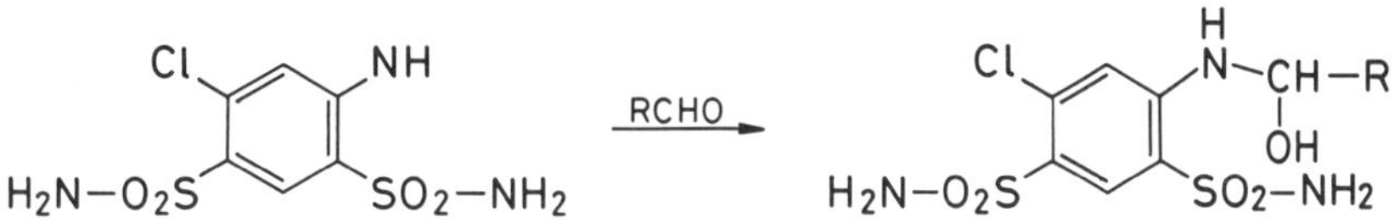

(13)

R = alkyl; reflux **13** with an excess of the aldehyde, with or without acidic (HCl) catalysis, R= aryl; fusion of **13** and the aldehyde (~200°C).

It is worth noting that in some cases different amounts of an anil derivative **14** were obtained[49] besides the main product, the 3-substituted hydrothiazide **15**:

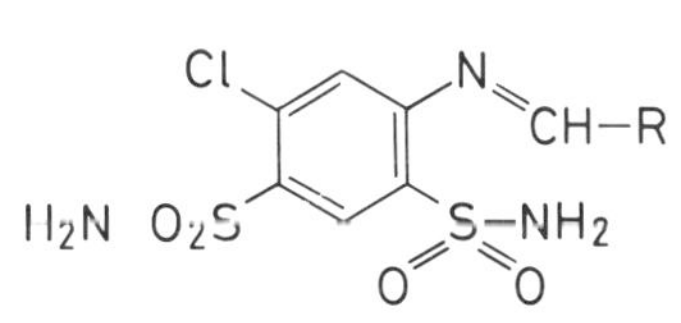

(14) 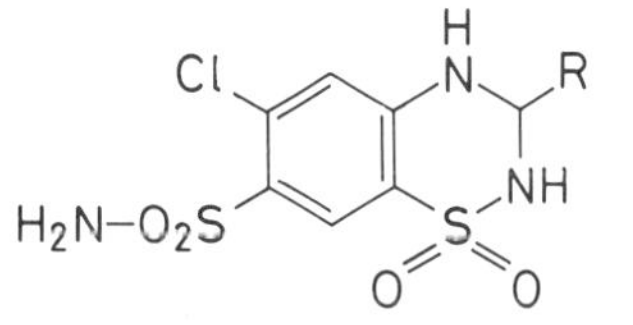 (15)

As examples, cyclopenthiazide, trichlormethiazide, and bendroflumethiazide may be mentioned here. The very high diuretic potency relative to that of hydrochlorothiazide and the low carbonic anhydrase-inhibitory activity are striking (Table 3).

The interpretation of the molecular basis of the effect of the 3-substituent in increasing the diuretic potency seems to be an unsolved problem.

Direct interaction between the substituent at C-3 and a receptor site must be neglected, and similarly, no mesomeric or inductive effect of the substituent at C-3 would be considered. It remains to assume some change in lipophilicity and/or steric influence of the substituent at C-3, either directly on the shape of ring A or through steric shielding of N-2 and N-4

TABLE 4
Some Properties of Thiazide Diuretics

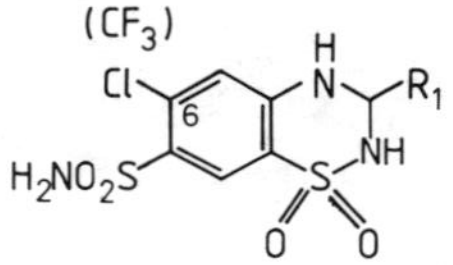

(CF3) Cl 6 H N R1 NH H2NO2S S O O

R_1	Name	Mp°C	pK_1	pK_2	Solubility in H_2O mol/1	Partition log P	Relative diuretic activity
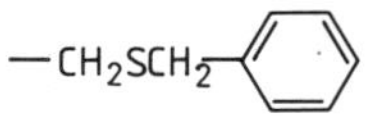 —CH_2SCH_2—	Benzthiazide	240				1.46	10
–H	Hydrochlorothiazide	269—270	7.0	9.2	$2.0 \cdot 10^{-3}$	0.0	10
–H	Hydroflumethiazide[a]		8.9	10.7			10
–$CHCl_2$	Trichlormethiazide	280—281[49]				0.62	100
–CH_2S–CH_2CF_3	Polythiazide					1.13	500
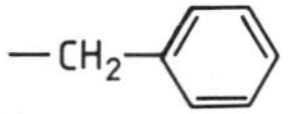 —CH_2—	Bendroflumethiazide[a]	228[49]	—	—	$9.4 \cdot 10^{-5}$	1.19	200
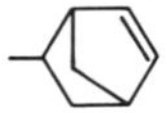	Cyclothiazide		—	—	—	1.80	250
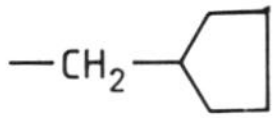 —CH_2—	Cyclopenthiazide	—	—	—	$1.30 \cdot 10^{-4}$	—	1000
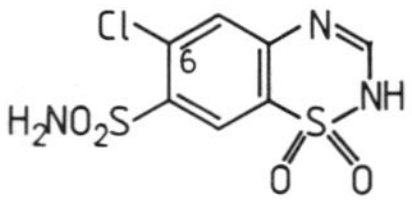 Cl 6 N NH H2NO2S S O O	Chlorothiazide	340	3.4	8.5	$7.4 \cdot 10^{-4}$	—	1

[a] CF_3 at position 6.

against metabolic processes. Some informative numerical data on the lipophilicity-influencing effects of substituents at C-3 may be seen in Table 4. The partition coefficient (log P) of the derivatives bearing apolar substituents at position 3 ranged between 1.50 and 2. Substitution at N-2 yielded very active and therapeutically useful compounds in the saturated series.[50,51] Methyclothiazide and polythiazide, two commercially available compounds, have a methyl group at position N-2.

Whether the substitutent effect is limited to the increase of lipophilicity (absorption) followed by metabolic demethylation is an unanswered question.

D. PROPERTIES

The therapeutically used chlorothiazide and hydrothiazide derivatives are white crystalline odorless powders. As for sulfonamides in general, their melting points are high (see Table 4). They are very slightly soluble in water and only somewhat better soluble in ethanol; some of them are sparingly soluble in methanol; most of them dissolve well in acetone, but are insoluble in ether, in chloroform, or in benzene. Due to their acidic character, they are soluble in alkaline solutions. The acidic property of thiazides stems from the endocyclic sulfonamide groups. The acidity of the free sulfonamido group is much weaker (see Table 4).

In connection with their acidic character, thiazide diuretics form complexes with copper(II) ions;[52] a variety of species have been identified in the equilibrium mixture. The relatively low values of the stability constants speak against chelate complex formation (which was assumed earlier), but rather indicate the role of the sulfonamide group in complex formation:

Log Stability Constants of Copper(II) Complexes of Hydrochlorothiazide (L)

CuL	CuHL	Cu(OH)L	$Cu(OH)_2L$	CuL_2	CuH_2L_2
4.09	5.72	6.82	7.50	6.76	12.72

Among the investigated sulfonamides, hydrochlorothiazide forms rather stable Ag(I) complexes.[53] The small differences between the stabilities of the sulfonamide derivatives suggest that the unsubstituted $-SO_2NH_2$ group takes part in the complex formation. It is an interesting observation that the diuretic activity and complex stability seem to be correlated.[53] The formation of the Co(II) complex of hydrochlorothiazide may be utilized for spectrophotometric determination of the compound.[54] The absorbance of the resulting blue-violet complex may be measured at 450 nm. Beer's law is obeyed for 0.2 to 1.12 $\mu g\ ml^{-1}$ hydrochlorothiazide.

The data available on the ultraviolet absorption spectra of thiazides are rather contradictory. Nevertheless, the hydrochlorothiazides and chlorothiazides can be differentiated via their UV spectra (Table 5 and Figure 2). Hydrochlorothiazides were found to have three absorption bands by Topliss et al.,[49] but only two peaks were reported by other authors.[53-56]

E. ANALYSIS

Several test-tube reactions are known for the identification of thiazides. These are based mainly on the (1) heteroatom (S, Cl, N) content, (2) substitution of the benzene ring, or (3) hydrolytic opening of the heterocycle.

For detection of the heteroatom content, the simple and effective procedures published by Rosenthaler[59] and Kala[60] should be mentioned. For the detection of nitrogen, 50 mg of

TABLE 5
Absorption Bands of Thiazides[a]

Name	Adsorption maxima, nm I	II	III	Ref.
Chlorothiazides	223—227 (25,000—30,000)	275—285 (8,000—12,000)		49 M
Chlorothiazide	224 (13,300)	280 (5,030)		55 E
	235	281		57 w
		280 (10,650)		59 M
Flumethiazide	223	279		57
Hydrochlorothiazides	223—228(m) (35,000—50,000)	266—275 (19,000—28,000)	310—320 (2,000—8,000)	49 M
Hydrochlorothiazide		271	318	56
	224 36,800)	270 (20,350)		55 E
	230	271	317	57 w
		269 (19,890)	315 (2,950)	58 M
Hydroflumethiazide		272 (19,880)	324 (3,910)	58 M
Cyclopenthiazide	229	272		57 w
		272 (22,220)	316 (3,040)	58 M
Cyclothiazide	224	272	312	59
	270 (21,180)	315 (2,790)		58 M
Bendroflumethiazide	221	274		57 w
		272 (23,600)	324 (4,170)	58 M
Methylclothiazide	228	272		57 w

[a] Solvents: M = methanol, E = ethanol, w = water.

the thiazide is heated in a dry test-tube with 150 mg of a 9:1 mixture of dry sodium carbonate and sodium thiosulfate until the mixture melts. The residue is extracted with 2 *M* nitric acid and filtered. On the addition of Fe^{3+}, a red color forms, $Fe(SCN)_3$. The chlorine and sulphur contents may be detected after fusion with a 9:1 MnO_2 + KOH mixture by the addition of nitric acid plus silver nitrate (Cl) or hydrochloric acid plus barium chloride to an aliquot of the filtrate.[59]

The aromatically bound chlorine can be split off for the purpose of quantitative determination by oxidation with potassium peroxodisulfate in aqueous potassium hydroxyde (30%) solution.[61] The solution is refluxed until the transitional color of the solution disappears. The excess of the oxidizing agent is eliminated with hydrazine sulfate and sodium sulfite. After the addition of nitric acid, chloride is titrated with 0.1 *N* silver nitrate solution. Nonoxidizing treatment (e.g., alkaline hydrolysis) does not result in the splitting-off of aromatically bonded chlorine, even on heating. This fact makes possible the selective determination of the aliphatic chlorine in methylclothiazide after alkaline hydrolysis. Reflux under full boiling for one hour results in quantitative cleavage of the chlorine from the

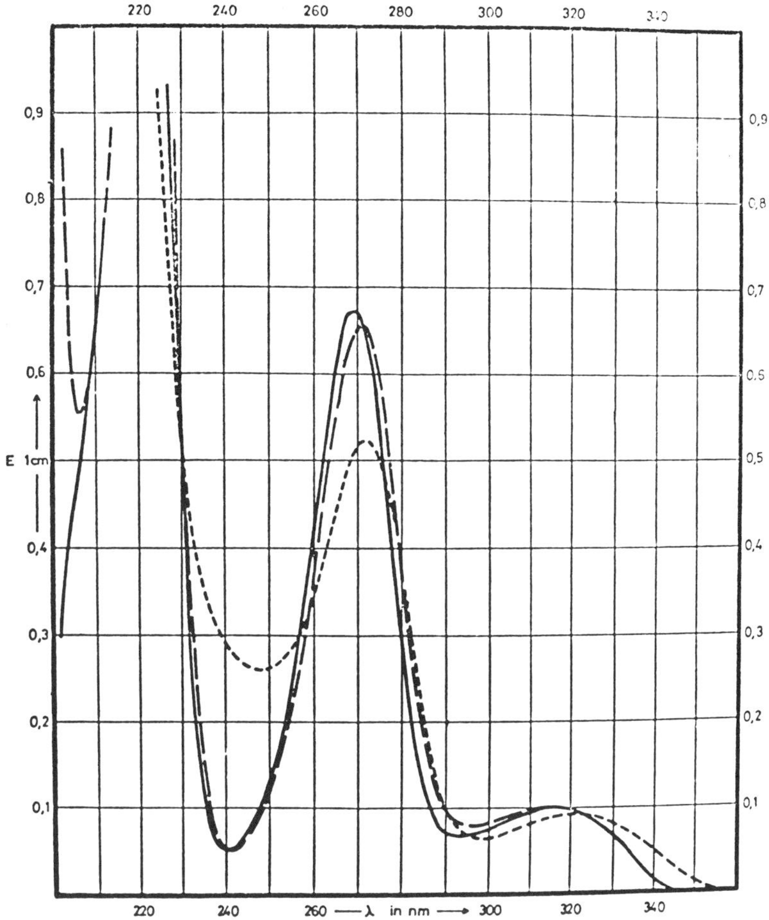

FIGURE 2. The UV spectrum of hydrochlorothiazide. Solvents: —— MeOH; ------ 0,1*N* NaOH; – – – 0.1 *N* Hcl. (From Dibbern, H.-W., *UV and IR-Spektren wichtiger pharmazeutischer Wirkstoffe,* Editio Cantor, Aulendorf, 1978. With permission.)

–CH_2Cl group without any change in the aromatic chlorine content. The one Cl^- anion/mol formed is titrated with 0.1 *N* silver nitrate volumetric solution:

heat, KOH (methanol)

+KCl
+HNO_3
+$AgNO_3$

Equiv. = M_r

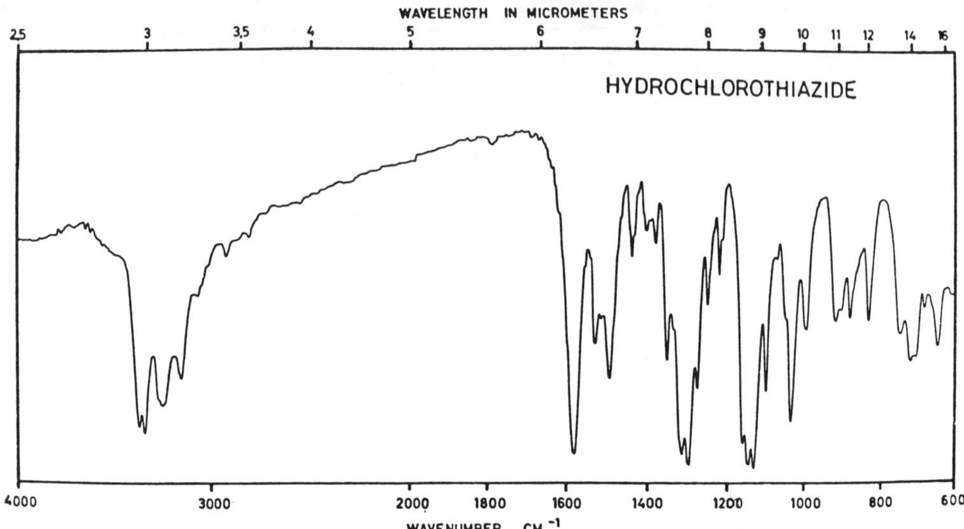

FIGURE 3. The IR spectrum of hydrochlorothiazide. (From Dibbern, H.-W., *UV and IR-Spektren wichtiger pharmazeutischer Wirkstoffe,* Editio Cantor, Aulendorf, 1978. With permission.)

The activated benzene ring of the thiazides may easily be substituted by bromine. The reaction can be used for the quantitative determination of hydrochlorothiazide.[62-64] Interestingly, bromochloride oxidizes hydrochlorothiazide quantitatively, with splitting of the sulfonamido side chain.[62]

Cl, H_2NO_2S, H, N, NH, S, O, O — Br_2 → Br, Cl, H_2NO_2S, + HBr

2BrCl, $2H_2O$ → Br, Cl, Br, $+2HCl + NH_4HSO_4$

Oxidation under milder conditions is also used for the quantitation of thiazides.[65] An excess of potassium hexacyanoferrate(III) or *N*-chlorosuccinimide in acidic medium oxidizes hydrochlorothiazide stochiometrically (at 50°C, or 20°C, with an incubation time of 35 min or 20 min, respectively). The excess of the reagents is measured iodometrically.

Thiazides can be identified with absolute certainty through their UV and IR spectra by comparison with spectra of reference standard substances. The bands generally present in the thiazide UV spectra are included in Table 5. Some characteristic features of the infrared spectra of thiazides include a band at 2.95 to 3.15 μ, which may be attributed to the N - H stretching vibration of the sulfonamide and a band at 7.5 μ assigned to the asymmetric S - O stretching vibration of S - O appears in the vicinity of 8.6 μ. As typical examples, ultraviolet and infrared spectra of hydrochlorothiazide is shown. (Figures 2 and 3).

Because of their weak acidic character, alkalimetry in aqueous solution cannot lead to good results in the quantitation of thiazides. However, under nonaqueous conditions in basic solvents, for example in *n*-butylamine (hydrochlorothiazide, USP XXI) or in pyridine (bendroflumethiazide, USP XXII), they may be determined by alkalimetric titration with 0.1 *N* sodium methoxide, with azo-violet as visual indicator. It is worth noting that under these conditions, 2,4-disulfanyl-5-chloro-(trifluoromethyl-, etc.)methylaniline, a possible acidic impurity in hydrochlorothiazide (bendroflumethiazide) will also be measured. Since the upper limit for the amount of such impurities is 1% or 1.5%, respectively, the official USP method may give somewhat higher results. (This is why the majority of thiazides are determined by a selective HPLC method, according to the USP).*

Undoubtedly, the best way to identify and quantify thiazides of pharmaceutical importance is to use the different chromatographic methods, mainly HPLC. In the reversed-phase system, the separation and determination for the widely used thiazide diuretics (10 to 12 derivatives) may be solved. A few studies have dealt with the identification of thiazide diuretics; most of the papers concentrate on the assay of a single or a small group of thiazides as known constituents in biological fluids.

The most informative publications in the last few years came from De Croo et al.[66,67] In their study of the main parameters of the chromatographic systems, an excellent separation of 12 thiazide compounds was achieved on a LiChrosorb RP 18 column.[66] The best mobile phase was found to be acetonitrile-water 40:60 as eluent, which proved suitable for the separation of all 13 diuretics within 10 min. The use of tetrahydrofuran as a third mobile phase component increased the selectivity in certain cases.

Similarly, a good separation could be obtained with the acetonitrile-1% aqueous acetic acid 35:65 mobile phase on an ODS Hypersil column when 10 thiazide diuretics were chromatographed.[68] Attention should be paid to the marked change in the retention when acetonitrile-water or acetonitrile-aqueous acetic acid was used as mobile phase.[66,68] Within the latter work, a detailed study of the retention reproducibility was carried out on diuretic compounds. Mainly such experimental conditions as the proportion of acetonitrile in the mobile phase, the column temperature, and the brand of column packing material were found to be important.

A good result was achieved by means of HPLC in the separation of thiazide steroisomers. Cyclothiazide could be separated by reversed-phase HPLC into four peaks.[69] The almost identical UV spectra of the eluted substances confirmed the conclusions drawn from earlier TLC and HPLC experiments[66,70] in which the presence of four steroisomers was postulated in cyclothiazide (USP).

HPLC also proved to be an efficient method for the resolution of racemic mixtures of thiazides[71] when a polyacrylamide stationary phase and toluene-dioxane as mobile phase were applied in a low-pressure liquid chromatographic system.

A comparative assay of the sensitivity of HPLC detection by UV absorptivity and by electrochemical (amperometric) means is to be found in the paper by Musch et al.[82]

A relatively small number of publications have appeared on the gas chromatographic assay of thiazides. However, the comparative data of Kim et al.[72] attested to the high sensitivity of flame photometric detection when thiazides and other sulfur-containing drugs were chromatographed on Chromosorb WAW-DMCS covered with 3% SE-30 and a 0.6% QF-1 layer. The resulting linear correlation ranged from 10^{-9} up to 2.10^{-8} mol.

Thin-layer chromatography (TLC) is often a method of choice when the identification of thiazide-containing pharmaceutical products is needed.[74-83] Optimized conditions of TLC separation may be found in the paper by Mistal et al.[78] who used an adsorptive mechanism (silica gel, alumina as stationary phase). In general, the strongest adsorptive activity on silica gel was displayed by chlorothiazide and hydrochlorothiazide owing to the relatively high polar character of these compounds. The sequence of the other compounds largely depended on the polarity of the substituent at C-3. The R_F value for chlorothiazide is lower than that for hydrochlorothiazide, a behavior which may be attributed to the presence of the active N(3) - H in chlorothiazide. Hydrochlorothiazide was recovered with good result by

* In USP XXII HPLC is used also for the determination of hydrochlorothiazide.

TLC with direct spectrometric evaluation from Esidrex™ tablet.[83] Linearity was found in the interval 0.2 to 2.0 μg ml^{-1}.

The great majority of the hydrochlorothiazide derivatives differ structurally in the substituents at C-3, with resulting small differences in polarity. Accordingly, a more reasonable way of TLC separation seems to use methods based on partitioning rather than on the adsorptive mechanism. However, 12 thiazide diuretics (together with several antihypertensive agents) were chromatographed in an adsorptive system by Stohs and Scratchley[81] using silica gel-G as stationary phase and methyl ethyl ketone-*n*-hexane and chloroform-acetone-triethanolamine as mobile phase. As expected, the sequence of the components and the drawbacks of the method were similar to those observed with other adsorptive TLC systems.[78] However, the retention values offer a possibility for the separation of pairs of thiazides and other antihypertensives from combinations.

Several spraying reagents have been used to determine the sensitivity of detection in the case of the most important diuretic and hypoglycemic sulfonamide drugs.[84] Earlier, a fairly detailed survey of the detection methods for thiazides and the other tested antihypertensives was published by Stohs and Scratchley.[81] The detection limits varied from 0.5 μg (Bratton-Marschal reagent) to 5 μg (Dragendorff reagent).

Hydrochlorothiazide may often be found combined with a hypotensive agent in formulated preparations. The HPLC method of Hitscherich et al.[85] for the determination of hydrochlorothiazide and propranolol in tablets seems to be highly sensitive and reproducible (limit of detection 0.1 and 0.4 μg ml^{-1}; recovery 100.5 and 100.2%, respectively; column: Ultrasphere-cyano, mobile phase: acetonitrile −0.05 *M* $NH_4H_2PO_4$ (pH 3.0)). A similar task and analytical solution was reported by Ramana et al.[86] and by Ficarra et al.,[87] in which hydrochlorothiazide and metoprolol were determined by HPLC (μBondapak amino column, 0.001 *N* sodium acetate in methanol as mobile phase; High Speed C_{18} (5 μm) Perkin-Elmer column, 1% acetic acid in acetonitrile as gradient elution) in dosage forms. No previous separation was needed for the spectrophotometric determination of hydrochlorothiazide (λ_{max} = 270, 293 nm) or amiloride (λ_{max} = 360 nm) in combination.[88]

Dual-wavelength spectrophotometry was also applied for the determination of hydrochlorothiazide and triamterene in two-component tablets.[89] The absorbance was measured at 271, 349 and 367 nm. Beer's law was obeyed for 1 to 6 μg ml^{-1} hydrochlorothiazide, and 2 to 12 μg ml^{-1} triamterene. Similarly, a three-wavelength (253, 274 and 323 nm in alkaline solution, pH = 10) spectrophotometric method was used for the determination of dihydrochlorothiazide in hypotensive tablets.[90] Orthogonal function spectrophotometry was applied effectively in the analysis of antihypertensive dosage forms containing hydrochlorothiazide as diuretic component.[91-93] Hydrochlorothiazide and baventolol were determined in feeds by reversed-phase HPLC.[94]

The stability of thiazides has been the subject of several investigations. The amino disulfonamide derivative was identified as generally occurring degradation product. However, this degradation proceeds very slowly in the case of bulk substances.[95] When the substances (hydrochlorothiazide and cyclopenthiazide) were stored at elevated temperature (90°C) and at high relative humidity (92%), exposed to UV irradiation, the amount of degradation products lay in the range 0.1 to 1%.[95,96]

Even in alkaline solution (0.1 *N* NaOH), the amount of amino disulfonamide formed from hydrochlorothiazide was less than 2%. The determination was performed by colorimetric[95] and HPLC[96] methods.

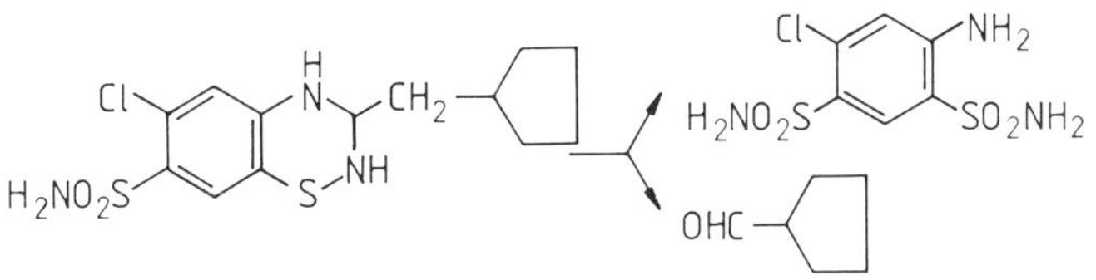

Degradation products of cyclopenthiazide in alkaline solution.

The most popular method for the determination of thiazides in biological fluids is HPLC. In the paper by Moyer and Anhalt[98] a short review is given on HPLC techniques for the analysis of diuretics in human urine. Abused diuretics (e.g., trichlormethiazide) were investigated in urine by means of TLC and HPLC.[99] A Nucleosil 5 (C_{18}) column and a Wakogel FM thin layer were used. The structures of the diuretics were established by mass spectrometry. The smallest concentration of trichlormethiazide that can be measured is 0.80 μg ml^{-1}. The urine is most often extracted with ethyl acetate.[99,100] However, the extraction does not eliminate the need for the use of a guard column[100] in the analysis of hydrochlorothiazide-containing urine. As a stationary phase, μBondapak Phenyl[100] or μBondapak C_{18}[101] has been applied. The mobile phases used are very rich in water, containing only 5 to 15% of organic solvent. A screening method for 12 diuretics (five thiazides) was published by Fullinfaw et al.[102] The TLC determination of hydrochlorothiazide is also preceded by extraction with ethyl acetate. Chromatography is carried out on silica gel, and moderately polar[103-105] mobile phases are used. Reversed-phase HPLC has been applied for the determination of eight thiazide diuretics on monomer and polymer columns.[106]

A highly sensitive photoconductivity detector has been applied for the analysis of sulfonamide derivatives in biological fluids. The method allows the detection of 0.5 to 1 ng amounts of hydrochlorothiazide.[107] Although the optimum procedure for the pretreatment of plasma is again extraction with ethyl acetate,[108-111] methods may be found in which the plasma proteins are precipitated with acetonitrile.[109] Several drugs that may be present together with hydrochlorothiazide (prednisolone, spironolactone, nifedipine, penicillin, etc.) did not interfere in the HPLC method of Yamazaki et al.[110] Hydrochlorothiazide was extracted from body fluids and blood cells into ethyl acetate. The HPLC column was Fine SIL-5, while the mobile phase consisted of ethanol-dichloroethane-hexane in different ratios for plasma, urine, blood cells, and bile. The detection limit was 5 μg ml^{-1}. Fully automated HPLC procedure was published for the determination of hydrochlorothiazide by Weinberger et al.,[111] using a C_8 column and an acetonitrile -0.1 *M* phosphate buffer (1:4) containing 5 m*M* tetrabutylammonium hydroxide. Methylclothiazid in human plasma was determined by HPLC.[112] In the reversed-phase system (μBondapak C_{18}/methanol-water 65:35) and with UV detection, a detection limit 1.5 ng/ml and a linearity range 10 to 100 ng/ml was achieved.

Hydrochlorothiazide may be measured in both urine and plasma with slight modifications (pH in the extraction) by HPLC by the potent method of Koopmans et al.[113] The method works on a LiChrosorb RP18 column with a mobile phase containing 10.24 m*M* tetrabutylammonium hydrogensulfate, 9.76 m*M* tris-(hydroxymethyl)-aminomethane and 20% methanol. Because of the mildly acidic pH of the mobile phase (pH = 5.5), it remains problematic whether ion-pair formation (between tetrabutylammonium hydrogensulfate and hydrochlorothiazide) plays a role during the chromatographic process. The method seems to be simple and fast. The calibration graphs were linear in a wide range (25 to 1000 μg ml^{-1} for plasma; 2.66 to 50 μg ml^{-1} for urine).

The chlorothiazide and hydrochlorothiazide contents of both plasma and urine can be assayed by the HPLC method of flow programming,[114] which offers some advantage over the traditional isocratic gradient elution method.

Capillary gas chromatography with electron capture detection has also been successfully used[115] for the determination of the plasma level of hydrochlorothiazide; 3 μg hydrochlorothiazide ml^{-1} plasma can be measured.

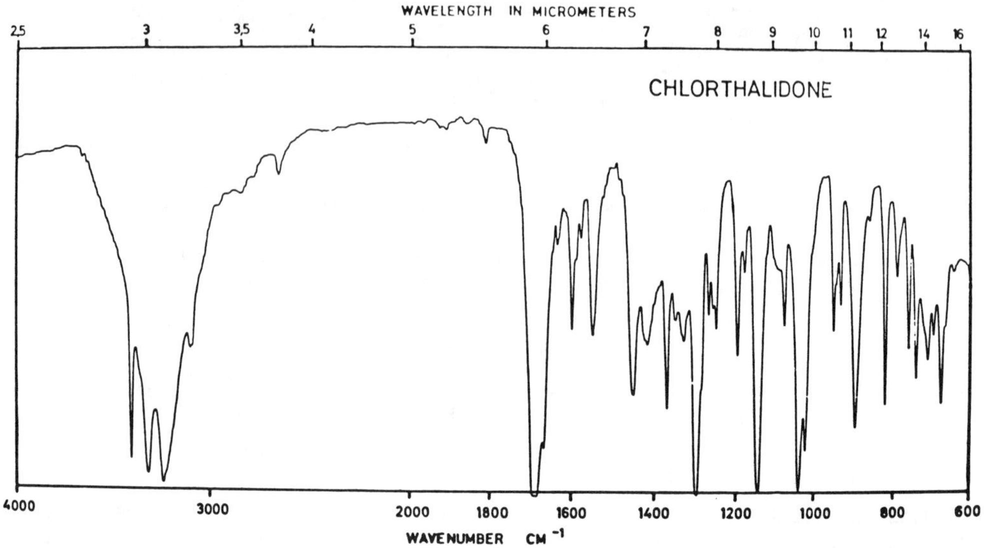

FIGURE 4. IR spectrum of chlorthalidone. (From Dibbern, H.-W., *UV and IR-Spektren wichtiger pharmazeutischer Wirkstoffe,* Editio Cantor, Aulendorf, 1978. With permission.)

III. CHLORTHALIDONE

Benzenesulfonamide, 2-chloro-5-(2,3-dihydro-1-hydroxy-3-oxo-1H-isoindol-l-yl)

2-Chloro-5-(1-hydroxy-3-oxo-1-isoindolinyl)benzenesulfonamide (77-36-1)

$C_{14}H_{11}N_2ClO_4S$ $M_r = 338.76$

A. PROPERTIES

A white or yellowish-white crystalline substance. Mp: above 215°C (with decomposition). Slightly soluble in water (1:500), insoluble in ether or in chloroform; soluble in methanol and moderately soluble in ethanol (1:150). Due to its weak acidic property, it is soluble in solutions of alkali metal hydroxides. A very significant increase in the water-solubility of chlorthalidone was achieved by solubilization with hydroxypropyl-β-cyclodextrin.[112] Identification through the IR and UV spectra is prescribed by USP. The IR spectrum (Figure 4) is characterized by a single carbonyl absorption band in the 5.8 μ region, indicative of the phthalamidine (hydroxylactam) structure **(1)** as opposed to the open-chain form **(2)**, which is expected to show two carbonyl absorptions. The UV spectrum of chlorthalidone exhibits maxima at 274 nm and 283 nm in methanol solution. In aqueous hydrochloric acid solution, almost the same maxima may be observed (Figure 5).

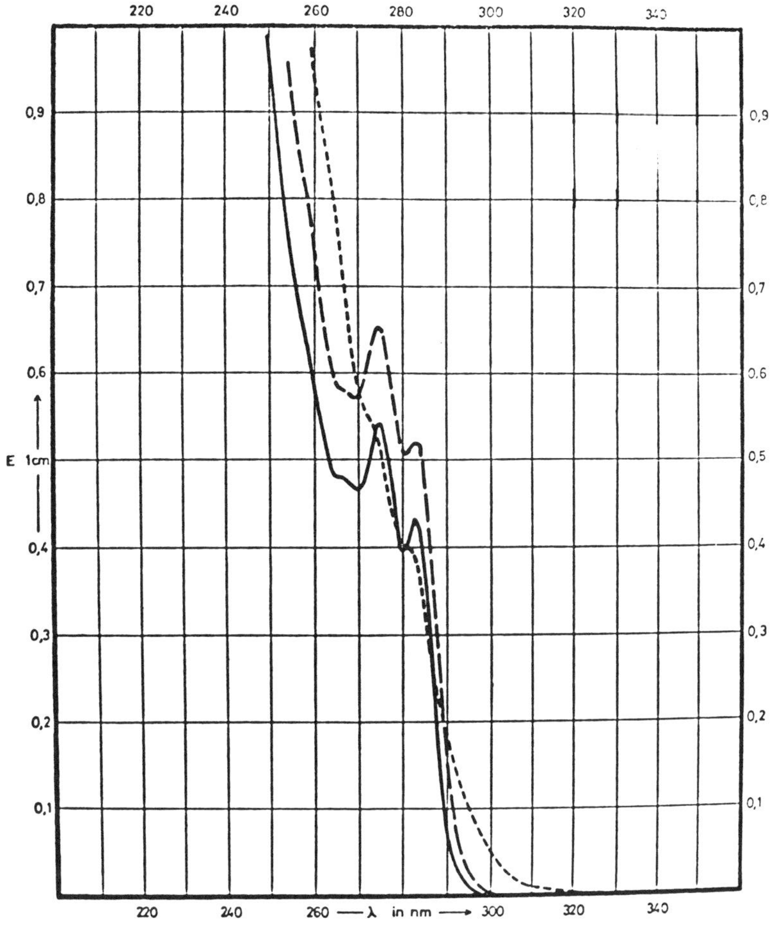

FIGURE 5. UV spectrum of chlorthalidone. Solvents: —— MeOH; ----- 0,1 *N* NaOH.; – – – 0,1 *N* HCl. (From Dibbern, H.-W., *UV and IR-Spektren wichtiger pharmazeutischer Wirkstoffe,* Editio Cantor, Aulendorf, 1978. With Permission.)

(1)

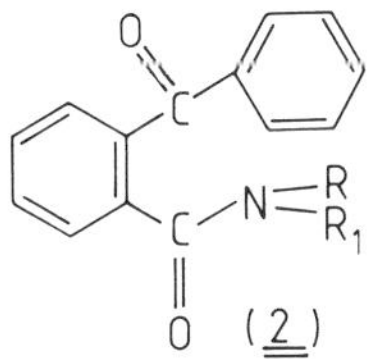

(2)

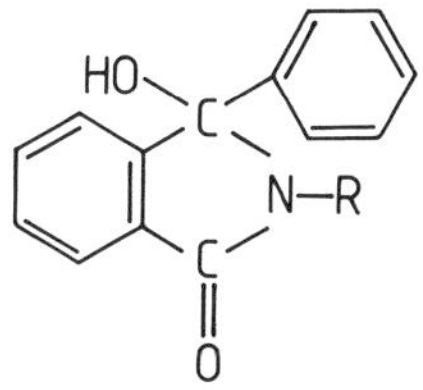

(3)

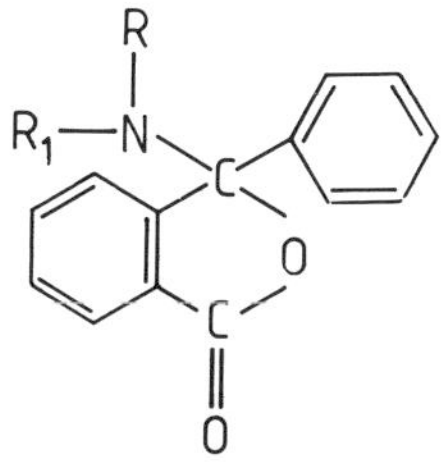

(4)

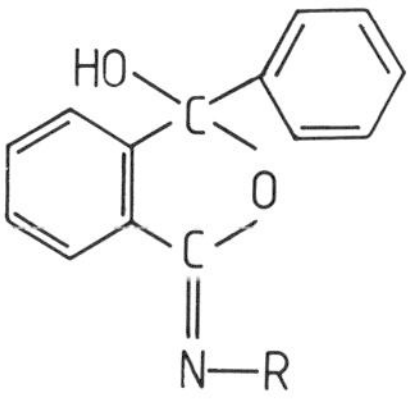

(5)

Due to the lactam structure, chlorthalidone reacts positively in the Zwikker reaction (violet color). Whether this behavior is due to the sulfamoyl or the cyclic carboxamide moiety, has not been established.

Together with numerous other drugs containing a secondary amino group, chlorthalidone has been examined *in vitro* for nitrosamine-forming ability.[117]

B. HISTORY AND STRUCTURE-ACTIVITY RELATIONSHIP

During the search for non-mercury but orally effective diuretics, the systematic investigation of benzophenone derivatives resulted in the finding of optimum saluretic-diuretic effect for the compounds with an *ortho*-substituted benzophenone structure. The method of synthesis of these compounds has been described in several patents, and also by Stoll and coworkers in a journal.[118] The optimum property was found for a chloro-sulfamoyl derivative of benzophenonecarboxylic acid, named chlorthalidone **(1)**. The amide derivatives of benzophenone-2-carboxylic acid may characterized by several structural forms. Apart from the normal (open-chain) structure **(2)**, the possible existence of the hydroxylactam **(3)**, the aminolactone **(4)**, and under certain conditions the cyclic iminoether **(5)** and iminocarboxylic acids **(6,7)** may be imagined:[118]

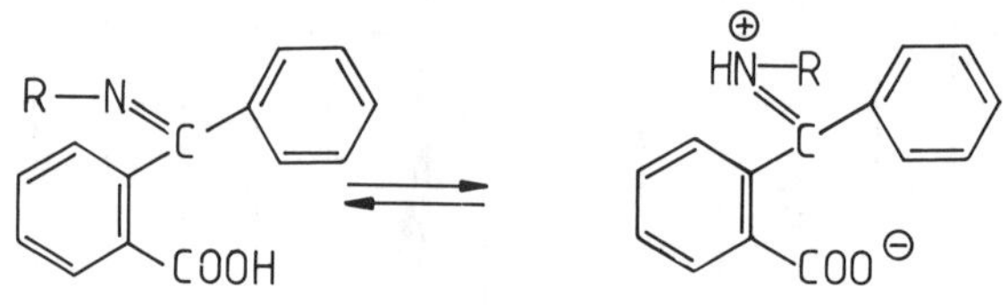

R—N=C ... COOH (6) ⇌ ⊕HN—R ... COO⊖ (7)

The observation that **(1)** could be transformed to simple ethers by treatment with aliphatic alcohols, together with some other experimental data, made only the hydroxylactam structure **(3)** plausible. The outstanding reactivity of C-3 could be explained by the existence of a resonance-stabilized carbonium ion structure **(8)**, which should play a role in different nucleophilic reactions (this may also be a cause of the reaction with sulfuric acid, when an intense yellow colour develops).

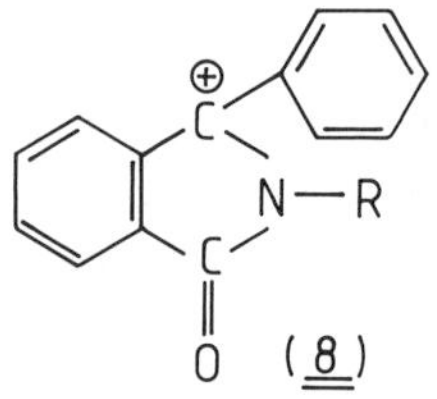

⊕C ... N—R ... C=O (8)

The synthesis of **(1)** starts with nitration of 2-(4-chlorobenzoyl)-benzoic acid **(9)** and reduction of the nitro compound **(10)** to the amino derivative **(11)**. The reaction of 3′-amino-4′-chlorobenzophenone-2-carboxylic acid **(11)** with sodium nitrite in hydrochloric acid solution results in a diazonium salt **(12)**, which is transformed to the sulfamoyl derivative **(13)** on the addition of sulfur dioxide and copper(II) chloride in acetic acid solution. On cyclization with thionyl chloride, crystallization from a chloroformic solution results in formation of the chlorophthalide **(14)**, a pseudo form of **(13)**. The latter is transformed to chlorthalidone **(1)** on the addition of ammonia in an ethanol-water mixture.

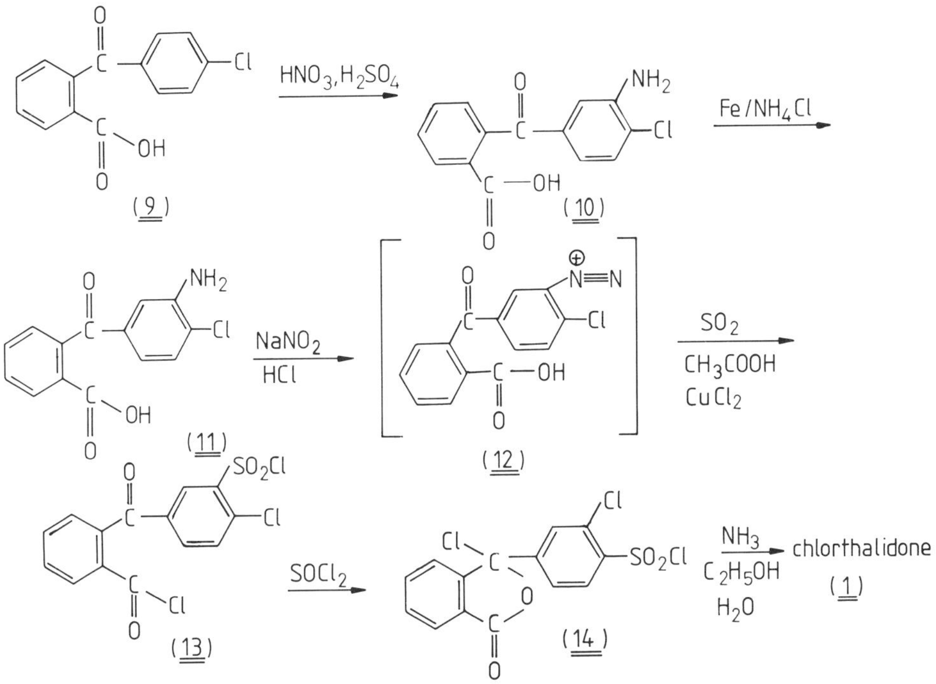

Chlorthalidone has good diuretic properties,[119-121] causing strong saluresis (Na^+, Cl^-) and a milder effect in increasing the output of K^+ and HCO_3^- ions. A noteworthy feature of chlorthalidone is its unusually long duration of action (~72 h) compared with that of thiazides. Just as for the latter, chlorthalidone exhibits a mild antihypertensive effect.[121] The site and mechanism of antihypertensive action are also similar to those for thiazides.[122] The open-chain analogs of chlorthalidone (sulfamoyl-benzoyl-benzoic acid derivatives, **15**) have a weaker diuretic activity than that of the parent compound, and also a shorter duration of action.

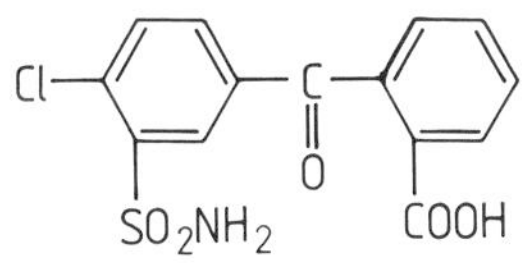

It is worth noting that removal of the sulfamoyl group from the chlorthalidone molecule resulted in a similar effect as in the thiazide series: elimination of the diuretic effect, with simultaneous enhancement of the antihypertensive activity could be observed.[123]

C. ANALYSIS

The lactam ring of chlorthalidone is subject to both alkaline and acidic degradation. In the stability indicating method of Fogel et al.,[124] the generally known hydrolysis product of chlorthalidone is the open-chain benzophenone carboxylic acid derivative **(16)**; an l-methyl ether derivative of chlorthalidone **(17)** has also been identified. Besides the carboxylic acid, another acidic degradation product **(18)** was detected by Bauer et al.[125] This may be regarded as a dehydrated derivative of chlorthalidone. The molecular absorptivity of each of the degradation products is greater than that of chlorthalidone itself, and the presence of these impurities could be a source of error if spectrophotometric assay is applied.

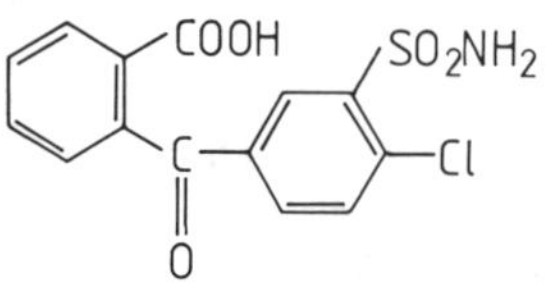

(16) 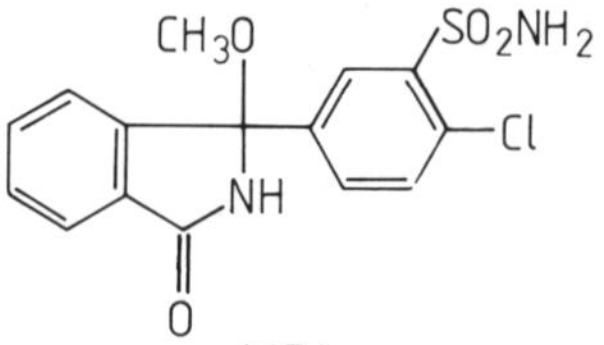 (17)

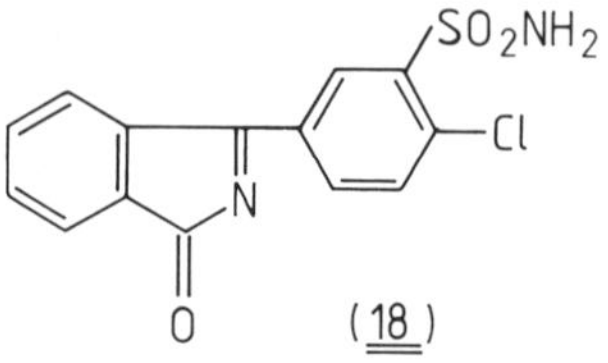

(18)

As a weak acid, chlorthalidone can be determined by simple alkalimetric titrimetry, which is a much easier, less expensive, and therefore in some cases a more reasonable method than the official pharmacopoeial procedures. Alkalimetry may be performed in dimethylformamide solution.[126] Catalytic thermometric titration in nonaqueous medium was also based on the weak acidic property of chlorthalidone.[127]

Similar to other sulfonamides, chlorthalidone exhibits anodic activity in alkaline solution.[128] This ability may serve for its sensitive electrometric determination. The electroanalytical parameters at rotating disc electrodes (Au, Pt) have been investigated.[128] The polarographic determination of chlorthalidone (and other medicinal compounds) in pharmaceuticals has been described, and the reduction products of chlorthalidone too have been discussed.[129]

Derivative spectrophotometry was used for the simultaneous determination of oxprenolol and chlorthalidone in pharmaceutical preparations.[130] Oxprenolol was selectively bound on a cation-exchange resin, while chlorthalidone was determined by measuring the absorbance of the eluate at 273 nm. Oxprenolol was estimated from the difference in absorbance before and after passage through the resin column at 273 nm, or from the second, third, and fourth-derivative spectra by means of regressional analysis. The polarimetric estimation of chlorthalidone in its formulations has been described.[131] The method is based on the "complex formation" between chlorthalidone and 4-dimethylaminobenzaldehyde. The absorbancy of the yellow reaction product is measured at 390 nm. Beer's law is obeyed for 2 to 7 ng ml^{-1} of the compound.

For the estimation of chlorthalidone in pharmaceutical formulations and in biological samples, HPLC is the most often used method. A representative compilation from the papers published during the last few years is given in Table 6. The N-functional group of chlorthalidone may be utilized for TLC detection with a versatile spray reagent consisting of ferric chloride in aqueous hydrochloric acid solution and potassium iodide in aqueous solution. A brown spot appears, indicating the oxidation of chlorthalidone.[139]

TABLE 6
HPLC Methods for the Determination of Chlorthalidone

Compounds present with chlorthalidone	Stationary phase	Mobile phase	Detection	Sensitivity/range of linearity	Note	Ref.
Clonidine	TC_{18}	50% aqueous methanol	UV		In tablets	132
Atenolol	C_{18}	Aqueous 1% acetic acid-acetonitrile (gradient elution)	275 nm		In tablets. Extraction with methanol	133
Reserpine	Bondapak NH_2	Chloroform-methanol (3:1)	UV 254 nm	From 20 to 125 μg ml^{-1} (for reserpine: 4,8—288 μg ml^{-1})	In tablets. Extraction into methanol (reserpine:chloroform)	134
—	Bondapak Phenyl	Aqueous phosphate buffer solution-acetonitrile		From 0.2 to 5.0 μg ml^{-1} in whole blood; from 4 to 190 μg ml^{-1} in urine	In whole blood and urine	135
—	Bondapak CN	Tetrahydrofuran-acetonitrile-water (2:0.5:97.5)	UV 214nm	0.20μg ml^{-1} 0; 20 to 5.27 μg ml^{-1}	In whole blood. Extraction into acetonitrile	136
—	LiChrosorb RP 18	0.01 *M* aqueous Na-acetate solution-acetonitrile (77:23)	UV 214nm	25μg ml^{-1}; 0.312 to 2.5 μg ml^{-1}	In whole blood. Extraction into t-butyl methyl ether	137
Probenicid (I.St.)	C_{18} (RCM-100)	0.01 *M* aqueous Na-acetate solution-acetonitrile (4:1)	UV 210 nm	From 100 to 4000 μg ml	In whole blood and urine. Column clean-up procedure: Bond-Elut BE 6001 column	138

IV. FUROSEMIDE

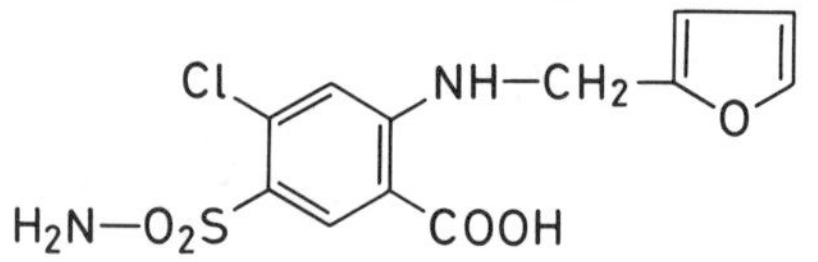

Benzoic acid, 5-(aminosulfonyl)-4-chloro-2-(2-furanylmethyl)-amino-
4-Chloro-*N*-furfuryl-5-sulfamoylanthranilic acid (54-31-9)
$C_{12}H_{11}ClN_2O_5S$ $M_r = 330.74$

A. PROPERTIES

Furosemide is a white or slightly yellow crystalline powder. It is odorless. Its X-ray powder diffraction data (Cu Kα radiation) are available,[140] and its crystal structure has also been determined.[141] The conformational lability of the furan ring and intermolecular interaction via NH$\cdots$O bonding have been established. The photochemical lability of furosemide was reported both for the solid form and for its solutions. On the action of light, first a yellow, then a yellow-brown, and finally a red-brown colour may be observed in furosemide solutions.[142] In the view of certain authors, the main products of photodegradation (or of heating at above 50°C) are identical with those of acidic hydrolysis[143] (see below).

Furosemide is insoluble in water or in chloroform. It is freely soluble in acetone, dimethylformamide, and alkali-metal hydroxide solutions. It is soluble in methanol, and slightly soluble in ether. From the aspect of pharmaceutical technology, important efforts have been made to increase the water-solubility and rate of dissolution of furosemide. It has been established that its solubility in water increases linearly in the range 0 to 50 m*M* as a function of the cyclodextrin concentration.[144]

The melting point of furosemide is 206°C; the eutectic melting point of its 1:1 mixture with salophen is 175 to 179°C.

The UV absorption spectrum of furosemide exhibits two or three maxima (Table 7 and Figure 6).

Due to its aromatic carboxy group, furosemide has a definite acidity; the proton of the sulfonamide group provides a second step of dissociation:

$$pK_1\text{: } 3.80 \text{ (-COOH)},^{145}$$

$$pK_2\text{: } 7.5 \text{ (-}SO_2NH_2\text{)}.$$

The *in vitro* stability of furosemide has been investigated by several authors. It was observed long ago by Sturm et al.[146] that furosemide undergoes proton-catalyzed hydrolysis, followed by splitting of the furylmethyl-amino bond. This observation was later confirmed and supplemented with further details by other authors.[147-149] Cruz et al.[150] investigated the kinetics and mechanism of hydrolysis of furosemide, and established the pH profile of the process. In the pH region above 7.2, the hydrolysis is extremely slow. After weeks of incubation at 70°C and pH = 8.5, hydrolytic degradation was not observed at all. The log K vs. pH function could be described by the equation

$$K_{obs} = k_H + (H^+)$$

and a plot of $K_{obs.}$ against (H^+) had a slope of 2.141 mol^{-1} $min.^{-1}$, which is the bimolecular rate constant, k_H+. The first-order rate constant was found to be over 500 times less when the pH was increased from 1.50 to 7.17. The knowledge of the Arrhenius parameters permitted estimation of the rate constant for furosemide hydrolysis under physiological conditions in the stomach: at pH = 1.0 and 37°C, the half-life of furosemide was 170 min (~20% degradation h^{-1}).

TABLE 7
Ultraviolet Absorbance of Furosemide

Solvent	Methanol	0.1 *N* HC1	0.1 *N* NaOH
$\lambda_{max,}$ nm	337	340	335
	273	274	270
ϵ	233	235	—
	5,260	5,590	4,330
	22,190	21,100	19,380
	45,970	47,290	—

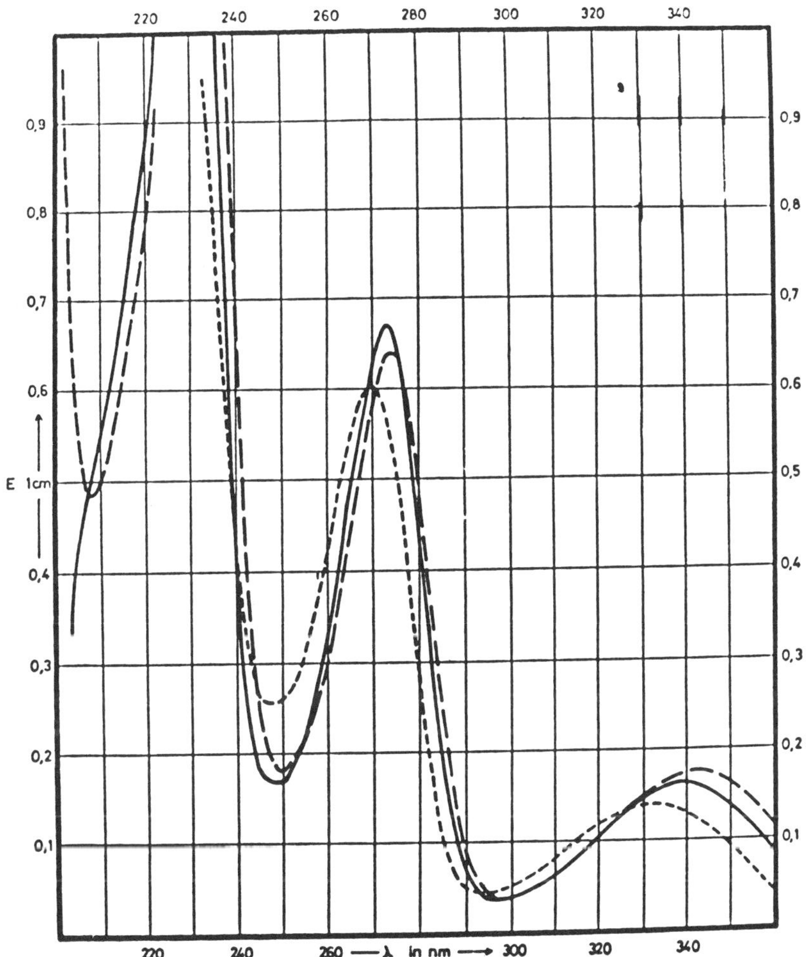

FIGURE 6. UV spectrum of furosemide. Solvents: ——— MeOH; ----- 0,1 *N* NaOH; – – – 0,1 *N* HCl. (From Dibbern, H.-W., *UV and IR-Spektren wichtiger pharmazeutischer Wirkstoffe*, Editio Cantor, Aulendorf, 1978. With permission.)

Bundgaard et al.[151] carried out very valuable experimental work on the photodegradation of furosemide. In acidic aqueous solution exposed to daylight, the disappearance of furosemide followed first-order kinetics. The shape of the pH vs. rate profile revealed that the un-ionized form of furosemide is the reactive species, and the overall kinetics can be described by the equation

$$k + k'\frac{a_H}{a_H + K_a}$$

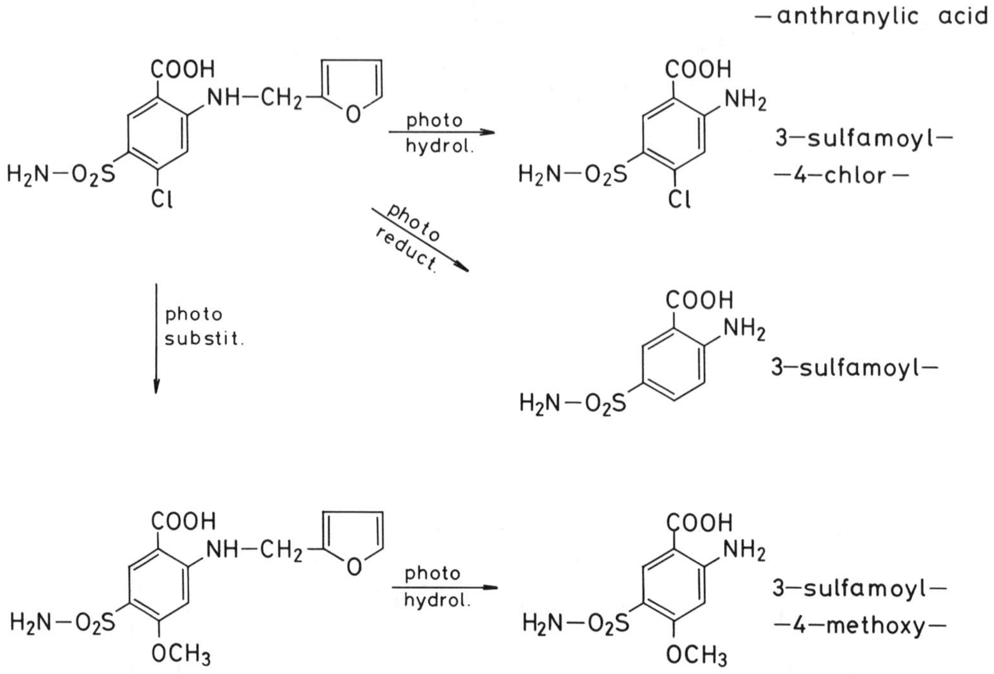

FIGURE 7. Photodegradation scheme of furosemide. (Modified from Reference 155.)

where k′ is the first-order rate constant for the photolytic degradation of the un-ionized form of furosemide, $a_H/a_H + K_a$ is the fraction of total furosemide in this form, and K_a is the apparent ionization constant of the carboxylic acid group in furosemide. Moore et al.[152-155] studied the photodecomposition of furosemide and paid special care to the problem of the photosensitizing ability of the compound. One of their significant findings was that the photolysis of furosemide (in methanolic oxygen-free solution, irradiated for 10 min) yields products with UV characteristics similar to those of furosemide. These products are therefore potential photosensitizers, and they may provide a source of error in the UV spectrophotometric determination of furosemide. The pathways of photodegradation of furosemide include both photoreductive and photohydrolytic reactions,[155] with the formation of different 3-sulfamoylanthranilic acid derivatives. (Figure 7).

Due to its acidic character, furosemide is present practically only in ionized form in the human organism. As a surface-active anion, it tends to displace lipid monolayers from the surface at positive polarizations, lowering their potential stability range.[156] The value of the partition coefficient of furosemide between lipid bilayers and water indicates that the drug facilitates the transport of ions.[157]

B. HISTORY AND STRUCTURE-ACTIVITY RELATIONSHIP

Furosemide belongs in the group of diuretics referred to as ''high-ceiling'' or ''loop'' diuretics as far as the intensity or the site of action is concerned. It is a common feature of high-ceiling diuretics that they cause strong saluresis and diuresis, but direct proof of the site and mode of their action is still lacking. In this respect, valuable information may be found in the paper by Schlatter et al.[158] Some details will be discussed later.

Although furosemide was synthetized by Siedel, Sturm et al.[146,159] in the late fifties, it was accepted as a highly effective saluretic and diuretic drug only a few years later. The discovery and development of furosemide have been reviewed by Muschawek.[160]

The discovery of furosemide was based on the synthesis of a series of *N*-substituted-5-

halogen-2,4-disulphamoylanilines.[159] The essence of this work was the condensation of 2,4-dihalogen-5-sulfamoylbenzoic acid **(1)** with amine derivatives **(2)**. Due to the high reactivity of the chlorine at position 2, activated by the *o*-carboxy group, the corresponding 2-anilide **(3)** was obtained as the main product. The great versatility of the method stems from the variability of the substituents of the two reacting compounds.

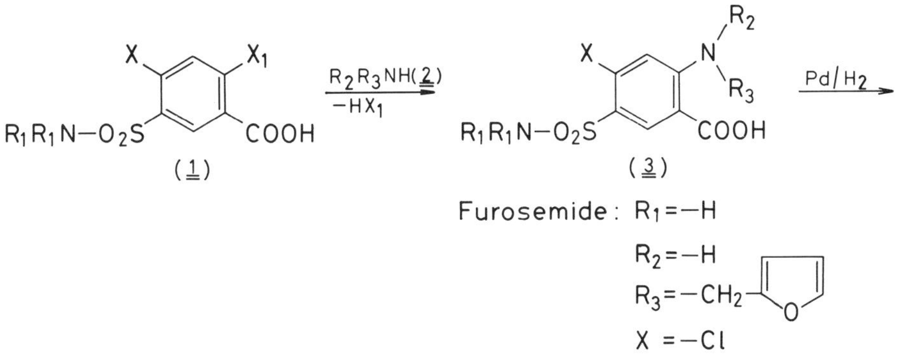

R_2R_3NH(2), $-HX_1$; (1); (3); Pd/H_2

Furosemide: R_1 = —H
R_2 = —H
R_3 = —CH_2—(2-furyl)
X = —Cl

Through variation of two of the experimental conditions, the yield of the condensation could be raised significantly: increasing the polarity (electronegativity) of X_1 by applying fluorine instead of chlorine, and decreasing the polarity of the halogen-activator carboxy group via its esterification or amidation. A necessary condition of the condensation is a temperature between 100 and 200°C, depending upon the structure of the amine. The optimum temperature for the synthesis of furosemide was found[159] to be 130°C. At this temperature, condensation with furfurylamine took place within 4 h with a yield of 40 to 50%.[159] Furfurylamine was taken in a 6 mol excess. Condensation under the same conditions, but using the 2-F derivative, resulted in a much higher yield: up to 85%. It is worthy of mention that the presence of a basic side chain stabilizes the 4-halogen and hinders the occurence of a diamination reaction.

On the other hand, a furosemide isomer **(6)** was obtained when 2,4-dichloro-5-sulfamoylbenzonitrile **(4)** and furylmethylamine **(5)** were condensed.[161]

An interesting feature that indicates the close relation between the structures of thiazides and furosemide is that the treatment of **(6)** with formic acid and alkaline hydrolysis resulted in the formation of 2-chloro-4-amino-5-sulfamoylbenzoic acid **(8)** via the formation of hydrochlorothiazide **(7)**:

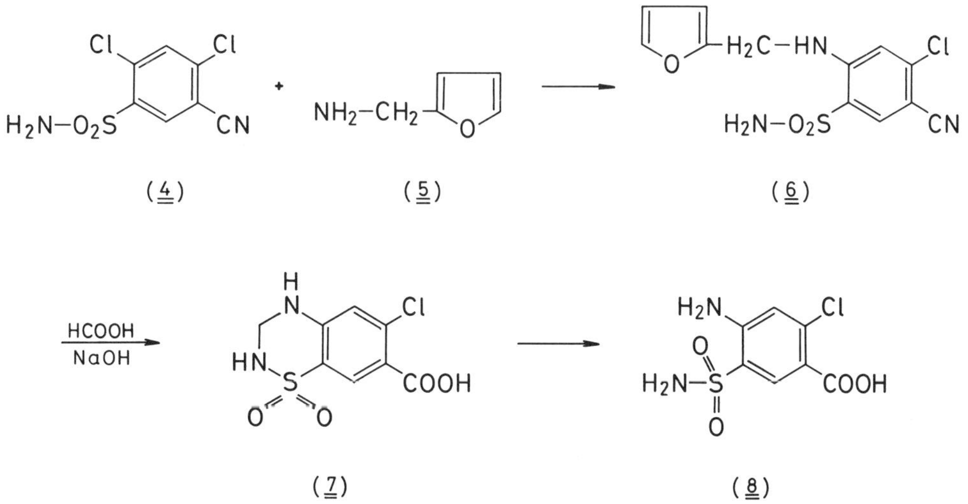

(4) + (5) → (6)

HCOOH / NaOH → (7) → (8)

Although furosemide has retained its dominant position among the anthranilic acid-derivative high-ceiling diuretics for 20 years, efforts to design and prepare even more potent and harmless congeners may be one of the aims of drug research. Almost 60 furosemide-related compounds were studied for diuretic effects (volume of urine, pH, chloride, potassium and sodium contents) by Shani et al.[162] The types of structural modifications are shown by the following scheme:

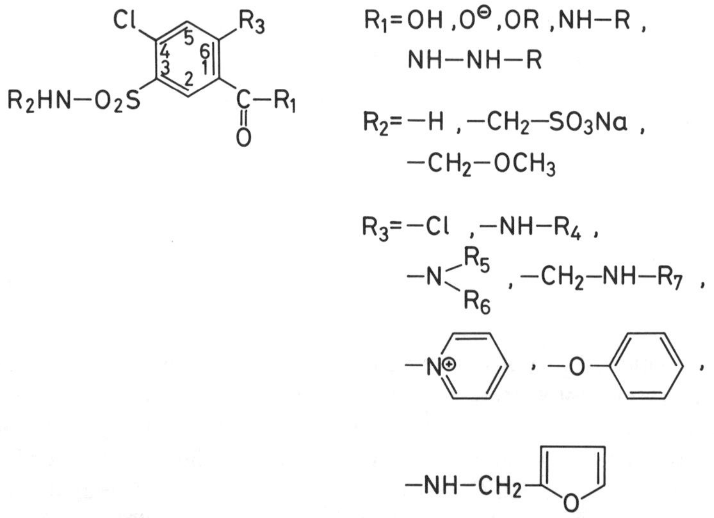

Cl, R_3, $R_2HN{-}O_2S$, $C(=O){-}R_1$; ring positions 1–6

R_1=OH, $O^{\ominus}$, OR, NH–R, NH–NH–R

R_2=–H, $-CH_2-SO_3Na$, $-CH_2-OCH_3$

R_3=–Cl, $-NH-R_4$, $-N(R_5)(R_6)$, $-CH_2-NH-R_7$, $-N^{\oplus}$(pyridinium), –O–(phenyl), $-NH-CH_2-$(furyl)

QSAR analysis has been accomplished by including in the function the biological activity vs. partition coefficient (calculated by the method of Hansch) and molar refraction. Non-inclusion of the group of 28 compounds with a substituent in the SO_2NH_2 group led to a significant correlation, but with a relatively low correlation coefficient ($r = 0.902$).[162] With regard to the numerous sources of error (for example, in the calculation of the log P values) this coefficient may be taken as good. It is interesting that the optimum biological activity was found at log P~5, representing a need for a high lipophilicity.

The mechanism of diuretic action of high-ceiling diuretics was studied at a molecular level in several works by Schlatter et al.[158,163] They provided proof that furosemide inhibits a Na^+-$2Cl^-$-K^+ cotransport system in the lumen membrane of the cortical thick ascending limb. They established that, of the several dozen compounds tested, only furosemide and certain structurally related compounds, but not several other high-ceiling diuretics, exerted the above-mentioned inhibitory effect. The measurement of the latter is based upon the continuous recording of the transepithelial electrical potential difference and transepithelial resistance. The ratio of these two values gives the equivalent short-circuit current. From this, in the knowledge of the ion fluxes, the direction and magnitude of the charge transfer are determined. The charge transfer stems mainly from the Cl^- diffusion from the cells. In this way, the *in vitro* activity of loop diuretics may be characterized. One of the conclusions relating to the structural requirements for K^+-$2Cl^-$-Na inhibitory activity is that each of the side chains influences the effect of the other, and thus the individual positions of the molecule cannot be examined separately. This latter is exemplified by the finding that position 1 may be substituted by an acidic group other than carboxyl (sulfinic acid, tetrazolyl residue) if position 2 is substituted by a furfurylamine residue; the same is true for positions 2 and 3: C-2 needs to bear a suitable substituent (this time the optimum is found with furfurylamine) if position 3 is not substituted. Substitution of position 2 is not necessary if position 3 is occupied by an amino group. Furfurylamine currently remains the most effective substituent at position 2, and even small alterations cause a dramatic reduction in inhibitory potency. Position 4 needs a bulky substituent at position 5, as exemplified by the most frequently occuring chlorine.

Only one of the hydrogens of the sulfonamide group may be substituted. In this case, the compound probably functions as a pro-drug, generating *in vivo* the molecule with unsubstituted 5-SO_2NH_2 residue.

C. ANALYSIS

The identification of furosemide as a drug substance includes the identity of its UV and IR spectra with those of the standard substance (USP). Additionally, a chemical reaction involving its acidic hydrolysis product can be performed. At the temperature of a steam bath, furosemide **(9)** can be hydrolyzed to yield 4-chloro-5-sulfamoylanthranilic acid **(10)** and furylmethanol **(11)**:

Cl, NH–CH2–(furyl), H2N–O2S, COOH (9) —H⊕ / H2O→ Cl, NH2, H2N–O2S, COOH (10) + HO–CH2–(furyl) (11)

On the addition of sodium nitrite in acidic medium, a diazonium salt **(12)** is formed, which is coupled to give an azo dye **(14)** by the addition of N-1-naphthylethylenediamine dihydrochloride **(13)**. The excess of sodium nitrite is bound by ammonium sulfamate. The formation of the red-violet reaction product gives a possibility for the spectrophotometric determination of furosemide and its potential impurity, 4-chloro-5-sulfamoylanthranilic acid (see above). In the latter case, the same reaction is carried out, but without preliminary hydrolysis (λ_{max} : 530 nm, solvent dimethylformamide). Furosemide should contain not more than 0.4% of **10**.

Cl, NH2, H2N–O2S, COOH (10) —NaNO2 / HCl→ ⊕N≡N, COOH (12) + NH–CH2–CH2–NH2 (13) → N=N–, COOH, NH–CH2–CH2–NH2 (14)

The color reaction with phloroglucinol has been suggested[164] for the spectrophotometric determination of furosemide. The reagent (0.05% phloroglucinol in 5.25 *M* hydrochloric acid) brings about rupture of the C - N bond, with liberation of furaldehyde. The latter condenses with phloroglucinol to yield a deep-coloured reaction product. The sensitivity of the method is 89 μM. The reaction goes to completion if the reaction mixture is kept at room temperature for 20 min (λ_{max} : 558 nm, log ϵ : 3.187). The method is suitable for the rapid and accurate determination of furosemide even in complex pharmaceutical mixtures.

Due to its rather strong acidic nature, furosemide can be titrated alkalimetrically in aqueous medium. Because of the poor solubility, volumetry is prescribed for nonaqueous (dimethylformamide) solution (USP).

In the analysis of furosemide in dosage forms, the method of choice is HPLC. In the case of tablets and injections, preliminary extraction into aqueous 0.1 *N* NaOH was found to be appropriate by Roth et al.[165] with subsequent linear gradient elution HPLC. Column: μBondapak C_{18}, mobile phase: an acetonitrile-aqueous 0.5% phosphoric acid mixture (3:17 to 2:3). The calibration graphs were linear for 75 to 500 ng. A similar HPLC procedure, but using a Bondapak NH_2 column and a mildly alkaline mobile phase, has been described[166] for the estimation of furosemide in dosage forms. The method allows identification of the preservatives in the formulations, and the detection of 4-chloro-5-sulfamoylanthranilic acid, a specific impurity of furosemide. Furosemide has been determined[167] in bulk and tablets by UV spectrophotometry of an ethanolic solution, with measurement of the drug maximum absorbances at 225, 273, and 326 nm and calculation of the absorbance ratio. Beer's law was obeyed in the concentration range 2.5 to 25 ng ml^{-1}.

The optimum conditions for the determination of furosemide in biological fluids have been studied repeatedly. A comparison of the extraction and precipitation methods applied in the HPLC determination of furosemide in plasma and urine was made by Bauza et al.[168] In frozen samples, furosemide was stable for several months. The extraction (from acidic solution with ether)[169,170] resulted in good recoveries from both of the fluids, but the precipitation method (addition of acetonitrile and centrifugation) was unsuitable for the testing of urine samples. A very small volume (50 μl) and very low amounts of furosemide can be prepared for further investigation by the microscale ultrafiltration technique devised by Wittfoht et al.[171] Further important details concerning the conditions of careful preparation of plasma samples for HPLC assay were reported by Rapaka and Roth.[172] Certain anticoagulants (e.g., Na-oxalate) interfere with the analysis of furosemide in plasma samples. Only EDTA (K-salt) did not result in any interfering peaks when either UV or fluorimetric HPLC detection was used. The preliminary investigation of sample containers before analysis is strongly recommended.

A small amount of plasma (100 μl) is required in the HPLC method of Nation et al.,[173] in which deproteinization with acetonitrile is followed by the application of a chromatographic system with a μBondapak C_{18} column, acetonitrile-0.05% phosphoric acid 3:7 as mobile phase and fluorimetric detection. Linearity was found in the range 0.1 to 10 μg ml^{-1}. A number of other drugs do not interfere. Surprisingly, a better sensitivity was reported by Lin et al.,[174] who used a precipitation and chromatographic system very similar to that in the method of Nation et al.,[173] but working with UV detection. The useful analytical range was found to be 0.08 to 0.245 μg ml^{-1}. Some further HPLC methods which operate under more or less the same conditions can be found in the literature.[175,176]

A special microscale HPLC method was described by Sord et al.[177] It requires only 25 μl of plasma or urine, and is therefore suitable for use in neonatal intensive care units. Nanogram amounts of furosemide can be measured by the procedures of Kubo et al.[178] and Guermouche et al.[179] It is interesting that the former authors added 5 m*M* Na hexane-1-sulfonate to the mobile phase (aqueous acetonitrile).

Ether, generally used for the extraction of furosemide, is replaced by dichloromethane to extract furosemide from plasma or wine by Lovett et al.[180] and Uchino et al.[181] The results obtained with the generally used detection method (UV detection) have been compared with those originating from the measurement of liquid scintillation.[182] Furosemide and one of its main metabolites, furosemide-glucuronide, were analysed by HPLC.[183] The amount of the latter in normal rabbits was 11%. Other furosemide metabolites, such as anthranilic acid and 4-chloro-5-sulfamoylanthranilic acid, were identified by TLC without any pretreatment by Kholodov et al.,[184] and after extraction with ether by Wesley-Hadzija and Mattocks.[185]

Attention should be paid to the warning of Berret et al.,[186] who demonstrated 4-chloro-5-sulfamoylanthranilic acid as an artifact of the HPLC process they used.

V. ETHACRYNIC ACID

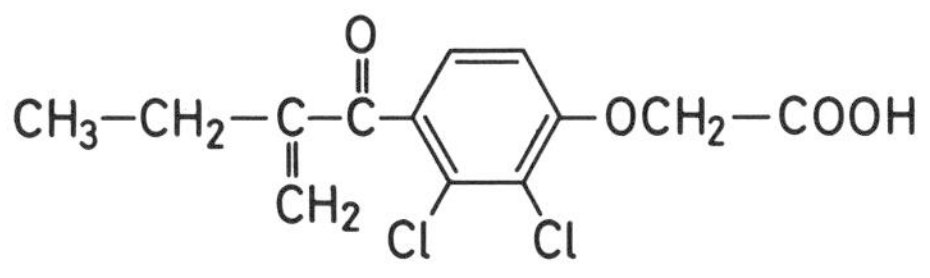

$$CH_3-CH_2-\underset{\underset{CH_2}{\|}}{C}-\overset{\overset{O}{\|}}{C}-C_6H_2Cl_2-OCH_2-COOH$$

2,3-Dichloro-4-(2-methylene-1-oxybutyl)-phenoxy acetic acid,
2,3-Dichloro-4-(2-methylenebutyryl)phenoxy acetic acid
$C_{13}H_{12}Cl_2O_4$ M_r = 303.2

A. PROPERTIES

Ethacrynic acid is a white or practically white, odorless crystalline powder. It is very slightly soluble in water, but freely soluble in alcohol (1 g in 1.6 ml), in chloroform (1 g/6 ml), and in ether (1 g/3.5 ml). The UV absorption maximum is at 271 nm. Melting point: 121 to 122°C. The pK_1 value is 3.5; pK_2 = 10.5. Partition coefficient (log P) = −1.15 (in ether).

B. HISTORY AND STRUCTURE-ACTIVITY RELATIONSHIP

The organomercuric compounds have strong diuretic properties, but they are very toxic molecules because of the release of mercury(II) ion in the kidney. The diuretic activity is proportional to the concentration of mercury.

These toxicological findings led to the search for non-mercury-containing compounds that would be just as effective and with a similar mechanism.

According to the present theory, the free mercury(II) ion from the organomercurials binds specifically to two sites of the receptor, one of which is a sulfhydryl group. This has prompted an intensive search for non-mercury-containing structures which also have the ability to interact with the functionally important sulfhydryl group.[187] The α,β-unsaturated ketones **(1)** were known to be substances which form addition products **(3)** with compounds containing a sulfhydryl group **(2)**:

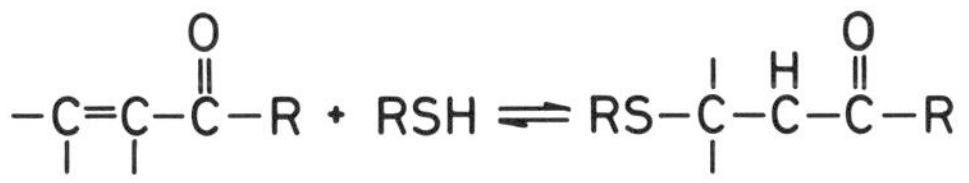

$$-\underset{|}{C}=\underset{|}{C}-\overset{\overset{O}{\|}}{C}-R + RSH \rightleftharpoons RS-\underset{|}{\overset{|}{C}}-\overset{H}{C}-\overset{\overset{O}{\|}}{C}-R$$

(1) (2) (3)

After a short time, it emerged that the search for non-mercury-containing sulfhydryl-active diuretics must be focused on the general structural type **(4)**:

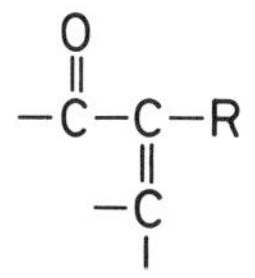

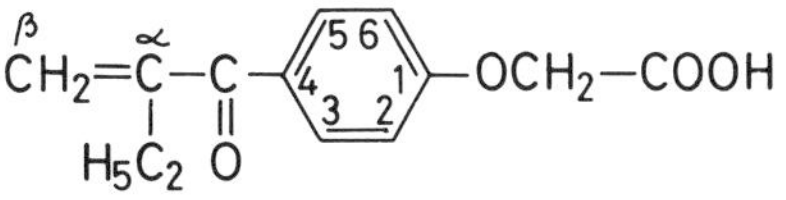

$$-\overset{\overset{O}{\|}}{C}-\underset{\underset{-\underset{|}{C}}{\|}}{C}-R \qquad \overset{\beta}{C}H_2=\underset{\underset{H_5C_2}{|}}{\overset{\alpha}{C}}-\underset{\underset{O}{\|}}{C}-C_6H_4-OCH_2-COOH$$

(4) (5)

Analysis of the effects of the subsequent structural modifications showed a parallelism with the conclusions drawn in connection with sulfonamide diuretics.

Substitution of the benzene ring with certain electronegative and ring-activating groups enhanced the diuretic activity in the sequence Cl>Br>CF_3, particularly when these substituents occupy both positions 2 and 3. The influence of these substituents is exerted via an enhancement of the electron flux from the ring (mesomeric effect) towards the α, β unsaturated carbonyl group, resulting in an increased activity of the latter. 2,3-dichloro substitution on the aromatic nucleus led to ethacrynic acid **(5)** as the most active derivative.[188,189] This compound has been synthetized in systematic research by Schultz et al.[187]

Substitution of molecules of the ethacrynic acid type in positions other than 2 and 3 reduces the diuretic activity. Methyl substituents in positions 2 and 3 cause somewhat less activation than Cl, Br or CF_3, and higher alkyl groups are much less effective.[190] Most of the related studies tend to support the notion that the diuretic-saluretic activity of ethacrynic acid is based on its sulfhydryl reactivity. Those derivatives of ethacrynic acid which are not able to bind to the sulfhydryl group also lack diuretic activity. For example, replacement of the two vinyl hydrogene with methyl groups yielded a compound **(6)** which lacks both sulfhydryl reactivity and diuretic efficacy. The same was observed when dihydroethacrynic acid **(7)** or the epoxy derivative **(8)** were tested:[191-195]

R–C(=O)–[2,3-dichlorophenyl]–O–CH_2–COOH

R = CH_2=C(C_2H_5)– ethacrynic acid

$(CH_3)_2$–C=C(C_2H_5)– (6)

CH_3–CH(C_2H_5)– (7)

CH_2–C(C_2H_5)– (epoxide, O bridging) (8)

It must be confessed that several compounds proved to have diuretic-saluretic action without displaying any *in vitro* sulfhydryl reactivity. A survey of this branch of research may be found in the excellent publication by Koechel.[196] Although this finding seems to suggest a mechanism of action different from that of ethacrynic acid, more exact data on these compounds can provide support for their *in vivo* reactivity.

Another important event in this field was the discovery of compounds serving as prodrugs, i.e., they release ethacrynic acid in the living organism. Certain sulfydryl adducts of **(9)** have been reported to have diuretic properties,[197] and it was found that the rate of ethacrynic acid release correlates with several components of the diuretic effect.[198]

C_2H_5–CH(CH_2–SR)–C(=O)–[2,3-dichlorophenyl]–OCH_2–COOH

(9)

C. SYNTHESIS[199]

2,3-Dichlorophenoxyacetic acid **(10)** is transformed in the Friedel-Crafts reaction with butyryl chloride **(11)** to the ketone **(12)**, which undergoes condensation with dimethylamine and formaldehyde to give a Mannich base **(13)**. When heated in vacuum, this splits off dimethylamine to yield ethacrynic acid:

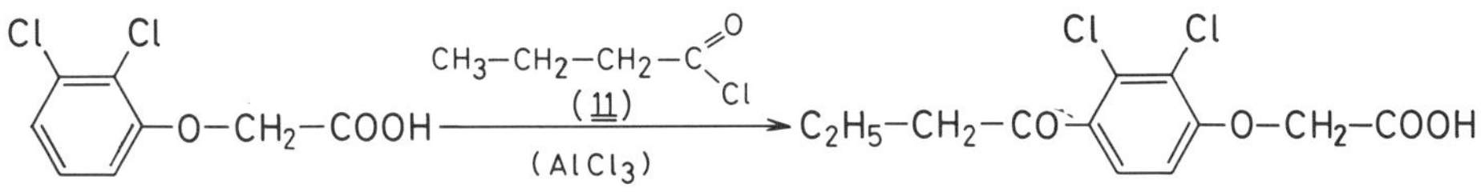

Cl Cl
$-O-CH_2-COOH$ (10) $\xrightarrow[(AlCl_3)]{CH_3-CH_2-CH_2-COCl\ (11)}$ $C_2H_5-CH_2-CO-$ (Cl, Cl) $-O-CH_2-COOH$ (12)

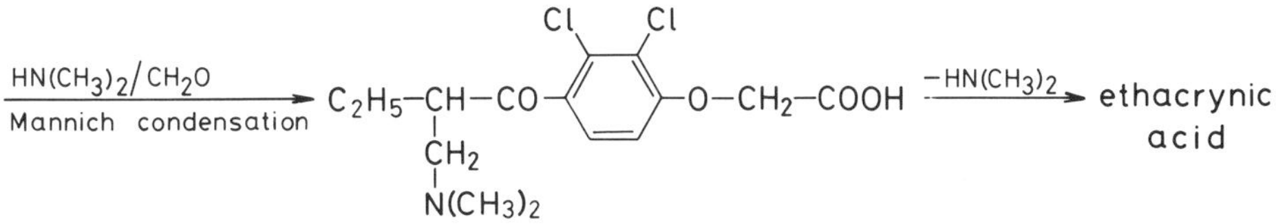

$\xrightarrow[\text{Mannich condensation}]{HN(CH_3)_2/CH_2O}$ $C_2H_5-CH(CH_2N(CH_3)_2)-CO-$ (Cl, Cl) $-O-CH_2-COOH$ (13) $\xrightarrow{-HN(CH_3)_2}$ ethacrynic acid

D. ABSORPTION AND METABOLISM

After absorption from the gastrointestinal tract, ethacrynic acid is bound to plasma proteins. Its biotransformation, like that of many other highly reactive alkylating agents, involves the formation of a glutathione conjugate which is subsequently degraded to the cysteine and mercapturic acid conjugates.[198,200,201] This conversion may occur in the liver or within the renal proximal tubular cells.[202,203] Accordingly, it is excreted in the urine, with larger quantities in the bile.

E. ANALYSIS

For the identification ethacrynic acid, USP specifies not only the IR and UV spectra, but also a color reaction.

When ethacrynic acid **(14)** is dissolved in sodium hydroxide, the solution is heated in a boiling water bath, then cooled and acidified, chromotropic acid and finally sulfuric acid are added, formalehyde is formed. This reacts with chromotropic acid to give a polymeric substance **(15)** containing a diarylmethane structural moiety.[204]

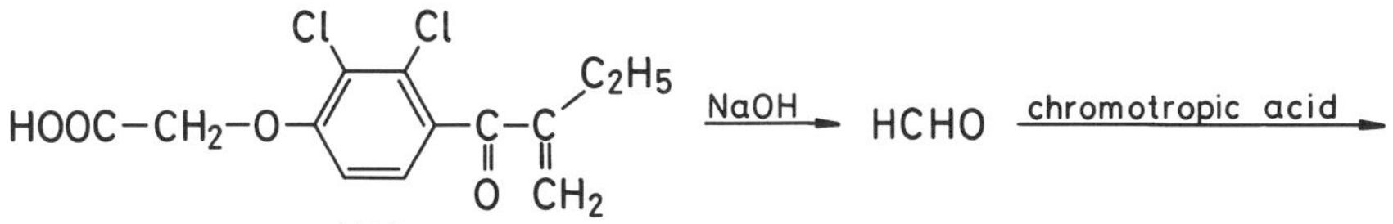

$HOOC-CH_2-O-$ (Cl, Cl) $-C(=O)-C(=CH_2)-C_2H_5$ (14) $\xrightarrow{NaOH}$ HCHO $\xrightarrow{\text{chromotropic acid}}$ (15)

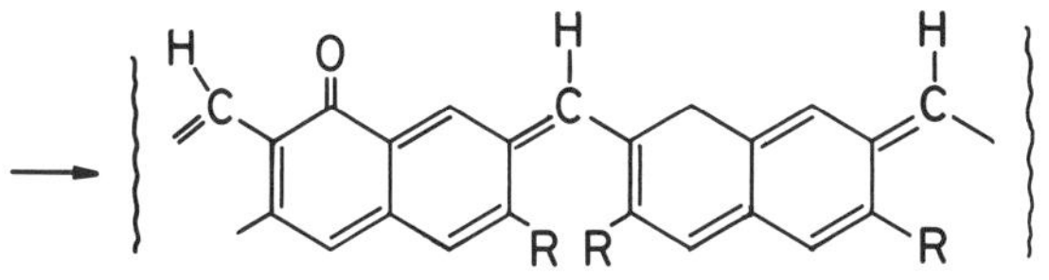

It is worth noting that the heating of ethacrynic acid in sodium hydroxide solution under suitable conditions[204] results in a dimeric derivative **(16)** of ethacrynic acid. This appears as a white precipitate, which may be extracted with ether from acidic medium.

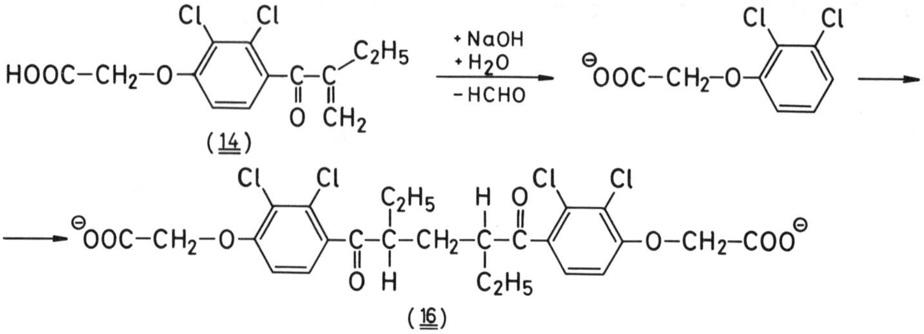

HPLC and GC assays have also been carried out for the determination of ethacrynic acid. Das Gupta[205] described a stability-indicating assay of ethacrynic acid in dosage forms by HPLC on a reversed phase (μBondapak C_{18}) with 0.01 *M* $(NH_4)_2HPO_4$-MeOH (11:9) as mobile phase. He stated that ethacrynic acid is stable in dosage form for about 1 year, and it undergoes degradation only in highly acidic or alkaline solution; however, NH_4 ions cause rapid decomposition.

Yarwood et al.[206] obtained the same result by reversed-phase HPLC with buffered aqueous solutions containing either Na^+ or NH_4^+ ions. The extent of degradation was influenced both by the species and by the concentration of the cation.

The GLC analysis of ethacrynic acid is possible after its derivatization. The approach involving its conversion to its methylester by treatment with diazomethan led to the formation of the following reaction product:

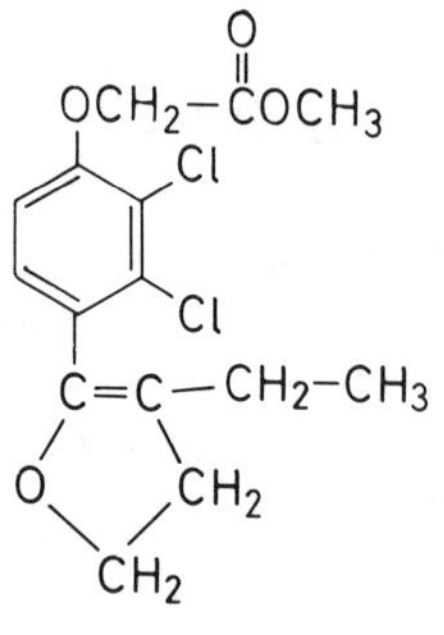

However, in methanol-hydrochloric acid solution the same reaction gave a quite different product: formation of the methyl ester of ethacrynic acid was observed. Both derivatives are suitable for the GLC determination of ethacrynic acid.[207]

In the volumetric determination of ethacrynic acid by the USP bromometric method, BrCl reacts by addition to the C=C double bond:

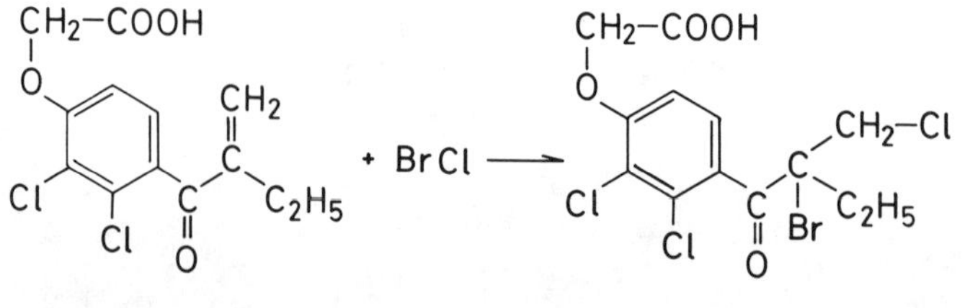

VI. SPIRONOLACTONE

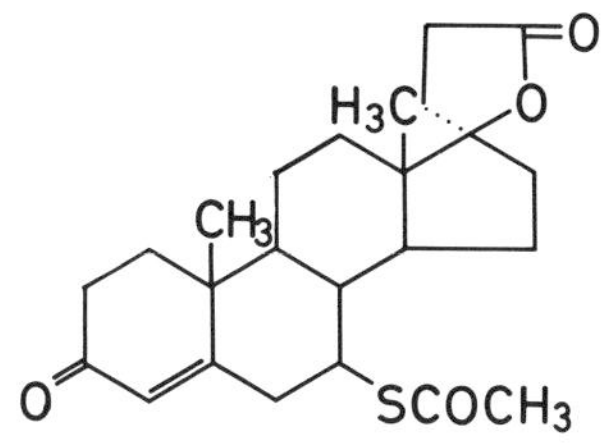

Pregn-4-ene-21-carboxylic acid, 7-Acetylthio(-17-hydroxy-3-oxo-
γ-lactone (7α, 17α)
17-Hydroxy-7α-mercapto-3-oxo-71α-pregn-4-ene-21-carboxylic
acid γ-lactone acetate

$C_{24}H_{32}O_4S$ M_r = 411.57

A. PROPERTIES

A light-cream to light-tan crystalline powder which has a faint mercaptan-like odor; it is stable in air.

The melting range of spironolactone is between 198 and 207°C (with decomposition). Occasionally, it displays preliminary melting at about 135°C, followed by resolidification. Because of the polymorphism, the solubility and melting point of spironolactone depend upon the conditions of crystallization and the solvent nature. El Dalsh et al.[208] found that spironolactone has four different modifications, as verified by X-ray analysis. None of them were soluble in water, but the extent of solubility was somewhat higher for the product obtained from EtOAc than for that crystallized from acetonitrile. At the same time, the melting point sequence was the reverse.

A similar observation was made by Salole and Sanay,[209] who synthetized three polymorphs and solvated forms of spironolactone with very different solubilities. The specific rotation is between −33° and −37°, calculated on the dried basis, determined in a chloroformic solution containing 10 mg spironolactone per ml.

Spironolactone is characterized by decreasing solubility with decreasing polarity of the solvent. It is practically insoluble in water, freely soluble in benzene and in chloroform, soluble in ethyl acetate and in alcohol, and slightly soluble in methanol and in fixed oils. Mention should be made of the efforts to stabilize spironolactone in water and in 0.1 *M* hydrochloric acid solutions by preparing its microcrystalline inclusion complexes with cyclodextrins. It was found that, of the different spironolactone *(S)* complexes formed — S α-cyclodextrin 1:1, *S*-γ-cyclodextrin 1:1, *S*-β-cyclodextrin 1:1, and S-β-cyclodextrin 1:3 — the last was the most stable, with good solubility properties.[210] This finding was confirmed by Vila et al.,[211] who identified the existence of the spironolactone-β-cyclodextrin 1:3 complex in the solid phase by means of IR, DSC, and X-ray diffraction methods. A bioavailability study revealed that approximately 45% of spironolactone is extracted as canrenone, 30% as the 1:3 S-β-cyclodextrin complex, and 25% as spironolactone itself.[211]

The partition coefficient (log P) is 2.78 (octanol).

The wavelength of maximum absorbance is about 238 nm, indicating the presence of the chromophore of the Δ^4-3-one moiety. This is also reflected by the IR spectrum of spironolactone; the conjugated carbonyl group of ring A exhibits a maximum in the IR spectrum at 1680 cm^{-1}, while the peak of the lactone carbonyl appears at 1750 cm^{-1} (Figure 8).

B. HISTORY AND STRUCTURE-ACTIVITY RELATIONSHIP

After the discovery of aldosterone **(1)** as the third genuine hormone of the adrenal cortex

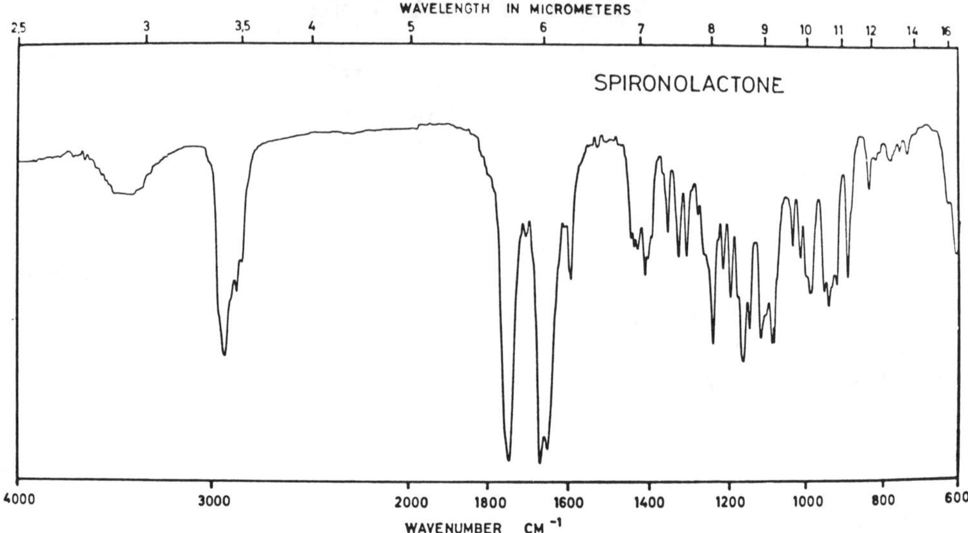

FIGURE 8. IR spectrum of spironolactone. (From Dibbern, H.-W., *UV and IR-Spektren wichtiger pharmazeutischer Wirkstoffe,* Editio Cantor, Aulendorf, 1978. With permission.)

and the demonstration of its powerful sodium-retaining activity,[212,213] the search could start for aldosterone antagonists and hence diuretically active compounds.

Aldosterone itself, which is formed in the zona glomerulosa of the adrenal cortex, causes sodium retention in the nephrons on the distal renal tubules.[214] The greatest efforts were made to synthesize active compounds which could antagonize the action of aldosterone at the site of action. The most important group of such agents is the group of steroidal spirolactones, in which the spirolactone ring is linked to ring D of the steroid skeleton by sharing C-17:

Cella and Kagawa reported the preparation and subsequent structure-activity study of potential aldosterone-inhibitor spirolactones.[215] The prototype of such compounds was a Δ^4-3-one derivative, denoted SC-5233 (**2**).

(1) (2)

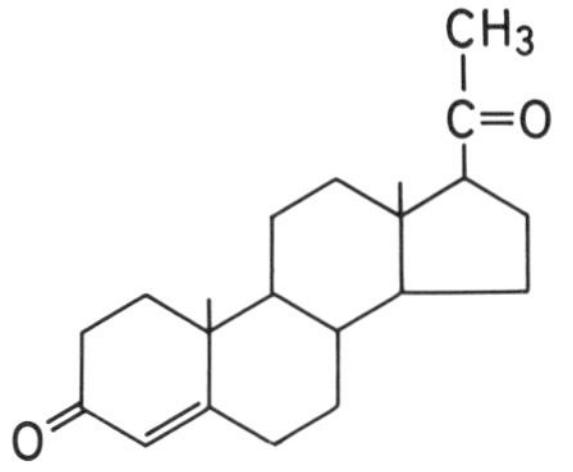

(3)

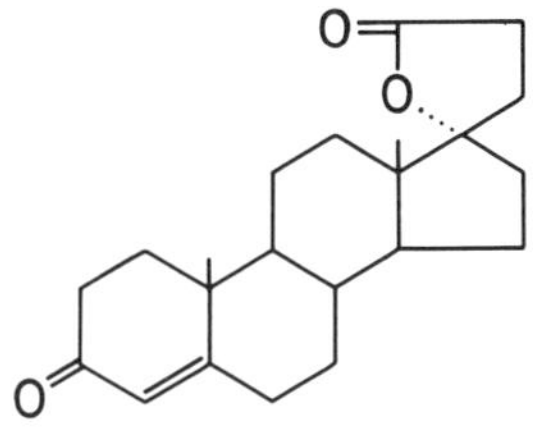

(4)

When administered subcutaneously to rats, SC-5233, the propanoic acid lactone of 17-β-hydroxyandrosterone, showed very intensive aldosterone-blocking activity.[215] In the course of the systematic work of Cella et al.[215-218] and other authors, many of the relations between the structure and the aldosterone inhibitory activity of 17-spirolactones have been elucidated. The outstanding result of this work was the preparation of spironolactone. The accumulated knowledge on the structure and action of aldosterone, and the antialdosterone effect of progesterone (**3**), may be taken as the basis of the drug design in this field. The most important conclusion of the structure-activity relationship studies may be summarized as follows. Methyl substitution in positions 2, 4, 6, 7, and 16 had relatively little positive influence on the diuretic effect[219,220] ; the experience was similar when 11-OH or 9-F substitution was performed.

Since the beginning of the search for spirolactone-type diuretics, the Δ^4-3-keto and the lactone-carbonyl groups have been regarded as the two most important (''receptor-active'') elements of the molecules. Consequently, the bulk of the research has been focused on the structural changes made on these most active sites of the compounds.

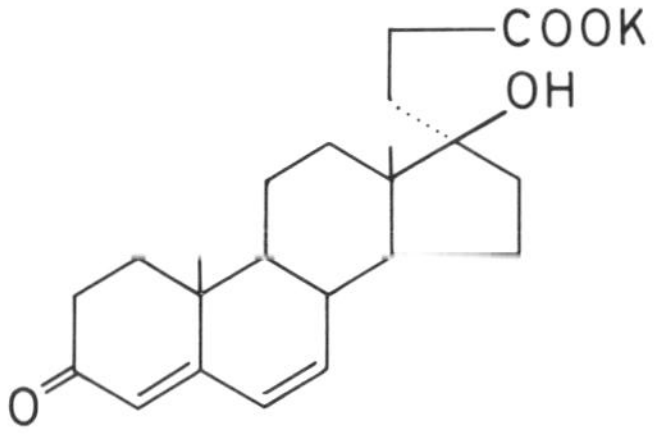

(5)

The isomer of SC 5233 with opposite configuration at position 17 (**4**) showed no electrolyte-regulating effect.[221] Potassium canrenoate (**5**), a water-soluble open lactone salt which can be prepared from SC 5233 (7-desthioacetylspironolactone) by saponification,[222] exhibited an effect that was equipotent with that of the corresponding spirolactone both orally and parenterally. The SC 5233 analog with a six-membered lactone ring at position 17 displayed diminished subcutaneous and oral activity.[213] The same is true for the corresponding spirolactam derivatives,[223] which have to be prepared through the corresponding androstane 17-oxime and nitro derivatives, bypassing the conventional synthesis route via the conversion of γ-lactones to γ-lactams by treatment of the lactone with an appropriate amine, resulting in this case in amide formation.[223-225]

Transfer of the spirolactone ring to position 16 completely eliminated the aldosterone-inhibiting activity. 16-β-Hydroxylation also caused a loss in activity of spironolactone derivatives.[226]

It should be noted that the preparation of 17-tetrahydrofuranyl-2′-spiro analog of spi-

TABLE 8
Antialdosterone Activity of Tetrahydrofuranyl Analogs of Spironolactone

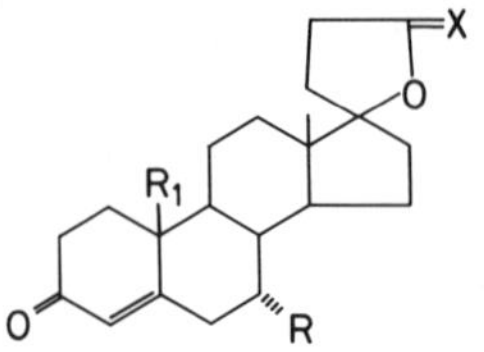

(6)

R	$-R_1$	=X	Activity: Subcutaneous	Activity: Oral
$SCOCH_3$ (spironolactone)	CH_3	O	1.0	1.0
$SCOCH_3$	CH_3	H_2	0.6—0.7	0.9—1.1
H	H	H_2	0.3—0.4	0.6—0.7
H	CH_3	H_2	0.2—0.3	0.2—0.3

rolactones (**6**) resulted in a somewhat lowered, but still very significant, antialdosterone activity.[227]

Table 8 seems to indicate that the lactone carbonyl is not an essential functional group in the spirolactone diuretic structure. Further, it shows the activity-enhancing effect of desmethylation at position 19.

As concerns the other (reactive) pole of the molecules, the carbonyl-oxygen at position 3 appears to be essential for the aldosterone-inhibitory effect of spirolactones. The relatively planar arrangement of rings A and B is ensured by the double bond at position 4-5. As in other biologically active steroid classes, further unsaturation of the rings, especially in the Δ^1 and Δ^6 positions, enhanced the oral activity.[228]

On the saturation of ring A, the activity practically disappeared in the spironolactone series. The same happened when the A/B junction was changed to *cis*.[216] Aromatization of ring A also abolished the antialdosterone activity.[216]

The route to the discovery of the outstanding importance of the 7-substitution in spirolactones led through the progesterone derivatives. Progesterone itself blocks aldosterone only in high doses. Oxygenation caused inactivity by producing the 15-keto derivative, but the 6-7 methylenation of the latter compound yielded a decreased antialdosterone activity.[229] However, progesterone showed an enhanced inhibitory effect when it was administered parenterally. On the basis of this observation, 7α-alkanethiolic acid adducts in the progesterone and spirolactone series have been prepared.[217] Similarly, the 7α-carbalkoxy group enhanced the activity. The most important compound in this series is mexrenoate-potassium (**8/a**), an open lactone derivative, which is two to four times as potent a diuretic as spironolactone.

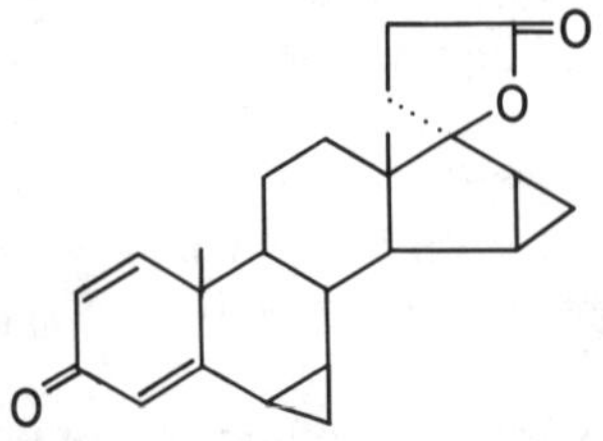

(7)

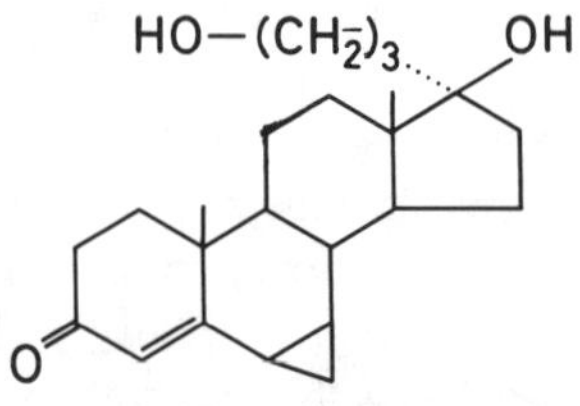

(8)

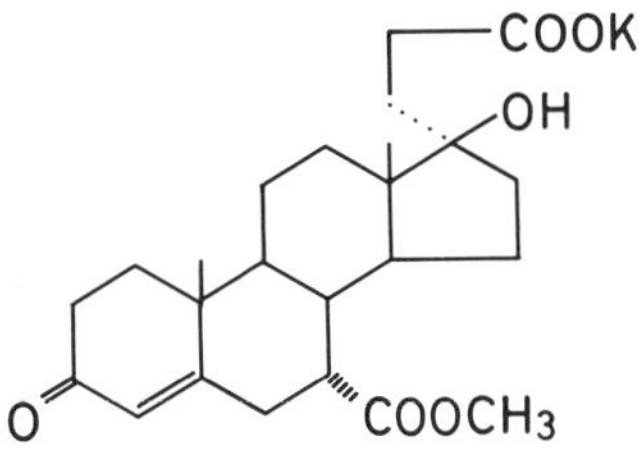

(8/a)

The observation that the antialdosterone activity of spironolactone is frequently accompanied by endocrine side-effects stimulated the search for new aldosterone antagonists. One result of this continued research was the synthesis of spironolactone dimethylene derivatives. One of the most promising compounds of this type, spirorenone, is a 3-(17β-hydroxy-6β, 7β; 15β, 16β-dimethylene-3-oxo)1,4-androstadien-17α-yl)propionic acid γ-lactone **(7)**. This compound proved 8.6 times as potent as spironolactone in inhibiting aldosterone *in vivo,* but showed a lower affinity for the mineralocorticoid receptor.[230]

The 17-open-chain analog of spirorenone **(8)** showed high *in vivo* activity. In the interpretation of this result, the transformation of compound **(8)** to an active metabolite was suggested.[228] The C-17 *(cis)* isomer of spirorenone had no biological activity and no affinity for the mineralocorticoid receptor, in accordance with the earlier results relating to the isomerism of compound SC 5233.[221]

Besides the diuretic effect, the nature and structure of the 17-substituents were suggested to have important roles in the hypotensive effect of spironolactone and its derivatives. The observation that spironolactone inhibits the excretion of the prostaglandin-E metabolite U-PGE-M resulted in the hypothesis of the possibility that aldosterone, angiotensin, spironolactone, and prostaglandin-E might act at the same locus in a highly associated mechanism related to the regulation of blood pressure.[231] Of the 52 compounds tested, only spironolactone and three of its derivatives **(9** and **10-12)** inhibited U-PGE-M excretion:

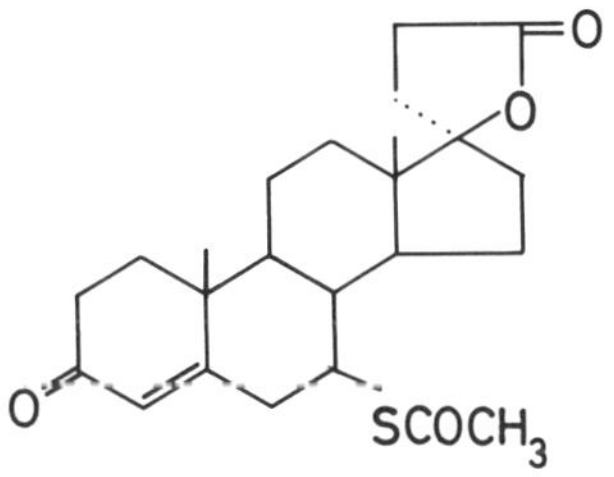

(9) 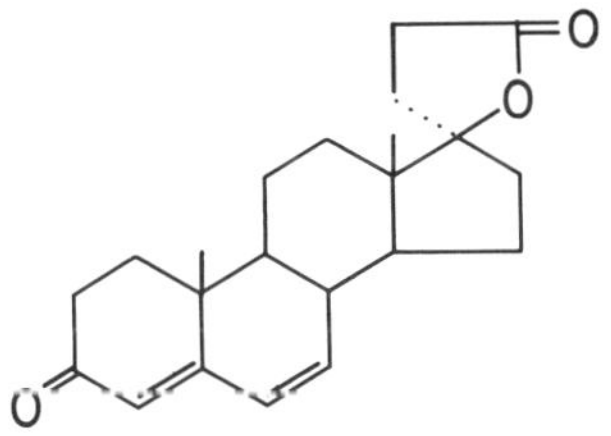 (10)

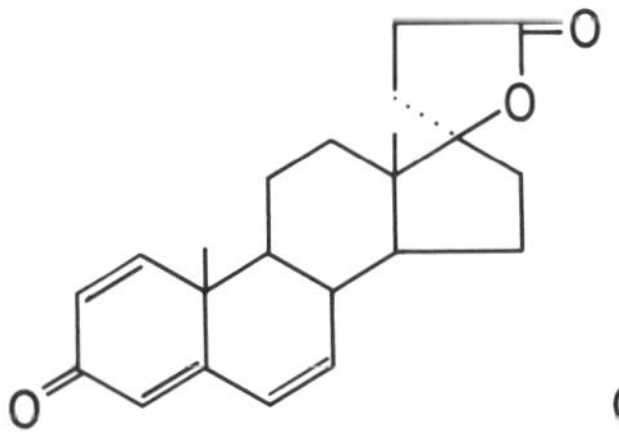

(11) 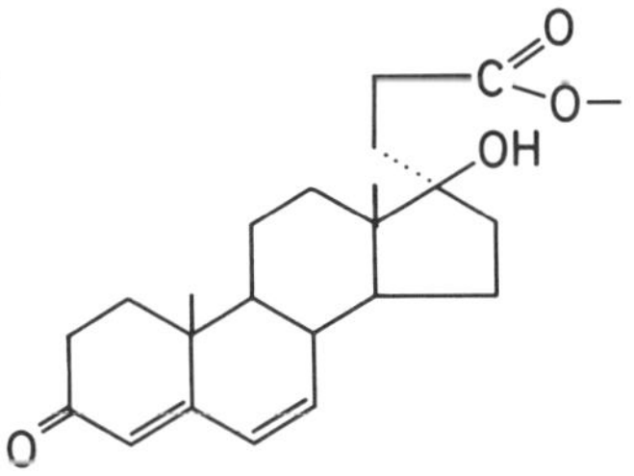 (12)

Examination of formulas *9-12* clearly demonstrates the close structural relation of *10-12* and spironolactone **(9)**. Moreover, the derivatives *10-12* are all metabolites (active?) of spironolactone. In the same work,[231] the surprisingly severe structural requirements for the inhibitory or stimulatory activity (blood pressure elevation or depression) of U-PGE-M were revealed.

The sexual endocrine side-effects of spironolactone seem to be successfully depressed in the epoxyspirolactone derivatives prepared by de Gasparo et al.[232] The introduction of a 9 α, 11α-epoxy group yielded the greatest selectivity; there was only a marginal lowering effect on the binding ability to the mineralocorticoid receptor, but it caused a definite decrease as regards the androgen and progesterone receptors (the latter should be involved in the usual sexual side-effects of spironolactone). Numerical data are given Table 9.

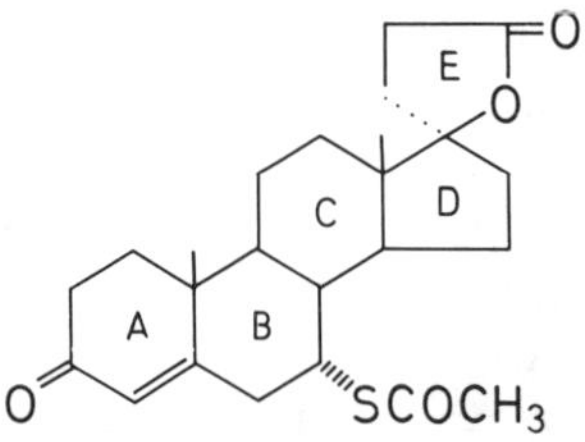

Spironolactone (A, B, C, D, E; O; $SCOCH_3$)

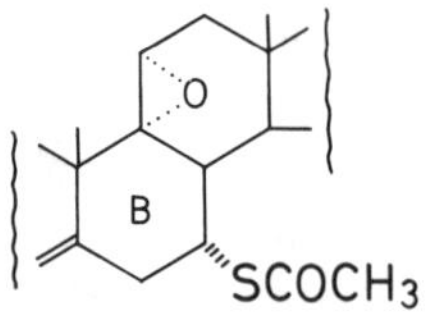

(13) (B; O; $SCOCH_3$)

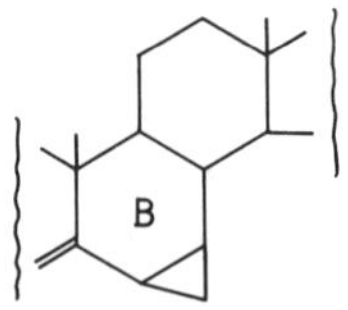

(14) (B)

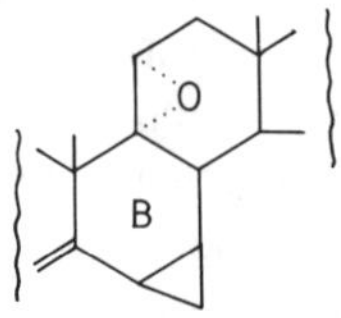

(15) (B; O)

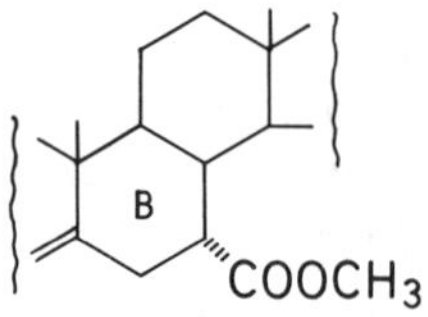

(16) (B; $COOCH_3$)

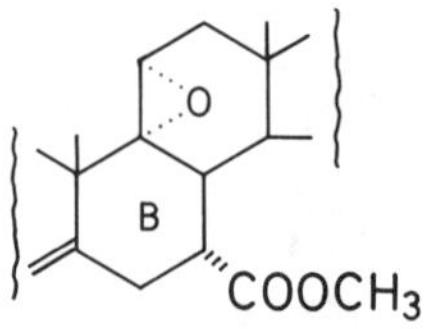

(17) (B; O; $COOCH_3$)

C. SYNTHESIS

In the reaction of 3β-hydroxy-5-androsten-17-one **(18)** and acetylene, 17α-ethynyl-3β, 17β-dihydroxy-5-androsterone **(19)** is formed, which is transformed to the 17α-propiolic acid **(20)** with methyl magnesium bromide and carbon dioxide. After selective reduction to the saturated acid, acidification afforded the unsaturated lactone, which was reduced catalytically to the lactone **(21)** immediately. Oppenauer oxidation yielded the 3-(17β-hydroxy-3-oxo-4-androstene-17α)propionic acid lactone **(22)**. On treatment with chloranil and thioacetic acid, the spironolactone was formed.[215]

TABLE 9
Relative Binding Affinities of Three Known Aldosterone Antagonists and Their Epoxy Derivative (Mean ± SE)[232]

	Mineralocorticoid receptor (aldosterone = 1)	Glucocorticoid receptor (dexamethasone = 1)	Androgen receptor (methyltrienolone = 1)	Progesterone receptor (progesterone = 1)
Spironolactone	1.1×10^{-1} ±36.0% (7)	1.8×10^{-3} ±16.1% (4)	9.1×10^{-3} ±16.5% (7)	7.0×10^{-3} ±11.0% (5)
Epoxyspironolactone (**13**)	7.5×10^{-2} ±4.3% (3)	7.5×10^{-3} ±9.5% (3)	1.1×10^{-5} ±16.4% (4)	7.4×10^{-4} ±16.8% (3)
Prorenone (**14**)	2.5×10^{-1} ±8.5% (10)	3.7×10^{-3} ±10.8% (6)	1.9×10^{-3} ±22.6% (4)	1.2×10^{-1} ±20.0% (5)
Epoxyprorenone (**15**)	1.2×10^{-1} ±8.3% (6)	5.0×10^{-4} ±8.0% (5)	2.4×10^{-4} ±14.6% (4)	8.3×10^{-3} ±9.8% (5)
Mexrenone (**16**)	7.5×10^{-2} ±2.6% (7)	3.1×10^{-3} ±9.7% (4)	1.1×10^{-3} ±12.7% (5)	1.2×10^{-3} ±21.7% (5)
Epoxymexrenone (**17**)	5.1×10^{-3} ±6.5% (7)	1.8×10^{-4} ±22.2% (4)	7.6×10^{-6} ±26.3% (6)	$<5 \times 10^{-5}$[a] (5)

Note: The SEs are expressed in percentage of the mean for easier comparison. Numbers in parentheses are number of determinations. (From de Gasparo et al., *J. Pharmacol. Exp. Ther.*, 240, 650, 1986. With permission.)

[a] Due to poor solubility in the assay conditions, this value is difficult to estimate.

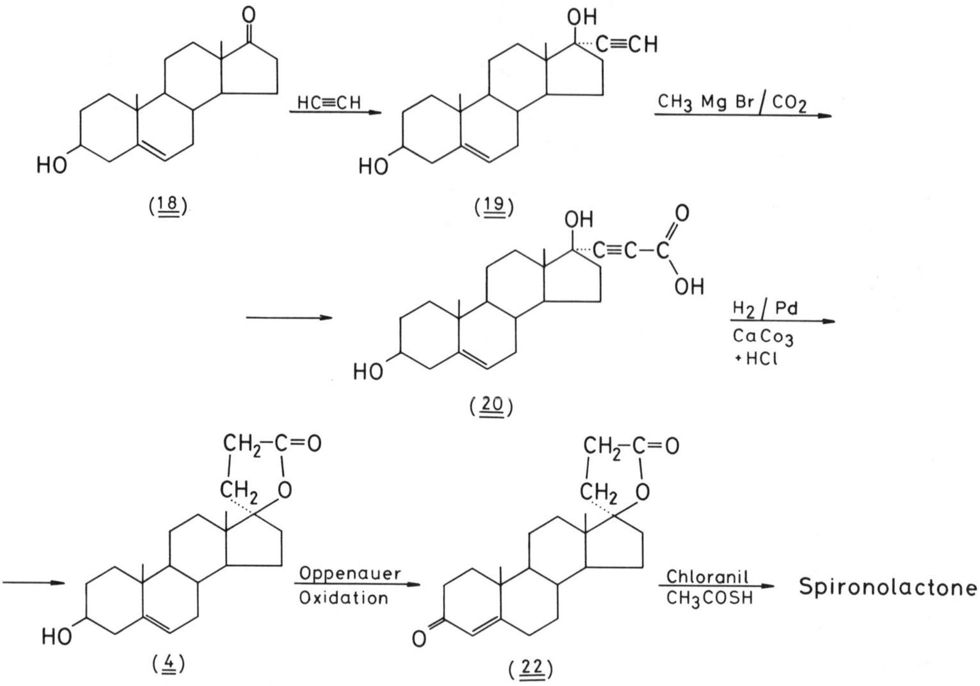

(18) —HC≡CH→ (19) —CH_3MgBr/CO_2→ (20) —H_2/Pd, $CaCO_3$ + HCl→ (4) —Oppenauer Oxidation→ (22) —Chloranil, CH_3COSH→ Spironolactone

D. ABSORPTION AND METABOLISM

Spironolactone is absorbed almost quantitatively following oral administration. 40 to 50% of it is excreted in the urine, and 20 to 40% in the feces. Its metabolites are formed mainly through the oxydative and hydrolytic splitting of the 7α-acetylthio group. Besides the major metabolite, canrenone **(23)**, other bioactive metabolites, a 7α-thiomethyl **(24)**, a 7α-thiol **(25)**, and a 6β-hydroxy-7α-methylthispirolactone **(26)** have been identified.

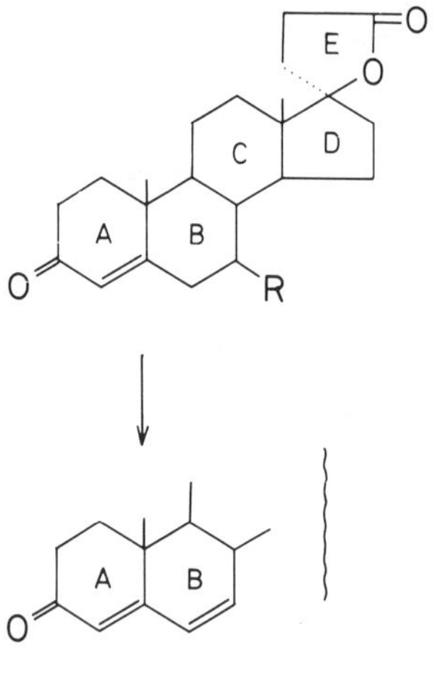

R	
SCH_3	(24)
SH	(25)
SCH_3 , 6β–OH	(26)
$SCOCH_3$	Spironolactone

(23)

E. ANALYSIS

The identification of spironolactone in USP is carried out through its IR and absorption spectra. Additionally, a chemical reaction is described. The sulfur content of the molecule can be detected by the addition of lead acetate (PbS formation).

Spectrophotometry is the most frequently used method for the determination of spironolactone in bulk as well as in pharmaceutical forms. The instructive data on the procedures are summarized in Table 10.

Fluorimetric measurements were carried out for the determination of canrenone by Neurath and Ambrosius.[239] On reaction with sulfuric acid, a fluorophore can be produced from canrenone and other metabolites, even those having an open ring:

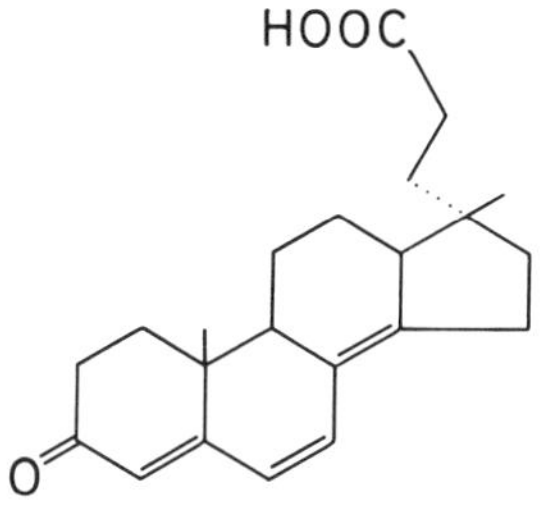

This reaction makes possible the determination of canrenone in ng amounts in serum samples: λ_{exc}: 457 nm, λ_{emiss}: 520 nm. The reaction between spironolactone and isoniazide in acidic methanol-chloroform medium has been used by others.

Spironolactone has been determined with acceptable accuracy by means of a polarographic and spectrophotometric routine analyzer for assaying content uniformity in pharmaceutical quality control.[240] HPLC is suitable for the estimation of spironolactone simultaneously with its metabolites or degradation products or analogs (Table 11).

VII. AMILORIDE HYDROCHLORIDE

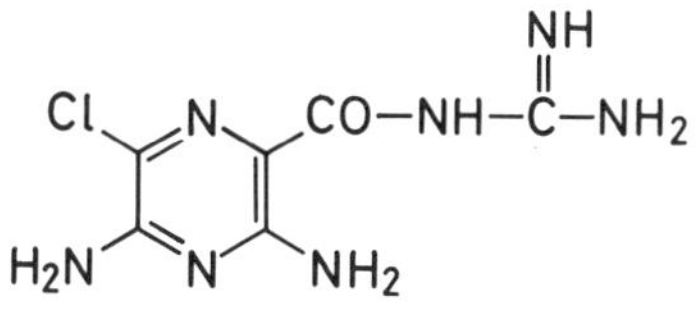

Pyrazinecarboxamide, 3,5-Diamino-*N*-(aminoiminomethyl)-6-chloro, monohydrochloride dihydrate

N-Amidino-3,5-diamino-6-chloropyrazinecarboxamide monohydrochloride dihydrate

$\underbrace{\overbrace{C_6H_8ClN_7O \cdot HCl}^{(b)}}_{(a)} \cdot 2H_2O\}(c)$ $\qquad$ M_r = 231.64(a), 266.1(b), 302.12(c)

A. PROPERTIES

A yellow to greenish-yellow, odorless or practically odorless powder. It is slightly soluble in water; insoluble in ether, in ethyl acetate, in acetone and in chloroform; freely soluble in dimethylsulfoxide; and sparingly soluble in methanol.

Its melting point is 240.5 to 241.5°C. The UV absorption maximum is at 361 nm. Through its amidine structure, amiloride is a definitely basic compound with a pK of 8.67.

B. HISTORY AND STRUCTURE-ACTIVITY RELATIONSHIP

The leading idea for the search for diuretics among pyrazine derivatives most probably arose when the diuretic activity of the pteridines was revealed at the beginning of the sixties. A series of pyrazinoylguanidine derivatives (**1**) proved to be highly effective natriuretic diuretics.[245-253] The outstanding feature of these substances is that their potassium-retaining ability is more marked than that of any other previous saluretics. They are also active in

TABLE 10
Spectrophotometric Methods for the Determination of Spironolactone

Reagent (medium)	Color of product	λ_{max}, nm	Sensitivity range of linearity	Note	Ref.
Na-nitroprusside (alkaline-solvent)	green	700	4 μg/ml	In pharm. formulations; other diuretics do not interfere.	233
Phosphotungstic acid	yellow	410	2 μg/ml		
2,3,5-Triphenyltetrazolium chloride + tetramethylammonium hydroxide	orange-yellow	480	10—70 μg/ml	In tablets; tartrazine amaranth and hydroflumethiazide do not interfere.	234
1,4-Dinitrobenzene (ethanol + ethanolic NaOH)	yellow	400	—	In tablets; common tablet excipients do not interfere.	235
Isoniazide (acidic methanol + chloroform)	yellow	375	5—40 μg/ml	In tablets; hydroflumethiazide does not interfere.	236
Folin-Ciocolteau reagent				Hydrochlorothiazide does not interfere.	237
Thiosemicarbazide (acidic medium)	orange	440	4—20 μg/ml	—	238

TABLE 11
HPLC Methods for the Estimation of Spironolactone and Its Metabolites

Compound(s) present together with spironolactone	Stationary phase	Mobile phase	Detection	Sensitivity/range of linearity	Note	Ref.
II, III, V Progesterone (I.St.)[a]	C_{18} Radialpak (5 μm)	Methanol-water (gradient elution methanol 65—100%)	UV 254 nm	25 μg/ml	In plasma, after extraction with ethyl acetate. Dry residue dissolved in acetonitrile.	241
III, IV, V	Partisil	Isopropyl ethermethanol (393:7)	UV 240 nm	5—20 μg/ml 50—400 μg/ml	In serum after extraction with CCl_4. Dry residue dissolved in the mobile phase.	242
VI, VII	RP	—	—	180-900 μg/ml (90 μg/ml VI) 250—900 μg/ml (180 μg/ml VII)	In human plasma.	243
Hydroflumethiazide	μBondapak-NH_2	Chloroform-methanol 45:55	—	—	In different dosage forms.	244
V	LiChrosorb Si-100 (10 μm)	Chloroform-methanol 45:55	UV 283 nm	5 μg/ml 1.55—20 μg	In serum and urine after extraction with chloroform.	239

Note: II: 7α-thiospironolactone; III: 7α-methylthiospironolactone; IV: 6 β-hydroxy-7α-methylthiospironolactone; V: Canrenone; VI: 7 α-methyl ethyl ester of spironolactone; VII: 6 β-Hydroxy-7α-methyl ethyl ester of spironolactone.

[a] I.St.: internal standard.

TABLE 12
DOC Inhibitory Effect of *N*-guanidinopyrazine Carboxamide Derivatives[254]

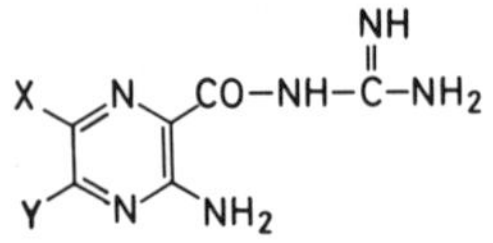

Y	X	Inhib. score
H	H	0
H	Cl	+3
H	Br	+2
H	I	+1
Cl	Cl	0
CH_3O	Cl	±
HO	Cl	±
CH_3S	Cl	+1
NH_2	H	+1
NH_2	Cl	+4 (amiloride)
NH_2	Br	+3
NH_2	I	+3
$(CH_3)_2N$	Cl	+3
	Spironolacetone	+1
	Triamterene	+2

the absence of mineralocorticoids, and they are therefore regarded as nonspecific inhibitors of the Na^+ uptake in the cells. Accordingly, their diuretic effect is based on reversal of the distal tubular handling of electrolytes, resulting in an enhanced extraction of both sodium and urine, and an inhibition of the renal loss of potassium.[246]

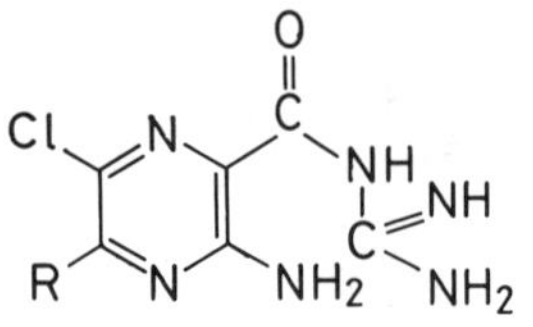

R = NH_2, amiloride

(1) (2)

The most potent and therapeutically most valuable compound of this series proved to be amiloride **(2)**, the guanidide of substituted pyrazinecarboxylic acid. The results of studies on the diuretic structure-activity relationship were reported in several papers by Cragoe and his team from the Merck, Sharp and Dohme Laboratories[247-250] in the sixties. Recently, a new wave of interest turned towards amiloride research, as concerns its Na^+ flux-inhibiting action in the different tissues and organs. Although this latter trend of amiloride research lies outside the diuretic field, some of the main results, together with the conclusions that hold for diuretics, are summarized below.

An early observation was made that the diuretic activity of pyrazinecarboxamides is markedly increased by the introduction of a halogen[247] or methyl[249] substituent at position 6. This observation was found to be valid for the substitution at position 5 too. Table 12 shows the inhibition scores for the compounds when 12 μg of desoxycorticosterone was administered to saline-loaded and adrenalectomized rats.

It may be seen that halo-substitution at position 6 is almost essential for amiloride-type diuretics to exert a marked activity. Whether this 6-substituent itself interacts with a receptor unit or merely has a modifying role on the electron density is a question that remains to be answered. An amino substituent at position 5 has a definite activating influence on the heteroaromatic ring, increasing the inhibitory activity of the molecule. The structure-activity relationships of amiloride and 38 of its analogs were studied by Vigne et al.[255] in relation to blockade of the Na^+/H^+ exchange system. One of their interesting findings suggests that the diuretic properties of amiloride and its derivatives bearing substituents on the terminal guanidino nitrogen are not due to direct inhibitory action on the Na^+/H^+ exchanger. This hypothesis[255] is based on the fact that amiloride derivatives with a substituent on the guanidino group are inactive as regards inhibition of Na^+/H^+ exchange, but they have the same diuretic ability as amiloride itself.[256] However, it must be confirmed that the isolated chick skeletal muscle (on which the experiments were performed)[255] and the living organism behave in the same manner metabolically. It is reasonably possible that the diuresis observed for the *N*-substituted guanidine derivatives of amiloride is due to their active metabolite amiloride.

It is also worthy of mention that the 6-F analog of amiloride is less active than the other haloderivatives; the latter have equipotent activity.[256] This fact seems to support the assumption of the ring-activating role of the substituent at position 6. The outstanding electron-attractive force of fluorine may counteract the positive mesomeric effect much more than in the case of the other halogen atoms. Substitution of the 5-amino group increased the inhibitory potency of amiloride derivatives. This effect was particularly strong for N-disubstituted compounds. With regard to the electron-withdrawing property of alkyl substituents, this observation again underlines the ring-activating role of the 5-amino group.

Although the function of the 3-amino group has not been completely elucidated, it seems to be less important than the other units of the amiloride molecule.

In spite of the fact that amiloride reversibly inhibits several Na^+ tranport mechanisms in living organisms, it undoubtedly works with the highest efficacy in the case of epithelial Na^+ channels. This ability of several amiloride analogs substituted on the terminal-N of the guanidine unit has been studied very thoroughly by Li et al.[257]

Chain elongation by terminal substitution on the guanidine nitrogens led to no surprises. All of the *N*-alkyl or aralkyl derivatives (pK_a ranging between 7.5 and 8.7) have very similar inhibitory activities. Elongation of the chain on C-2 by means of –O– or –NH– group insertion, and replacement of the amidino terminal by some other nitrogen-containing group, were also studied. The results of this work[258] confirm the earlier assumptions[259] relating to the binding of amiloride derivatives at the sites of the epithelial Na^+ channels, which takes place in two steps:

1. The positively charged guanidinium side chain invades the outward-facing channel entrance, which has a fixed negative charge. The stability of the encounter complex formed codetermines the value of the "on rate constant", K_{on}. When the channel contains Na^+, formation of the complex is impossible, i.e., Na^+ exerts a very strong competitive effect.
2. In a second step, the complex is transformed into a more stable blocking position, which may be aided by an adaptive change of the receptor conformation.[259] The negative charge (electronegativity) on the ligand at position 6 strongly influences the complex stability in the "blockade state" (blocking complex stability).

It has been suggested[258] that the ligand on C-6 binds to a suitable (positively charged) area of the receptor. In the latter respect, recent review papers may provide further valuable information.[260-262]

From the rather broad range of Na^+-transporting sites blocked by amiloride, the most known and well characterized are the apical Na^+ channels which mediate the luminal entry of the Na^+ ion. These channels have a high affinity for amiloride and some related test compounds (benzamil, phenamil), which form encounter complexes with a constant less than 10^{-6} *M*.[263] The relationship between the structure and blocking activity of amiloride-type compounds was elucidated to a large extent during the research on the diuretic activity of amiloride. Much less information is available on another pathway of Na^+ cross-membrane transport. This pathway has a relatively low but definite affinity for amiloride.[264] The structure-activity relationships for this pathway are quite different from those established for the Na^+ channels[265,266] or Na^+/H^+ exchangers.[267,268] The structural variation among the amiloride derivatives caused only small changes in the inhibitory constants for this electrogenic pathway of Na^+ ions.[264] Structure-activity relationships were also revealed when the inhibitory property of a series of amiloride derivatives on Na^+/H^+ exchange in A431 cells was studied by Zhuang et al.[271] A number of 5-N-alkyl-substituted amiloride derivatives displayed a much higher affinity than that of the parent compound for the Na^+/H^+ antiporter in A431 cells. In accordance with this finding, 5-N-alkyl and dialkyl amiloride derivatives showed a 10 to 500-fold increased potency in the inhibition of chemotactic factor-activated Na^+/H^+ exchange in human neutrophils.[272] The data of Rocco et al.[273] suggest that the 2-carbonylguanidinium moiety and the 6-chloro atom are important for the binding of amiloride and its derivatives to sites at or near the rabbit renal Na^+/H^+ antiporter.

It has been found[274] that amiloride reversibily inhibits Na^+/Ca^{2+} exchange in rat brain synaptosomal plasmalemme vesicles. The inhibitory ability proved to be a pH-dependent property, and amiloride is more effective at lower than at higher pH values; in other words, amiloride and its derivatives must be present in protonated form. As a result of this work,[274] it can be stated that the Na^+/Ca^{2+} exchange system is distinct from the epithelial Na^+ system in several respects. The differences are mirrored by the different structural requirements for increasing inhibitory effect. For example, although a protonated guanidino side-chain is required by both inhibitory systems, the elongation of this side-chain (by the insertion of an NH group between the carbonyl and guanidino moieties) exerts quite different effects: where the elongated derivative is comparable to amiloride as an inhibitor of epithelial Na^+ transport, it has no inhibitory effect on Na^+/Ca^{2+} exchange.[274]

C. SYNTHESIS

The systematic structure-activity relationship study among pyrazine diuretics was prompted by the strong effects of the N-amidine-3-amino-6-halopyrazine carboxamides.[248] The amiloride-type substances have been synthetized by the reaction of the appropriate ester **(3)** with a guanidine derivative **(4)**. The reaction can usually be carried out by heating the ester with a methanolic solution of guanidine or the N-substituted guanidine derivative:

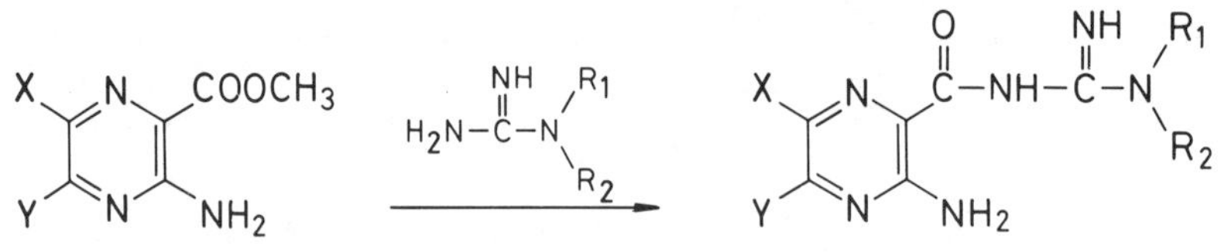

X, Y, N, N, $COOCH_3$, NH_2 (3) — $H_2N{-}C({=}NH){-}N(R_1)(R_2)$ → X, Y, N, N, $C({=}O){-}NH{-}C({=}NH){-}N(R_1)(R_2)$, NH_2 (4)

(3) (4)

Treatment of methyl 3-amino-5,6-dichloropyrazinecarboxylate **(5)** with guanidine afforded acylguanidine, which reacted with dimethylamine in dimethylformamide solution to give the dimethylamino derivative of amiloride. In the analogous reaction, but when liquid ammonia was used instead of dimethylamine and the reaction was conducted in a highly polar solvent (dimethylsulfoxide, dimethylformamide, dimethylsulfone, etc.), methyl-3,5-diamino-6-chloropyrazinecarboxylate **(6)** was formed in good yields.[248] The latter reacts with guanidine to give amiloride **(7)**.

$$\text{(5)} \xrightarrow[(CH_3)_2SO_2]{NH_3\ (liqu)} \text{(6)} \xrightarrow{H_2N-C(=NH)-NH_2} \text{(7)}$$

(5) Cl, Cl, $COOCH_3$, NH_2 — (6) Cl, H_2N, $COOCH_3$, NH_2 — (7) Cl, H_2N, NH_2, $C(=O)-NH-C(=NH)-NH_2$

D. ANALYSIS

The chromatographic methods, and especially HPLC, are of outstanding importance in the analysis of amiloride in dosage forms and in biological samples.

USP prescribes a TLC purity test on silica gel, with tetrahydrofuran-3 N ammonium hydroxide as mobile phase. Being a rather strong base (pK_a = 8.67), amiloride can be measured as its hydrochloride salt by acidimetric titration in nonaqueous medium, after the addition of mercury(II) acetate and crystal violet indicator (USP).

The combination of HPLC and spectrofluorimetry proved highly effective in the analysis of amiloride in biological samples. After extraction into a silica cartridge and perchloric acid elution, HPLC with a μBondapak C_{18} column and a methanol-0.1 *M* sodium perchlorate (pH = 4.0) 40:60 mixture as mobile phase was used for the determination of amiloride in plasma and urine by Vincek et al.[275] The fluorimetric detection (λ_{exc}: 368 nm, λ_{emiss}: 415 nm) was suitable for the determination of ng amounts of amiloride by this[275] and the similar procedure of Yip et al.[276] and, more recently, Forrest et al.[277] Amiloride has been assayed[278] in a pharmacokinetic study on a Spherisorb ODS column with acetonitrile-methanol-ammonium phosphate buffer (pH = 2.8) as mobile phase, again with fluorescence detection. Amiloride hydrochloride was detected and quantitated in tablets by spectrofluorimetry,[279] on the basis of measurement of the fluorescence of amiloride in acetic acid medium (λ_{exc}: 364.5 nm, λ_{emiss}: 440 nm). Beer's law was obeyed in the interval 0.1 to 1 $\mu g\ ml^{-1}$. Hydrochlorothiazide and common excipients used in the tablet formulations did not interfere in the determination. A spectrophotometric determination of amiloride was described by Sastry et al.,[280] who added sodium hydroxide and Folin-Ciocolteau reagent and measured the absorbance at 740 to 750 nm. Beer's law was obeyed in the range 1 to 12 $\mu g\ ml^{-1}$. Another interesting method of amiloride determination was reported by Sastry et al.,[281] in which 3-methylbenzothiazolin-2-one hydrazone was used as spectrophotometric reagent. The thiazide diuretics were hydrolysed in alkali with Ce(IV), Fe(III) or IO_4^- as oxidant before the reaction. Absorbance was measured at 545 nm. The color was stable for 2 h. The methods described were applied to dosage forms.

VIII. TRIAMTERENE

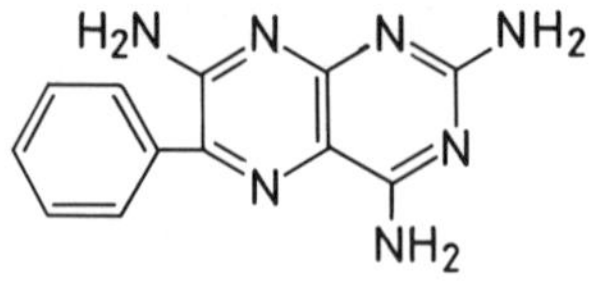

2,4,7-Pteridinediamine, 6-phenyl-
2,4,7-Triamino-6-phenylpteridine
$C_{12}H_{11}N_7$ $M_r = 253.27$

A. PROPERTIES

This is a yellow, odorless, crystalline powder. It is practically insoluble in water, in benzene, in chloroform, in ether, and in dilute alkali metal hydroxides. It is soluble in formic acid (1 g/30 ml), sparingly soluble in methoxyethanol (1 g/85 ml), and very slightly soluble in acetic acid, in alcohol and in dilute mineral acids. Its melting point is 316°C. The partition coefficient (log P) is 0.98 (octanol).

Triamterene is a much weaker base than amiloride (pK_a = 6.2 and 8.7, respectively) because of the aromatic nature of its amino groups.

Crystallographic measurements have revealed that the molecules of triamterene are linked into ribbons by H-bonds between both H atoms of the amino group and N(1) and N(3) of adjacent rings. The ribbons are connected by paired N(7)-H....N(8)H bonds around centres of symmetry.[282]

Triamterene is identified through its IR spectrum and the reaction in formic acid (1 in 1000) which produces an intense bluish fluorescence (USP).

A TLC method is offered for identification and also for testing of the purity of triamterene.

On Silufol plates with a methyl ethyl ketone-methanol-ammonia mixture as mobile phase,[283] or on a silica gel 60 G layer containing fluorescence indicator, with an ethyl acetate-methanol-ammonia 80:10:10 mixture as mobile phase,[284] triamterene can be detected via the bright-blue fluorescence or by spraying the plate with Meyer's reagent.

B. HISTORY, STRUCTURE-ACTIVITY RELATIONSHIP, AND SYNTHESIS

The variety of the biological action of pteridine derivatives has been known for several decades. The systematic research into pteridine diuretics was prompted by the observation of renally active xanthopterine. It may be assumed that the activating property of the phenolic moiety served as leading idea for the synthesis of pteridine compounds containing other ring-activating substituents. The chain of events began with the report by Wiebelhaus, Weinstock et al.[285] about the synthesis of triamterene, a compound which proved to be a very effective diuretic agent.[286,287] The research was extended to a rather broad circle of pteridine derivatives; synthetic methods and structure-activity relationships were described in a number of papers by Pachter et al.[288-291] and by the team of Weinstock[292-299] from the Smith, Kline and French laboratories (U.S.).

The main structural types of pteridines prepared in the above-mentioned research programs and found to be compounds with strong diuretic activity may be seen in Table 13.

Besides the types of compounds detailed in this Table, some other classes of aminopteridines have been prepared,[288-299] but their diuretic-natriuretic activity and/or potassium-retaining property are far poorer than those of triamterene.

To comment on the structure-activity relationship data in Table 13, it must be realized that the aminopteridines are weak organic bases; their electronegativity should be localized around the pyrimidine unit of the pteridine skeleton. Another point to be considered is that at the physiological pH of the human organism (pH$\sim$7) triamterene and its derivatives ($K_a \sim 10^{-6}$) are present mainly ($\sim$90%) as free bases. It is very plausible to assume that the main point of interaction of dimethyl pteridines should be this Lewis base centre of the

TABLE 13
Synthetic Diuretic Pteridines

Structural type (number of tested compound)	References	Comments		
Diaminopteridines	294,299—303	Maximum diuretic activity:		
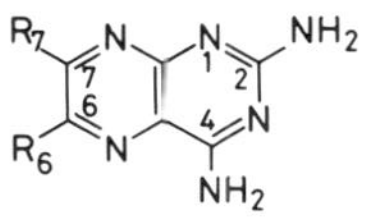		R_6		R_7
		H		H
		CH_3		H
		CH_3		CH_3
			$CH_2CH_2CH_3$	
(22)		$COOCH_3$		H
		$COOCH(CH_3)_2$		H
		C_6H_5		OH
		Inactivity:		
		CH_3		OH
		$CONH_2$		OH
		COOH		H
		CH_3		C_6H_5
			n-C_5H_{11}	
			n-C_3H_7	
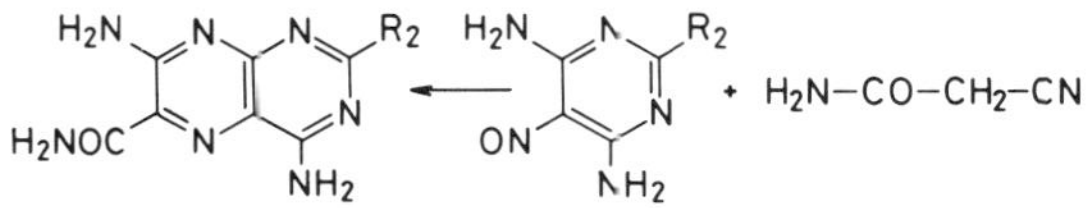	292	Maximum activity:		
			R_2	
			C_6H_5	
			p-$CH_3C_6H_5$	
			m-$CH_3C_6H_5$	
(16)			$N(CH_3)_2$	
		Inactivity:		
		H, CH_3, $CH_2C_6H_5$, NH_2, etc.		
Triaminopteridines	293		R_6	
		Maximum activity:		
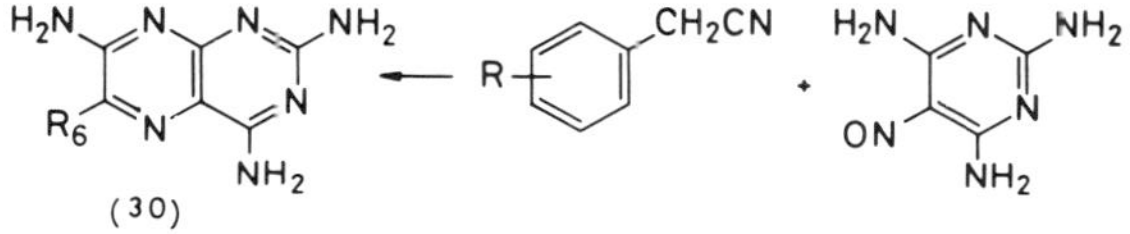			C_6H_5, CH_3, C_2H_5	
		Inactivity:		
			p-OHC_6H_4	
			p-$H_2NC_6H_4$	
(30)			COOH	
			$CONH_2$, etc.	

TABLE 13 (continued)
Synthetic Diuretic Pteridines

Structural type (number of tested compound)	References	Comments	
Dimethylpteridine	294, 301, 304—310	R_2	R_4
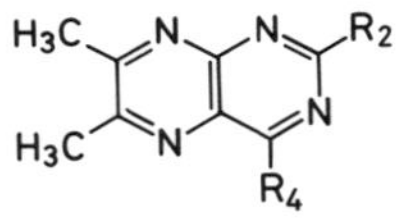 (15)		Maximum activity:	
		NH_2	NH_2
		Inactivity:	
		H	NH_2
		CH_3	NH_2
		C_6H_5	NH_2
		OH	NH_2
		NH_2	CH_3, etc.

molecules; hence, a rather bulky electron-deficient site with a hydrophobic nature must be involved on the receptor surface. The amino substituent, which is present in almost all active compounds, can play only a role of minor importance in the receptor-pteridine interaction; it may either activate the heteroaromatic rings or/and form H bonds with the appropriate unit of the receptor. It should be noted that, of the 2,4-diaminopteridines, only the 6,7-dimethyl derivative exhibited a high activity by both methods of rating. Although the carboxamide moiety at position 6 provides a good basis for the diuretic activity in 4,7-diaminopteridines, only the 2-phenyl and the *p*- and *m*-methylphenyl derivatives exerted effects that might be strong enough for therapeutic use. It may be concluded that very few of the hundreds of pteridines can rival triamterene in diuretic-natriuretic and potassium-retaining ("sparing") action. When this fact is considered together with the relatively simple structure of triamterene, the difficulties involved in the more precise elucidation of the receptor site, e.g., receptor-pharmacon interactions, can be realized.

Although triamterene may be regarded as a good potassium-retaining diuretic, which is also successfully used in the therapy of hypertension, it must be taken into consideration that the compound is only poorly soluble in water, and thus it is not commercially available for parenteral administration. The potassium-retaining effects of hydroxytriamterenesulfuric acid ester, one of the main metabolites of triamterene,[311,312] stimulated the systematic synthesis of acidic triamterene esters.[313] These compounds display good solubility properties and can be combined with high-ceiling diuretics; they should be potential candidates for the preparation of parenteral medicines. The optimum balance for natriuretic and antisaluretic properties was found[313] when the 3-side-chain contained five carbon atoms **(1)**. The water solubility is higher for branched-chain (iso-) derivatives **(2)**, but this effect coincides with a weakening of the antipotassuretic ability.

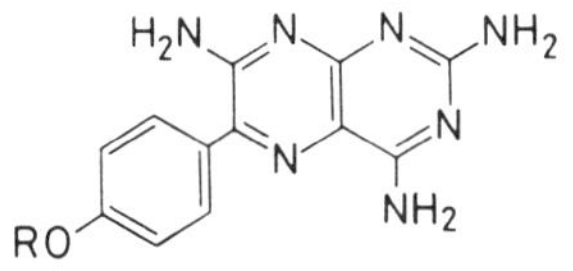

H_2N, N, N, NH_2, N, N, NH_2, RO

R = $-(CH_2)_4-COOH$ (1)

$-H\overset{*}{C}(-(CH_2)_2-CH_3)-COOH$ (2)

$O\ (CH_2)_n-N(CH_3)_2$ (3)

$-O-CH_2-CH(OH)-CH_2-N(R_1)(R_2)$ (4)

n = 2–3

$R_1, R_2 - -CH_3, -C_2H_5$

The acidic side-chain may represent a second binding moiety in these substances.[313] Whether the changes in biological activity accompanying the changes in the number of side-chain carbon atoms are related to the differences in pharmacon-receptor interactions, or whether the absorption and transport are influenced by the changes in lipophilicity, is a question that remains to be answered. The metabolism may also lead to new interpretations of the mechanism of action of the acidic pteridines.

The above-mentioned problems are complicated by the finding[314] that the potassium-retaining property of hydroxytriamterene ethers is greatly increased when the ω-carboxy group of **1** or **2** is substituted by a basic unit **(3)**. A further increase in anti-potassuretic ability was observed on OH-substitution in the basic side-chain **(4)**.

C. ANALYSIS

For the assay of triamterene, USP prescribes acidimetry in a nonaqueous medium in a previously prepared mixture of strong levelling solvents: formic acid-acetic anhydride-glacial acetic acid (1:1:2).

TABLE 14
HPLC Methods for the Estimation of Triamterene

Compound(s) present together with triamterene	Stationary phase	Mobile phase	Detection	Sensitivity/range of linearity	Note	Ref.
6-(4-Methoxyphenyl) analog (I.St.)	μBondapak C_{18} (10 μm)	Methanol-water 0.1% KH_2PO_4 (pH = 3.8) 9:11	(1) F 440 nm emiss, 365 nm exc (2) UV, 230 nm	1 μg/ml 20—40 μg/ml in plasma 10 μg/ml UV 100—800 μg/ml in urine	In plasma and in urine, after extraction with ethyl acetate. Dry residue dissolved in methanol	314
—	Lichrosorb Si 60	Methylene chloride-hexane-methanol-70% H_3PO_4 57:35:8:0.1	F 470 nm emiss, 320 nm exc	2—1000 μg/ml	In plasma or urine after extraction with $HClO_4$ + isobutyl methyl ketone; directly injected	315
p-Hydroxy-T	μBondapak C_{18}	Acetonitrile-water-acetic acid 120:79:1	UV, 365 nm	20 μg/ml 40—240 μg/ml	In plasma, after extraction with ether-isopropyl alcohol 19:10	316
Hydrochlorothiazide, 3′-hydroxy acetophenone (I.St.)	μBondapak C_{18} (10 μm)	0.2 *M* Na-acetate (pH = 5.0) water-acetonitrile-methanol 2:78:15:5	UV, 273 nm	15—30 mg/capsule hydrochlorothiazide 30—70 mg/capsule triamterene	In powdered sample of capsule, after extraction with acetonitrile-acetic acid	317
Benetizid, Bayranolol	Zorbax CN	Methanol-acetonitrile-tetrahydrofuran-water 1:1:1:7	—	—	In multicomponent pharmaceuticals	318

Hydrochlorothiazide	LiChrosorb RP-2 (10 μm)	Tetrahydrofuran-propan-2-ol-water 62:11:127; pH = 3.8(H_3PO_4)	UV, 266 nm	—	In tablets, after extraction with 2-methoxyethanol. The major related impurities are also detectable: 5-chloro-2,4-disulphamoyl aniline; 2,7 diamino-4 hydroxy-6-phenylpteridine; 2,4-diamino-7-hydroxy-6-phenylpteridine	
4-Hydroxytriamterene, 4-hydroxytriamterene sulfate	Novapak C_{18}	Acetonitrile-methanol	F	1 ng/ml 4-100 ng/ml	In plasma and urine	320

Note: F = fluorimetry; I.St. = internal standard.

The selective estimation of triamterene in the presence of its degradation products or other compounds in biological fluids and dosage forms is carried out mainly by HPLC. Table 14 summarizes some methods from this field. Fluorimetric[322] and UV spectrophotometric[322] methods are also known for the determination of triamterene. Both types of methods provide very sensitive ways to estimate triamterene and its metabolites in blood[322] and tablets.[323]

REFERENCES

1. **Taylor, S. H.,** *Z. Kardiol.* 74. Suppl., 2, 2, 1985.
2. **Gifford, R. W., Jr.,** *Am. J. Med.,* 75, 102, 1984.
3. **Berglund, G. and Andersson, O.,** *Lancet,* 744, 1981.
4. **Freis, E. D.,** *Am. J. Med.,* 75, 107, 1984.
5. **Cragoe, E. J.,** *Diuretics, Chemistry, Pharmacology and Medicine,* John Wiley & Sons, New York, 1983.
6. **Kokko, J. P.,** *Am. J. Med.,* 77, 11, 1984. (Suppl.)
7. **Jacobson, H. R. and Kokko, J. P.,** *Annu. Rev. Pharmacol. Toxicol.,* 16, 201, 1976.
8. **Freis, E. D.,** *Am. Heart J.,* 106, 185, 1983.
9. **Southworth, H.,** *Proc. Soc. Exp. Biol. Med.,* 36, 58, 1937.
10. **Mann, T. and Keilin, D.,** *Nature (London),* 146, 164, 1940.
11. **Davenport, H. W. and Wilhelmi, A. E.,** *Proc. Soc. Exp. Biol. Med.,* 48, 53, 1941.
12. **Pitts, R. F. and Alexander, R. S.,** *Am. J. Physiol.,* 144, 239, 1945.
13. **Roblin, R. O., Jr. and Clapp, J. W.,** *J. Am. Chem. Soc.,* 72, 4890, 1950.
14. **Miller, W. H., Dessert, A. M., and Roblin, R. O., Jr.,** *J. Am. Chem. Soc.,* 72, 4893, 1950.
15. **Maren, T. H., Mayer, E., and Wadsworth, B. C.,** *Bull. Johns Hopkins Hosp.,* 95, 199, 1954.
16. **Ford, R. V., Spurk, C. L., and Moyer, J. H.,** *Circulation,* 16, 394, 1957.
17. **Vaughan, J. R., Jr., Eichler, J. A., and Anderson, G. W.,** *J. Org. Chem.,* 21, 700, 1956.
18. **Young, R. W., Wood, K. H., Eichler, J. A., Vaughan, J. R., Jr., and Anderson, G. W.,** *J. Am. Chem. Soc.,* 78, 4649, 1956.
19. **Travis, D. M.,** *J. Pharmacol. Exp. Ther.,* 167, 253, 1969.
20. **Kunan, R. T., Jr.,** *J. Clin. Invest.,* 51, 294, 1972.
21. **Posner, A.,** *Am. J. Ophthalmol.,* 45, 225, 1958.
22. **Hill, T. W. K. and Randall, P. J.,** *J. Pharm. Pharmacol.,* 28, 552, 1976.
23. **Gussin, R. Z.,** in *Cardiovascular Pharmacology,* Antonaccio, M., Ed., Raven Press, New York, 1977.
24. **Beyer, K. H.,** *Arch. Intern. Pharmacodyn.,* 98, 97, 1954.
25. **Sprague, J. M.,** *Ann. N. Y. Acad. Sci.,* 71, 328, 1958.
26. **Beyer, K. H. and Baer, J. E.,** *Pharmacol. Rev.,* 13, 517, 1961.
27. **Beyer, K. H., Jr.,** *Perspect. Biol. Med.,* 19, 500, 1976.
28. **Baggiano, B. G., Condon, S., Davies, M. T., Jackman, J. B., Overell, B. G., Petrow, V., Stephenson, O., and Wild, A. M.,** *J. Pharm. Pharmacol.,* 12, 419, 1960.
29. **Jackman, J. B., Petrow, V., Stephenson, O., and Wild, A. M.,** *J. Pharm. Pharmacol.,* 12, 648, 1960.
30. **David, A. and Fellowes, K. P.** *J. Pharm. Pharmacol.,* 12, 65, 1960.
31. **Novello, F. C., Bell, S. C., Abrams, E. L. A., and Sprague, J. M.,** *J. Org. Chem.,* 25, 965, 1960.
32. **Topliss, J. G., Daly, M. C., Lepski, J., Saphiro, E. P., and Sperber, W.,** *J. Med. Chem.,* 6, 312, 1963.
33. **Bourdais, J. and Meyer, F.,** *Bull. Soc. Chem. France,* 550, 1961.
34. **Krebs, H. A.,** *Biochem. J.,* 43, 525, 1948.
35. **Schrader, E.,** *J. Prakt. Chem.,* 95, 392, 1917.
36. **Freeman, J. H. and Wagner, E. C.,** *J. Org. Chem.,* 16, 815, 1951.
37. **Novello, F. C. and Sprague, J. M.,** *J. Am. Chem. Soc.,* 79, 2528, 1957.
38. **De Stevens, G., Werner, L. H., Halamandaris, A., and Rica, A., Jr.,** *Experientia,* 14, 463, 1958.
39. **De Stevens, G.,** *Diuretics, Chemistry and Pharmacology,* Academic Press, New York, 1963.
40. **Orita, Y., Ando, A., Yamabe, S., Nakanishi, T., Arakawa, N., and Abe, H.,** *Arzneim. Forsch.,* 33, 688, 1983.
41. **Orita, Y., Ando, A., Takamitsu, Y., Urakabe, S., Furukawa, T., and Abe, H.,** *Jpn. Circul. J.,* 32, 547, 1968.
42. **Ando, A., Orita, Y., Takamitsu, Y., Urakabe, S., Shirai, D., Abe, H.,** *Jpn. Circul. J.,* 34, 609, 1970.

43. **Orita, Y., Ando, A., Takamitsu, Y., Shirai, D., Urakabe, S., and Abe, H.,** *Jpn. Circul. J.*, 36, 187, 1972.
44. **Dupont, L. and Diedberg, U.,** *Acta Cryst.*, B 28, 2240, 1972.
45. **Novello, F. C., Bell, S. C., Abrams, E. L. A., Ziegler, C., and Sprague, J. M.,** *J. Org. Chem.*, 25, 970, 1960.
46. **P'An, S. Y., Scriabine, A., McKersie, D. E., McLamore, W.,** *J. Pharmacol. Exptl.*, 128, 122, 1960.
47. **Scriabine, A., P'An, S. Y., Kondratas, B., McManus, J., McLamore, W.** *Chemotheraóia,* 4, 405, 1962.
48. **Cragoe, W. J., Woltersdorf, O. W., Jr., Baer, J. E., and Sprague, J. M.,** *J. Med. Pharm. Chem.*, 5, 896, 1962.
49. **Topliss, J. G., Sherlock, M. H., Clarke, F. H., Daly, M. C., Pettersen, B. W., Lipski, J., and Sperber, N.,** *J. Org. Chem.*, 26, 3842, 1961.
50. **Scriabine, A., Korol, B., Kondratas, B., Yu, M., P'An, S. Y., and Schneider, J. A.,** *Proc. Soc. Exptl. Biol. Med.*, 107, 804, 1961.
51. **Close, W. J., Swett, L. R., and Nordeen, C. W.,** *J. Org. Chem.*, 26, 3423, 1961.
52. **Malecki, F., Staroscik, R., and Weiss-Gdanska, W.,** *Pharmazie,* 34, 158, 1984.
53. **Malecki, F., Staroscik, R., and Weiss-Gdanska, W.,** *Pharmazie,* 38, 174, 1983.
54. **Moussa, B. A. and El Konsy, N. M.,** *Pharm. Weekbl. Sci. Ed.*, 7, 79, 1985.
55. **Láng, L., Ed.,** Absorption Spectra in the ultraviolet and visible region, Publ. of Hung. Acad. Sci., Budapest, 1963.
56. **Roth, H. J., Eger, K., and Troschütz, R.,** *Arzneistoffanalyse,* Georg Thieme Verlag., Stuttgart, 1981.
57. **Agrawal, D. K. and Deshpande, A. V.,** *Pharmazie,* 37, 150, 1982.
58. **Dibbern, H.-W.,** *UV und IR-Spektren wichtiger pharmazeutischer Wirkstoffe,* Editio Cantor, Aulendorf, 1978.
59. **Rosenthaler, L.,** *Pharm. Ztg. verein. Apoth. Ztg.*, 104, 378, 1960.
60. **Kala, H.,** *Pharmazie,* 20, 82, 1960.
61. **Friedrich, F., Kottke, K.,** *Zbl. Pharm.*, 115, 235, 1976.
62. **Kertész, P.,** *Acta Pharm. Hung.*, 33, 150, 1963.
63. **Kertész, P.,** *Acta Pharm. Hung.*, 39, 127, 1969.
64. *Pharmacopoeia Hungarica,* 7th ed., Medicina, Budapest, 1986.
65. **Shukla, I. C., Ahmad, S., Singh, D., and Srivastava, M. K.,** *Ind. J. Pharm. Sci.*, 45, 249, 1983.
66. **De Croo, F., Van den Bossche, W., and De Moerloose, P.,** *J. Chromatogr.*, 325, 395, 1985.
67. **De Croo, F., Van den Bossche, W., and De Moerloose, P.,** 349, 301, 1985.
68. **Smith, R. M., Murilla, G. A., Hurdley, T. G., Gill, R., and Moffat, A. C.,** *J. Chromatogr.*, 384, 259, 1987.
69. **Davies, G. E. and Williamson, M. J.,** *J. Chromatogr.*, 361, 407, 1986.
70. **Duchene, M. and Lapiere, C. L.,** *J. Pharm. Belg.*, 20, 275, 1965.
71. **Blaschke, G. and Maibaum, J.,** *J. Pharm. Sci.*, 74, 438, 1985.
72. **Kim, K., Cho, Y. H., Park, M. K., and Lee, W. K.,** *Arch. Pharmacol. Res.* 6, 103, 1983.
73. **Osborne, G. B.,** *J. Chromatogr.*, 70, 190, 1972.
74. **Schaefer, M., Geissler, H. E., and Mutschler, E.,** *J. Chromatogr.*, 143, 615, 1977.
75. **Smith, P. J. and Hermann, T. S.,** *Analyt. Biochem.*, 22, 134, 1968.
76. **Sohn, D., Simon, J., Hanna, M. A., Ghali, G., and Tolba, R.,** *J. Chromatogr.*, 87, 570, 1974.
77. **Thielemann, H. and Paepke, M.,** *Z. analyt. Chem.*, 266, 128, 1973.
78. **Misztal, G., Przyborowska, M., and Przyborowska, L.,** *Pharmazie,* 38, 67, 1983.
79. **Adam, R. and Lapiere, C. L.,** *J. Pharm. Belg.*, 19, 79, 1964.
80. **Duchene, M. and Lapiere, C. L.,** *J. Pharm. Belg.*, 20, 275, 1965.
81. **Stohs, S. J. and Scratchley, G. A.,** *J. Chromatogr.*, 114, 329, 1975.
82. **Musch, G., De Smet, M., and Massart, D. L.,** *J. Chromatogr.*, 348, 97, 1985.
83. **Székely, G.,** *Mitt. Geb. Lebensmittelunters. Hyg.*, 73, 155, 1982.
84. **Agarwal, S. P. and Nwaiwu, J.,** *J. Chromatogr.*, 383, 35, 1986.
85. **Hitscherich, M. E., Rydberg, E. M., Tsilifonis, D. C., and Daly, R. E.,** *J. Liq. Chromatogr.*, 10, 1011, 1987.
86. **Ramana, R., Raghuveer, S., and Khadgapathi, P.,** *Indian Drugs,* 23, 39, 1985.
87. **Ficarra, P., Ficarra, R., Tommasini, A., Calabro, M. L., and Feneh, C. G.,** *Il Farmaco* ed. pr., 41, 332, 1986.
88. **Bulut, P., Turelli, F., and Turk H. J.,** *Denejsel Biyol. Derg.*, 40, 206, 1983.
89. **Qui, Y., Zheng, L., and Shen, Q.,** *Zhongguo Yaoke Dexiee Xiebao,* 18, 130, 1987; ref. A. A., 50, 436, 1988.
90. **Chu, J. and Yu, R.,** *Nanjing Yaoxueyuan Xuebao,* 22, 19, 1984; ref.: C. A., 102, 32421d, 1985.
91. **Xu, J., Yang, Q., Dong, S., and Yu, R.,** *Nanjing Yaoxueyuan Xuebao,* 20, 9, 1982. ref.: C. A., 99, 93843, 1983.

92. **Xu, J., Yang, Q., Dong, S., and Yu, R.,** *Nanjing Yaoxueyuan Xuebao,* 20, 15, 1982. ref.: C. A., 99, 93844, 1983.
93. **Duan, L., Xiang, D., Sheng, L., Wu, R., and An, D.,** *Yaoxie Xuebao,* 22, 761, 1987. ref.: A. A., 50, 656, 1988.
94. **Spurlock, C. H. and Schneider, H. G.,** *J. Assoc. Off. Anal. Chem.,* 67, 321, 1984.
95. **Daniels, S. L. and Wanderwielen, A. J.,** *J. Pharm. Sci.,* 70, 211, 1981.
96. **Hennig, B.,** *Pharmazie,* 39, 779, 1984.
97. **Ings, R. M. and Stevens, L. A.,** *Prog. Drug Metab.,* 7, 57, 1983.
98. **Moyer, T. P. and Anhalt, J. P.,** *Methodol. Anal. Toxicol.,* 3, 79, 1985.
99. **Uchino, K., Yamamura, Y., Saitoh, Y., Isozaki, S., Tamura, Z., Nakagawa, F., Sakina, K., and Kojima, I.,** *Yakugaku Zasshi,* 104, 1101, 1984. ref.: C. A., 102, 41073n, 198.
100. **Shiu, G. K., Prasad, V. K., Lin, J., and Worsley, W.,** *J. Chromatogr. Biomed. Appl.,* 50, 430, 1986.
101. **Hessey, G. A., Constanzer, M. L., and Bayne, W. F.,** *J. Chromatogr. Biomed. Appl.* 53, 450, 1986.
102. **Fullinfaw, R. O., Bury, R. N., Moulds, R. F. W.,** *J. Chromatogr. Biomed. Appl.* 347, 1987.
103. **Chen, J., Jiao, Zh., Wang, D.,** *Yaou Fenxi Zazhi,* 3, 357, 1983. ref.: C. A., 100, 79365, 19
104. **Lu, M., Win, G., Chen, S., and Ruan, Z.,** *Nanjing Yaoxueyuan Xuebao,* 1, 1983, ref. : A. A., 46, 739, 1984.
105. **Stumph, M. J., and Noall, M. W.,** *J. Anal. Toxicol.,* 8, 170, 1984.
106. **Shah, V. P., Lee, J., and Prasad, V. K.,** *Anal. Lett.,* 15, 529, 1982.
107. **McKinley, W. A.,** *J. Anal. Toxicol.,* 5, 209, 1981.
108. **Alton, K. B., Desrivieres, D., and Patrick, J. E.,** *J. Chromatogr.,* 374, 103, 1986.
109. **Lin, E. T.,** *Clin. Liq. Chrom.,* 1, 115, 1984.
110. **Yamazaki, M., Ito, Y., Suzuka, T., Yaginuma, H., Itoh, S., Kamada, A., Orito, Y., Nakama, H., Nakamishi, T., and Ando, A.,** *Chem. Pharm. Bull.,* 32, 2387, 1984.
111. **Weinberger, R. and Pietrantonio, T.,** *Anal. Chim. Acta,* 146, 219, 1983.
112. **Hartman, C. A., Kucharczyk, N., Duane, S. R., and Perhach, J. L.,** *J. Chromatogr. Biomed. Appl.,* 15, 510, 1981.
113. **Koopmans, P. P., Tan, Y., Van Ginneken, C. A. M., and Gribnau, F. W. J.,** *J. Chromatogr. Biomed. Appl.,* 32, 445, 1984.
114. **Shah, V., Walzer, M. A., Prasad, V. K.,** *J. Liq. Chromatogr.,* 6, 1949, 1983.
115. **Lutz, D., Ilias, E., and Jaeger, H.,** *Glass Capillary Chromatography in Clinical and Medical Pharmacology,* Marcel Dekker, New York, 1985, 453.
116. **Pitha, J., Milecki, J., and Fales, H.,** *Int. J. Pharm.,* 29, 73, 1986.
117. **Gillatt, P. N., Hart, R. J., Walters, C. L., and Reed, P. I.,** *Food Chem. Toxicol.,* 22, 269, 1984.
118. **Graf, W., Girod, E., and Stoll, W. G.,** *Helv. Chim. Acta,* 42, 1085, 1959.
119. **Stenger, E. G., Wirz, H., and Pulver, R.,** *Schweiz. Med. Wochenschr.,* 89, 1130, 1959.
120. **Veyrat, R., Arnold, E. F., Duckert, A.,** 89, 1133, 1959.
121. **Reutter, F. and Schaub, F.,** 89, 1158, 1959.
122. **Mudge, G. H.,** Diuretics, in *Pharmacological Basis of Therapeutics,* Gilman, A. G., Goodman, L. S., Gilman, A. Eds., 6th ed., Macmillan, New York, 1980, 899.
123. **Topliss, J. G., Konzelman, L. M., Sperber, M., and Roth, F. E.,** *J. Med. Chem.,* 7, 453, 1964.
124. **Fogel, J., Sisco, J., and Hess, F.,** *J. Assoc. Off. Anal. Chem.,* 68, 96, 1985.
125. **Bauer, J., Quick, J., Krogh, S., and Shada, B.,** *J. Pharm. Sci.,* 72, 924, 1983.
126. **Schnekenburger, J. and Quade-Henkel, M.,** *Dtsch. Apoth. Ztg.,* 124, 1167, 1984.
127. **Wu, Q., Yang, Q., and Yu, R.,** *Vanjing Yaoxueyuan Xuebao,* 63, 1983.
128. **Bishop, E. and Hussein, W.,** *Analyst (London),* 109, 913, 1984.
129. **Boneva, A. S., Loginova, N. F., Filippova, P. M., Mischenko, V. V., Bekker, A. R., Starostina, A. K., Nino, N., and Mairanovskii, V. G.,** *Khim. Farm. Zh.,* 19, 1494, 1985.
130. **Vetuschi, C., Ragno, G., Mazzeo, P., and Mazzeo-Farina, A.,** *Farmaco Ed. Prat.,* 40, 215, 1985.
131. **Shinghal, D. M. and Prabhudesai, J. S.,** *Indian Drugs,* 21, 466, 1984.
132. **Walters, S. M. and Stonys, D. B.,** *J. Chromatogr. Sci.,* 21, 43, 1983.
133. **Ficarra, R., Ficarra, P., Tommasini, A., Calabro, M. L., and Guarniera, F. C.,** *Farmaco* ed. Prat., 40, 307, 1985.
134. **Prasad, T. N. V., Rao, E. D., Sastry, C. S. P., and Rao, G. R.,** *Indian Drugs,* 24, 398, 1987.
135. **Lin, E. T.,** *Clin. Liq. Chromatogr.,* 1, 107, 1984.
136. **Rosenberg, M. J., Lam, K. K., and Dorsey, T. H. E.,** *J. Chromatogr. Biomed. Appl.,* 48, 438, 1986.
137. **Muirhead, D. C. and Christie, R. B.,** *J. Chromatogr. Biomed. Appl.,* 60, 420, 1987.
138. **MacGregor, T. R., Farina, P. R., Hagopian, M., Hay, N., Esber, H. J., and Keirus, J. J.,** *Ther. Drug Monit.,* 6, 83, 1984.
139. **Fiorese, F., Vermuelen, G., and Turcotte, C.,** *Subst. Alcohol Actions Misuse,* 3, 47, 1982.
140. **De Camp., W. H.,** *J. Assoc. Off. Anal. Chem.,* 67, 927, 1984.
141. **Shiu, W. and Jeon, G. S.,** *Chayosi Kwahak Talhak Nomunyip (Soul),* 8, 45, 1983.

142. **Roth, H. J., Eger, U., and Troschütz, R.,** *Arzneistoffanalyse,* Georg Thieme Verlag., Stuttgart, 1981.
143. **Constantinescu, T.,** *Rev. Chim. (Bucarest),* 34, 749, 1983.
144. **Kata, M.,** *Pharmazie,* 39, 856, 1984.
145. **Orita, Y., Ando, A., Urakabe, S., and Abe, H.,** *Arzneim. Forsch.,* 26, 11, 1976.
146. **Sturm, K., Siedel, W., Weyer, R., and Ruschig, H.,** *Chem. Ber.,* 99, 328, 1966.
147. **Kovar, K. A., Wojtovicz, G. P., and Auterhoff, H.,** *Arch. Pharm.,* 307, 657, 1974.
148. **Toyooka, T., Sano, A., Kuziki, T., and Suzuki, N.,** *J. Pharm. Soc. Jpn.,* 101, 489, 1981.
149. **Andreasen, F., Bøtker, H. E., and Lorentzen, K.,** *Br. J. Clin. Pharmacol.,* 14, 306, 1982.
150. **Cruz, J. E., Maness, D. D., and Yakatan, G. J.,** *Int. J. Pharm.,* 2, 275, 1979.
151. **Bundgard, H., Nørgaard, T., and Mielsen, N. M.,** *Int. J. Pharm.,* 42, 217, 1988.
152. **Moore, D. E.,** *J. Pharm. Sci.,* 66, 1282, 1977.
153. **Moore, D. E. and Burt, C. D.,** *Photochem. Photobiol.,* 34, 431, 1981.
154. **Moore, D. E. and Tamat, S. R.,** *J. Pharm. Pharmacol.,* 32, 172, 1980.
155. **Moore, D. E. and Sithipitkas, V.,** *J. Pharm. Pharmacol.,* 35, 489, 1983.
156. **Miller, I. R.,** *J. Membr. Biol.,* 101, 113, 1988.
157. **Bach, D., Vinkler, C. H., Miller, I. R., and Caplan, S. R.,** *J. Membr. Biol,* 101, 103, 1988.
158. **Schlatter, E., Greger, R., and Weidtke, C.,** *Pflügers Archiv.,* 396, 210, 1983.
159. **Sturm, K., Siedel, W., and Weyer, R.,** *Deutsch. Bundespat.,* 1, 119290, 1959.
160. **Muschawek, R.,** *Diuretics: Chemistry, Pharmacology, Clinical Application,* Proc. Int. Conf. Diuretics, Elsevier, New York, 1984.
161. **Siedel, W., Sturm, K., and Scheurich, W.,** *Chem. Ber.,* 99, 345, 1966.
162. **Shani, J., Schoenberg, S., Lien, E. J., Cherkez, S., Pfeifel, N., Schonberger, C., and Yellin, H.,** *Pharmacology,* 26, 172, 1983.
163. **Greger, R., Schlatter, E., and Lang, F.,** *Pflügers Arch.,* 396, 1983.
164. **Casassas, E. and Faleregas, J. L.,** *Anal. Chim. Acta,* 106, 151, 1979.
165. **Roth, J., Rapaka, R. S., and Prasad, V. K.,** *Anal. Lett.* Part B., 14, Lo13, 1981.
166. **Rao, G. R. and Raghuveer, S.,** *Indian Drugs,* 22, 217, 1985.
167. **Moustafa, A. A., Abdel-Moetyins, E. N.,** *Farmaco* ed. Prat., 42, 51, 1987.
168. **Bauza, M. T., Lesser, C. L., Johnston, J. T., Smith, R. V.,** *J. Pharm. Biomed. Anal.,* 3, 459, 1985.
169. **Carr, K., Rane, A., and Froelich, J. C.,** *J. Chromatogr. Biomed. Appl.,* 2, 421, 1978.
170. **MacDougall, M. L., Shoeman, D. W., and Azarnoff, D. L.,** *Res. Commun. Pathol. Pharmacol.,* 10, 285, 1975.
171. **Wittfoht, W., Duwe, K., Kuhnz, W., and Nau, H.,** *Clin. Chem. (Winston-Salem N.C.)* 30, 878, 1984.
172. **Rapaka, R. S. and Roth, J.,** *Clin. Chem. (Winston-Salem N.C.),* 27, 1470, 1981.
173. **Nation, R. L., Peng, G. W., and Chion, W. L.,** *J. Chromatogr. Biomed. Appl.,* 4, 88, 1979.
174. **Lin, E. T., Smith, G. E., Benet, L. Z., and Hoener, B. A.,** *J. Chromatogr. Biomed. Appl.,* 5, 315, 1979.
175. **Snedden, W., Sharma, J. N., and Fernandez, P. G.,** *Drug Monit.,* 4, 381, 1982.
176. **Kerremans, A. L. M., Tan, Y., Van Ginneken, C. A. M., and Gribnau, F. W. J.,** *J. Chromatogr. Biomed. Appl.,* 18, 129, 1982.
177. **Sord, S. P., Green, V. I., and Norton, Z. N.,** *Drug Monit.,* 9, 484, 1985.
178. **Kubo, H., Li, H., Kobayashi, Y., and Kinoshita, T.,** *Bunzeki Kagaku,* 35, 259, 1986.
179. **Guermouche, S., Guermouche, M. H., Mansouri, M., Boukhari, D., and Sassard, J.,** *Analysis,* 12, 438, 1984.
180. **Lovett, L. J., Nygard, G., Dura, P., and Khalil, S. K. W.,** *J. Liq. Chromatogr.,* 8, 1611, 1985.
181. **Uchino, K., Isozaki, S., Saitoh, Y., Nakagawa, F., Tamura, Z., and Tanaka, N.,** *J. Chromatogr. Biomed. Appl.,* 33, 241, 1984.
182. **Guermouche, S., Guermouche, M. H., Mansouri, M., and Abdel, L.,** *J. Pharm. Biomed. Anal.,* 3, 453, 1985.
183. **Yoshitomi, H., Ikega, K., and Goto, S.,** *Yakugaku Zasshi,* 102, 1171, 1982.
184. **Kholodov, L. E., Tishchenkova, I. F., and Glezer, M. G.,** *Khim. Farm. Zh.,* 18, 626, 1984.
185. **Wesley-Hadzija, B. and Mattocks, A. M.,** *J. Chromatogr. Biomed. Appl.,* 18, 425, 1982.
186. **Berret, L. Z., Smith, D. E., Lin, E. T., Vincenti, F., and Gambertoglio, J. G.,** *Fed. Proc. Fed. Am. Soc. Exp. Biol.,* 42, 1695, 1983.
187. **Schultz, E. M., Cragoe, E. J., Jr., Bicking, J. B., Bolhofer, W. A., and Sprague, J. M.,** *J. Med. Pharm. Chem.,* 5, 660, 1962.
188. **Baer, J. E., Michaelson, J. K., McKinstry, D. N., and Beyer, K. H.,** *Proc. Soc. Exp. Biol. Med.,* 115, 87, 1964.
189. **Beyer, K. H., Baer, J. E., Michaelson, J. K., and Russo, H. F.,** *J. Pharmacol. Exptl. Therap.,* 147, 1, 1965.
190. **Sprague, J. M.,** Diuretics, in *Topics in Medicinal Chemistry,* Vol. 2., Rabinowitz, J. L. and Myerson, R. M., Eds., Interscience, New York, 1968.

191. **Cragoe, E. J., Jr. and Schultz, E. M., Schneeberg, J. D., Stokker, G. E., Woltersdorf, O. W., Jr., Fanelli, G. M., Jr., and Waltson, L. S.,** *J. Med. Chem.*, 18, 225, 1975.
192. **Woltersdorf., O. W., Jr., de Solms, S. J., Shultz, E. M., and Cragoe, E. J., Jr.,** *J. Med. Chem.*, 20, 1400, 1977.
193. **de Solms, S. J., Woltersdorf, O. W., Jr., Cragoe, E. J., Jr., Watson, L. J., and Fanelli, G. M.,** *J. Med. Chem.*, 21, 437, 1978.
194. **Thuillier, G., Laforest, J., Cariou, B., Bessin, P., Bonnet, J., and Thuillier, J.,** *Eur. J. Med. Chem.*, 9, 625, 1974.
195. **Woltersdorf, O. W., Jr., de Solms, J. S., Cragoe, E. J., Jr.,** in *Diuretic Agents,* Cragoe, E. J., Jr., Ed., ACS Symp. Ser. 83, 1978.
196. **Koechel, D. A.,** *Am. Rev. Pharmacol. Toxicol.*, 21, 265, 1981.
197. **Sprague, J. M. and Schultz, E. M.,** US Patent 3.453.312, 1969.
198. **Koechel, D. A. and Cafruny, E. J.,** *J. Med. Chem.*, 16, 1147, 1973.
199. **Schultz, D. E. and Schnekenburger, J.,** *Einführung in die Pharmazeutische Chemie,* Verlag Chemie, Weinheim, West Germany, 1984, 409.
200. **Beyer, K. H., Baer, I. E., Michaelson, J. K., and Russo, H. F.,** *J. Pharmacol. Exp. Ther.* 147, 1, 1965.
201. **Klaassen, C. D. and Fitzgerald, T. I.,** *J. Pharmacol. Exp. Ther.*, 191, 548, 1974.
202. **Fine, L. G., Goldstein, E. J., Trizna, W., Rosmaryn, L., and Arias, I. M.,** *Proc. Soc. Exp. Biol. Med.*, 157, 189, 1978.
203. **Duggan, D. E. and Noll, R. M.,** *Proc. Soc. Exp. Biol. Med.*, 139, 762, 1972.
204. **Auterhoff, H. and Hinnes, J.,** *Arch. Pharm.*, 312, 1037, 1979.
205. **Das Gupta, V.,** *Drug Dev. Ind. Pharm.*, 8, 869, 1982.
206. **Yarwood, R. J., Moore, W. D., and Collett, J. H.,** *J. Pharm. Biomed. Anal.*, 5, 364, 1987.
207. **Midha, K. K., Hubbard, J. W., Charette, C., and Jun, H. W.,** *J. Pharm. Sci.*, 67, 975, 1978.
208. **El Dalsh, S. S., El Sayed, A. A., Badawi, A. A., Khattab, F. I., and Pouli, A.,** *Drug Dev. Ind. Pharm.*, 9, 877, 1983.
209. **Salole, E. G. and Al-Sanaj, F. A.,** *Drug Dev. Ind. Pharm.*, 11, 855, 1985.
210. **Moellgaard, A. F. and Bundgaard, H.,** *Arch. Pharm. Chem. Sci. Ed.*, 11, 7, 1983.
211. **Vila, J., Jose, L., Blanco, J., and Bilar, A.,** *Acta Pharm. Technol.*, 32, 82, 1986.
212. **Luetscher, A., Jr., Deming, Q. B., and Johnson, B. B.,** Ciba Foundation Colloquia on Endocrinology, Wolstenholm, G. E. W. and Cameron, M. P., Eds., Churchill Livingstone, London, 1952.
213. **Simpson, S. A. and Tait, J. F.,** *Endocrinology,* 50, 150, 1952.
214. **Laragh, J. H.,** *J. Chronic Diseases,* 11, 292, 1960.
215. **Cella, J. A. and Kagawa, C. M.,** *J. Am. Chem. Soc.*, 79, 4808, 1957.
216. **Cella, J. A., Brown, E. A., and Burtner, R. R.,** *J. Org. Chem.*, 24, 243, 1959.
217. **Cella, J. A. and Tweit, R. C.,** *J. Org. Chem.*, 24, 1109, 1959.
218. **Brown, E. A., Muir, R. D., and Cella, J. A.,** *J. Org. Chem.*, 25, 96, 1960.
219. **Cella, J. A.,** in *Edema,* Moyer, J. H. and Fuchs, M., Eds., W. B. Saunders, Philadelphia, 1960.
220. **Atwater, N. W., Bible, R. H., Jr., Brown, E. A., Burtner, R. R., Mihina, J. S., Nysted, L. N., and Sollmann, P. B.,** *J. Org. Chem.*, 26, 3077, 1961.
221. **Hess, H. J.,** *J. Org. Chem.*, 27, 1096, 1962.
222. **Kagawa, C. M., Bouska, B. J., Anderson, M. I., and Krol, W. F.,** *Arch. Int. Pharmacodyn.*, 149, 8, 1964.
223. **Nysted, R. L. and Burtner, R. R.,** *J. Org. Chem.*, 27, 3175, 1962.
224. **Patchett, A. A., Hoffman, F., Giarusso, F. F., Schwam, H., and Arth, G. E.,** *J. Org. Chem.*, 27, 3822, 1962.
225. **de Stevens, G.,** *Diuretics, Chemistry and Pharmacology,* Academic Press, New York, 1963.
226. **Chinn, L. J. and Hofmann, L. M.,** *J. Med. Chem.*, 16, 839, 1973.
227. **Arth, E. G., Schwam, H., Saratt, L. H., and Glitzer, M.,** *J. Med. Chem.*, 6, 618, 1963.
228. **Kagawa, C. M., Cella, J. A., and van Arman, C. G.,** *Science,* 126, 1015, 1957.
229. **Chinn, L. J. and Desai, B. N.,** *J. Med. Chem.*, 18, 268, 1975.
230. **Casals-Stenzel, J., Buse, M., Wambach, G., and Lovert, W.,** *Arzneim. Forsch. (Drug. Res.),* 34, 241, 1984.
231. **Fretland, D. J., Brown, E. A., and Cammarata, P. S.,** *J. Steroid Biochem.*, 22, 305, 1985.
232. **de Gasparo, M., Joss, U., Ramjoue, S. E., Whitebread, S. E., Haenni, H., Schenkel, L., Kraehenbuehl, C., Biollaz, M., Grob, J., Schmidlin, J., Wieland, P., and Wehrli, H. U.,** *J. Pharmacol. Exp. Ther.*, 240, 650, 1986.
233. **Sastry, C., Prasad, T., and Rao, E.,** *Indian J. Pharm. Sci.*, 47, 190, 1985.
234. **Nevrekar, V.,** *Indian Drugs,* 21, 349, 1984.
235. **Shingbal, D. M. and Rao, V. R.,** *Indian Drugs,* 23, 232, 1986.
236. **Shingbal, D. M. and Prabhudesai, I. S.,** *Indian Drugs,* 21, 306, 1984.

237. **Anon.,** *Indian Drugs,* 24, 54, 1986.
238. **Shingbal, D. M.,** *Indian Drugs,* 24, 450, 1987.
239. **Neurath, G. B. and Ambrosius, D.,** *J. Chromatogr. Biomed. Appl.,* 163, 230, 1979.
240. **Fehér, Zs., Horvai, Gy., Nagy, G., Niegreisz, Zs., Tóth, K., and Pungor, E.,** *Anal. Chim. Acta,* 145, 41, 1983.
241. **Sherry, J. H., O'Donnel, J., and Colby, H.,** *J. Chromatogr.,* 374, 183, 1986.
242. **Overdick, J. W., Hermens, W. A., and Merkus, F. W.,** *J. Chromatogr. Biomed. Appl.,* 42, 279, 1985.
243. **Jackson, L. S. and Stafford, J. E.,** *J. Chromatogr.,* 428, 377, 1988.
244. **Prasad, T. N., Rao, E. D., and Sastry, C. S.,** *Indian Drugs,* 24, 346, 1987.
245. **Glitzer, M. S. and Steelman, S. L.,** *Proc. Soc. Exptl. Biol. Med.,* 120, 364, 1965.
246. **Glitzer, M. S. and Steelman, S. L.,** *Nature,* 212, 191, 1966.
247. **Bicking, J. B., Mason, J. W., Woltersdorf, O. W., Jr., Jones, J. H., Kwong, S. F., Robb, C. M., and Cragoe, E. J., Jr.,** *J. Med. Chem.,* 8, 638, 1965.
248. **Cragoe, E. J., Jr., Woltersdorf, O. W., Jr., Bicking, J. B., Kwong, S. F., and Jones, J. H.,** *J. Med. Chem.,* 10, 66, 1967.
249. **Bicking, J. B., Robb, C. M., Kwong, S. F., and Cragoe, E. J., Jr.,** *J. Med. Chem.,* 10, 598, 1967.
250. **Jones, J. H., Bicking, J. B., and Cragoe, E. J., Jr.,** *J. Med. Chem.,* 10, 899, 1967.
251. **Moukheibir, N. W. and Kirkendall, W. M.,** *Clin. Res.,* 13, 425, 1965.
252. **Reynolds, T. B. and Pelle, H. C.,** *Clin. Res.,* 14, 184, 1966.
253. **Alter, S., Cushman, P., and Hilton, J. G.,** *Clin. Pharmacol. Therap.,* 8, 243, 1967.
254. **Sprague, J. M.,** Diuretics, in *Topics in Medicinal Chemistry,* Vol. 2., Rabinowitz, J. L. and Myerson, R. M., Eds., Interscience, New York, 1968.
255. **Vigne, P., Frelin, C., Cragoe, E. J., Jr., and Lazdunski, M.,** *Mol. Pharmacol.,* 25, 131, 1984.
256. **Cragoe, E. J., Jr.,** *Diuretics, Chemistry, Pharmacology and Medicine,* John Wiley & Sons, New York, 1983.
257. **Li, H. J. Y., Cragoe, E. J., Jr., and Lindemann, B.,** *J. Membrane Biol.,* 95, 171, 1987.
258. **Li, H. J. Y., Cragoe, E. J., Jr., and Lindemann, B.,** *J. Membrane Biol.,* 83, 45, 1985.
259. **Cuthbert, A. W.,** *Mol. Pharmacol.,* 12, 945, 1976.
260. **Kaczorowski, G. J., Slaughter, R. S., Garcia, M. L., and King, V. F.,** *Progr. Clin. Biol. Res.,* 252, 261, 1988.
261. **Villereal, M. L.,** *Curr. Top. Membr. Transp.,* 27, 55, 1986.
262. **Benos, D. J.,** Ionic channels Cells Model Syst., 401, 1986.
263. **Benos, D. J.,** *Am. J. Physiol.,* 242, C131, 1982.
264. **Asher, C., Cragoe, E. J., Jr., and Garty, H.,** *J. Biol. Chem.,* 262, 8566, 1987.
265. **Cuthbert, A. W. and Fanelli, G. M.,** *Br. J. Pharmacol.,* 63, 139, 1978.
266. **Li, J. H. I., Cragoe, E. J., Jr., and Lindemann, B.,** *J. Membrane Biol.,* 83, 45, 1985.
267. **Vigne, P., Prelin, C., Cragoe, E. J., Jr., and Lazdunski, M.,** *Mol. Pharmacol.,* 25, 131, 1984.
268. **Zhuang, Y. X., Cragoe, E. J., Jr., Shaikewitz, T., Glaser, L., and Cassel, D.,** *Biochemistry,* 23, 4481, 1984.
269. **Jurkowitz, M. S., Altschuld, R. A., Brierley, G. P., and Cragoe, E. J., Jr.,** *FEBS Lett.,* 162, 262, 1983.
270. **Kaczorowski, G. J., Barros, F., Dethmers, J. K., Trumble, M. J., and Cragoe, E. J., Jr.,** *Biochemistry,* 24, 1394, 1985.
271. **Zhuang, Y. X., Cragoe, E. J., Jr., Shaikewitz, T., Glaser, L., and Cassel, D.,** *Biochemistry,* 23, 4481, 1984.
272. **Simchowitz, L. and Cragoe, E. J., Jr.,** *Mol. Pharmacol.,* 30, 112, 1986.
273. **Rocco, V. K., Cragoe, E. J., Jr., and Warnock, D. G.,** *Am. J. Physiol.* 252 /Renal Fluid Electrolyte Physiol. 21./ F517, 1987.
274. **Schellenberg, G. D., Anderson, L., Cragoe, E. J., Jr., Swanson, Ph.D.,** *Mol. Pharmacol.,* 27, 537, 1985.
275. **Vincek, W. C., Hessay, G. A., Constamer, M. L., and Bayne, W. F.,** *Pharm. Res.,* 143, 1985.
276. **Yip, M. S., Coates, P. E., and Thiessen, J. J.,** *J. Chromatogr.,* 307, 343, 1984.
277. **Forrest, G., McIanes, G. T., Forshead, A. P., Thompson, G. G., and Brodie, M. J.,** *J. Chromatogr.,* 428, 123, 1988.
278. **Van der Meer, R. J. and Brown, L. W.,** *J. Chromatogr.,* 423, 350, 1987.
279. **Kurani, S., Desai, D., and Seshadrinathan, A.,** *Indian Drugs,* 23, 230, 1986.
280. **Sastry, C. S. P., Prashad, T. N. V., Rao, A. R. M., and Rao, E. V.,** *Indian Drugs,* 25, 206, 1988.
281. **Sastry, C. S. P., Prashad, T. N. V., Sastry, B. S., and Rao, E. V.,** *Analyst,* 113, 255, 1988.
282. **Schwalbe, C. and Williams, G.,** *Acta Crystallogr. Sect. C.: Cryst. Struct. Commun.,* C43, 109, 1987.
283. **Vogel, I.,** *Pharmazie,* 42, 165, 1987.
284. **Molling, J., Leùschner, U., Meyer, F., and Walther, H.,** *Pharmazie,* 41, 301, 1986.

285. **Wiebelhaus, V. D., Weinstock, J., Brennan, F. T., Sosnowski, G., and Larsen, T. J.,** *Fed. Proc.*, 20, 409, 1961.
286. **Crosley, A. P., Jr., Ronquiilo, L. M., Strickland, W. H., and Alexander, F.,** *Ann. Internal. Med.*, 56, 241, 1962.
287. **Donnelly, R. J., Turner, P., and Sowry, G. S. C.,** *Lancet*, 1, 245, 1962.
288. **Pachter, I. J. and Nemeth, P. E.,** *J. Org. Chem.*, 28, 1187, 1963.
289. **Pachter, I. J.,** *J. Org. Chem.*, 28, 1191, 1963.
290. **Pachter, I. J., Nemeth, P. E., and Villani, A. J.,** *J. Org. Chem.*, 28, 1197, 1963.
291. **Pachter, I. J. and Nemeth, P. E.,** *J. Org. Chem.*, 28, 1203, 1963.
292. **Weinstock, J., Dunoff, R. Y., and Williams, J. G.,** *J. Med. Chem.*, 11, 542, 1968.
293. **Weinstock, J., Dunoff, R. Y., Sutton, B., Trost, B., Kirkpatrick, J., Farina, F., and Straub, A. S.,** *J. Med. Chem.*, 11, 549, 1968.
294. **Weinstock, J., Pachter, I. J., Nemeth, P. E., and Jaffe, G.,** *J. Med. Chem.*, 11, 557, 1968.
295. **Weinstock, J., Graboyes, H., Jaffe, G., Pachter, I. J., Snader, K., Karash, C. B., and Dunoff, R. Y.,** *J. Med. Chem.*, 11, 560, 1968.
296. **Weinstock, J. and Dunoff, R. Y.,** *J. Med. Chem.*, 11, 565, 1968.
297. **Graboyes, H., Jaffe, G. E., Pachter, I. J., Rosenbloom, J. T., Villani, A. J., Wilson, J. W., and Weinstock, J.,** *J. Med. Chem.*, 11, 568, 1968.
298. **Weinstock, J., Wilson, W. W., Wiebelhaus, V. D., Maass, A. R., Brennan, F. T., and Sosnowski, G.,** *J. Med. Chem.*, 11, 573, 1968.
299. **Weinstock, J., Dunoff, R. Y., Carevic, J. E., Williams, J. G., and Villani, A. J.,** *J. Med. Chem.*, 11, 618, 1968.
300. **Campbell, N. R., Dunsmuir, J. H., and Fitzgerald, M. E. H.,** *J. Chem. Soc.*, 2743, 1950.
301. **Mallette, M. F., Taylor, E. C., and Cain, C. K.,** *J. Am. Chem. Soc.*, 69, 1814, 1947.
302. **Seeger, D. R., Cosulich, D. B., Smith, J. M., and Holtquist, M. E.,** *J. Am. Chem. Soc.*, 71, 1753, 1949.
303. **Renfrew, A. G., Piatt, P. C., and Cretcher, L. H.,** *J. Org. Chem.*, 17, 469, 1952.
304. **Daly, J. W. and Christensen, B. E.,** *J. Am. Chem. Soc.*, 78, 225, 1956.
305. **Roth, B., Smith, J. M., and Holtquist, M. E.,** *J. Am. Chem. Soc.*, 73, 2864, 1951.
306. **Taylor, E. C. and Cain, C. K.,** *J. Am. Chem. Soc.*, 74, 1644, 1952.
307. **Brown, D. J. and Jacobsen, N. W.,** *J. Chem. Soc.*, 4413, 1961.
308. **Cain, C. K., Mallette, M. F., and Taylor, E. C.,** *J. Am. Chem. Soc.*, 68, 1996, 1946.
309. **Albert, A., Brown, D. J., and Cheeseman, G.,** *J. Chem. Soc.*, 474, 1951.
310. **Blicke, F. F. and Godt, H. C.,** *J. Am. Chem. Soc.*, 76, 2798, 1954.
311. **Knauf, H., Mutschler, E., Völger, K. D., and Wais, U.,** *Arzneim. Forsch. (Drug. Res.)*, 28(2), 1417, 1978.
312. **Leilich, G., Knauf, H., Mutschler, E., and Völger, K. D.,** *Arzneim. Forsch.*, 30(1), 949, 1980.
313. **Priewer, H., Kraft, H., Mutschler, E.,** 36(1), 213, 1986.
314. **Pushett, J. B. and Greenberg, A. Eds.,** *Chemistry, Pharmacology, Clinical Application of Diuretics*, Proc. Int. Conf. Diuretics, Elsevier, New York, 1984.
315. **Brodie, R. E., Chasseaud, L. F., Taylor, T., and Walmsley, L. M.,** *J. Chromatogr. Biomed. Appl.*, 6, 527, 1979.
316. **Sved, S., Sertie, J. A. A., and McGilverany, I. J.,** *J. Chromatogr. Biomed. Appl.*, 4, 474, 1979.
317. **Yakatan, G. J. and Cruz, J. E.,** *J. Pharm. Sci.*, 70, 949, 1981.
318. **Menon, G. N. and White, L. B.,** *J. Pharm. Sci.*, 70, 1083, 1981.
319. **Waechter, W., Sczepanik, B., Gottmann, G., and Cordes, G.,** *Pharm. In.*, 45, 1000, 1983. ref.: Anal. A., 46, 12E36, 1984.
320. **Korany, M. A. and Franzky, H. J.,** *Sci. Pharm.*, 51, 291, 1983.
321. **Swart, K. J. and Botha, H.,** *J. Chromatogr.*, 413, 315, 1987.
322. **Lebedev, A. A., Kuznetsov, G. P., and Lebedev, P. A.,** USSR SV, 1, 076, 833, 29. Febr. 1984.
323. **Qin, Y., Zheng, L., and Shen, Q.,** *Zhonggno Yaoke Daxne Xuebao*, 18, 130, 1987.

Chapter 2

RENIN INHIBITORS

I. HISTORY AND STRUCTURE-ACTIVITY RELATIONSHIP

The inhibitors of angiotensin converting enzyme (ACE) have proved to be important modulators of the blood pressure. In parallel, it has been discovered that renin inhibitors should exhibit antihypertensive activity, so they may be therapeutically important in the development of novel antihypertensive agents. Most of the potent renin inhibitors found so far are peptides, and their metabolic lability and poor absorbability make them unsuitable for oral administration, i.e., for the therapy of hypertension.

An early step in the designing of renin inhibitors was the construction of hypothetical renin structures (the X-ray crystallographic structure of human renin not having been determined yet), on the basis of which new active inhibitory compounds could be synthetized. Three-dimensional models were therefore constructed in the knowledge of the structures of proteins of biological interest, e.g., penicillopepsin, rhisopuspepsin, endothiapepsin, and pepsin.

The main criteria required of a potent renin inhibitor were as follows:

1. A high specificity for renin
2. A high metabolic stability
3. Water solubility
4. A prolonged duration of action

Another quite reasonable step in the designing of renin inhibitory agents was to prepare peptides involving modified residues from the amino acid sequence of angiotensinogen (ANG) (Figure 1) around the site of cleavage by renin.[1] ANG 6-13, as the minimal octapeptide sequence of the renin substrate, was chosen (Figure 2). Szelke et al.[2] synthetized compounds in which the scissile bond grouping Leu P-Val P′ was replaced by the isosteric secondary amine (CH_2-NH-) (reduced amide)[3] unit. The change from sp^2 to sp^3 is thought to allow the carbon of the "reduced" peptide bond to assume the tetrahedral configuration of the transition state and thereby increase the binding energy significantly.

```
                              6                    10    11
H—Asp—Arg—Val—Tyr—Ile—His—Pro—Phe—His—Leu—Val—Ile—His ~
```

FIGURE 1. N-terminus of angiotensinogen.

```
                                  | Renin
                                  ↓
His—Pro—Phe—His—Leu—Val—Ile—His
P5   P4   P3   P2   P1   P1'  P2'  P3'
```

FIGURE 2. ANG 6—13 with the bonding site code.

A systematic study of the ANG-based structure-activity relationship was performed by Burton et al.[4-8] between 1973 and 1980. They synthetized a decapeptide, renin inhibitor peptide (RIP)[6] (H-Pro-His-Pro-Phe-His-Phe-Phe-Val-Tyr-Lys-OH), which possessed Phe-Phe in the position P-P′, with Pro as N-terminal and Lys as C-terminal residue. This compound was also found to be highly active *in vivo*. It is soluble in water, and for a long time served as a basis of comparison in the design of peptides with similar structures and modes of action.

Thaisrivongs et al.[9] reported the synthesis of the ANG analog peptides **(1)** and **(2)**, containing the dipeptide isostere (2R,3R,4R,5S)-5-amino-3,4-dihydroxy-2-isopropyl-7-methyloctanoic acid residue at the scissile site, as potent inhibitors of human plasma renin.

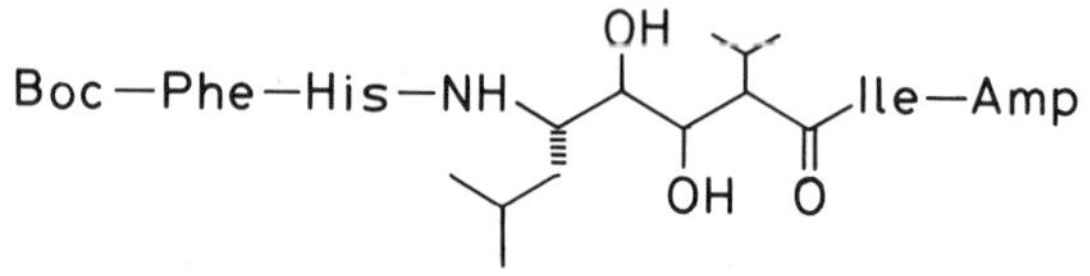

Boc—Phe—His—NH … Ile—Amp

(1)

Amp = 2-amino-methyl pyridine, Boc = butoxycarbonic acid.

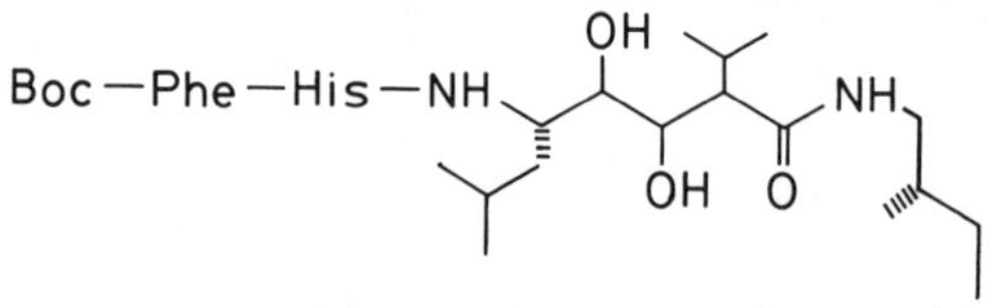

Boc—Phe—His—NH … NH

(2)

It has been discovered that an unusual γ-amino acid, statine (3*S*,4*S*-4-amino-3-hydroxy-6-methylheptanoic acid), is suitable for replacement of the Leu-Val hydroxyethylene as an isostere, even though it is one atom shorter and lacks the Val side-chain.[10] Rosenberg et al.[11] prepared substituted 1,3- and 1,4-diamines from epoxides derived from Boc-Leu or Boc-cyclohexylamine, and these structural units were incorporated into small peptide analogues of ANG, replacing the Leu-Val scissile bond. They described renin inhibitors based upon statine and its analog extended by one carbon atom homostatine, retro-inverted at the C-terminus. It is worth noting that replacement of the isopropyl portion of the statine side-chain with a cyclohexyl group significantly enhances the activity. A comparative analysis of the results of structure-activity relationship studies on renin substrate analogue inhibitory peptides has been published by Hui et al.[12] They stated that the renin inhibitory potency varied over a wide range (from inactivity to IC_{50} = 3 n*M*) when the basic decapeptide sequence (RIP) was substituted with statine or AHPPA, (3*S*,4*S*)-amino-3-hydroxy-5-phenylpentanoic acid, at the P-P′ position, or was modified at other positions too.

When the position P_5, P_3, or P'_3 contained Ftr (N^{in}-formyltryptophan), there was little effect on the inhibitory potency (in the case of AHPPH-substituted RIP), or a reduced activity (in the case of statine-substituted RIP) was observed.

However, RIP with Ftr at position P'_3 is more potent than the corresponding inhibitor with tryptophan or tyrosine. The best result was obtained with a modified statine substrate analog containing Ftr at the N- and C-terminals, acetyl-Trp(For)-Pro-Phe-His-statine-Val-Trp(For)-NH_2 (IC_{50} = 0.38 n*M*).

The potency was reduced at least tenfold when the peptide was shortened by two residues at either the amino or carboxy terminus. The AHPPA-containing inhibitors were severalfold less potent than the statine-containing inhibitors, due to the steric hindrance between the phenyl ring of the AHPPA residue and the S_1 active subsite of the receptor.

Substitution of the His at position P_2 by other basic amino acids caused a great loss in potency, possibly due to the disruption of H-bond formation, which is suggested to be important by molecular modeling. All these findings hold for *in vitro* circumstances.

Sawyer et al.[13] have explored the structure-activity relationships for ANG-based inhibitory peptides involving P_5 His substitution by Ftr; P-P′, Leu-Val replacement by Phe-Phe, Sta, Leu Ψ [CH-(OH)-CH_2] Val, or by Leu Ψ [CH_2NH] Val, and C-terminal modification. The potency of the basic compound Ac-Ftr-Pro-Phe-His-Phe-Phe-NH_2 (RIP derivative), when it was substituted at P-P′, Phe-Phe by Sta, Leu Ψ[CH(OH)CH_2] Val, or Leu Ψ [CH_2NH] Val, increased more than 100,000-fold.

The N-terminal Boc-protected compounds containing Pro Ψ [CH_2O] Phe in positions P_4-P_3 were potent inhibitors of renin. When the P_4-P_3 "ether" pseudopeptide moiety was combined with the hydroxyethylene isostere at the P-P′ positions, renin inhibitors with activities in the nanomolar range were obtained.[14]

Yizuka et al.[15] reported that KRI-1230, an orally active compound which contains norstatine (nor-Sta = 2*R*, 3*S*-3-amino-2-hydroxy-5-methylhexanoic acid), a nonnatural amino acid, is a highly potent renin inhibitor, meeting all the requirements demanded of an active renin inhibitor. They propose the following features for a potent renin inhibitor structure:

1. The presence of a large hydrophobic residue in positions P_1 and P_3 of the inhibitor. (The binding sites S_1 and S_2 located under the "flop" are wide and hydrophobic.)
2. Hydrogen-bonding possibilities (as many as possible) can contribute to the control of inhibitor orientation on the enzyme surface and also stabilize the complex formed.
3. The third requirement for a long-lived inhibitor is the minimum number of natural peptide bonds.
4. Hydroxy groups contribute to the binding energy by substituting a water molecule bound to the active site; a statine or norstatine residue is therefore very important for the inhibition.

Several nonpeptide replacements corresponding to the dipeptide Leu-Val of human ANG have been prepared and coupled to a protected dipeptide (3).[16]

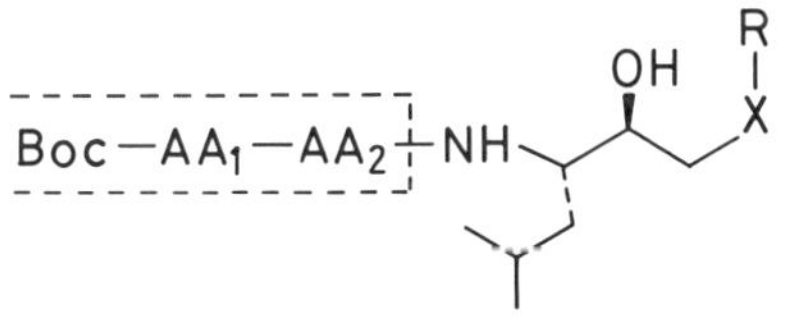

Boc—AA_1—AA_2—NH ... OH ... X—R

(3)

AA_1 were mainly Phe; AA_2 was His or Ala. R groups that closely resemble the Val side-chain of ANG were found to be preferable; thus, isopropyl ≥ higher alkyl > substituted phenyl. Group X included the following moieties: NH, S, SO, SO_2, O, and CH_2. Sulfur was found to be the best X moiety. All the compounds with some other amino acid than His in position AA_2 had a lower activity and a reduced selectivity. This attempt was aimed at minimizing the peptide feature in the compounds, and the molecular weight (water solubility, absorption).

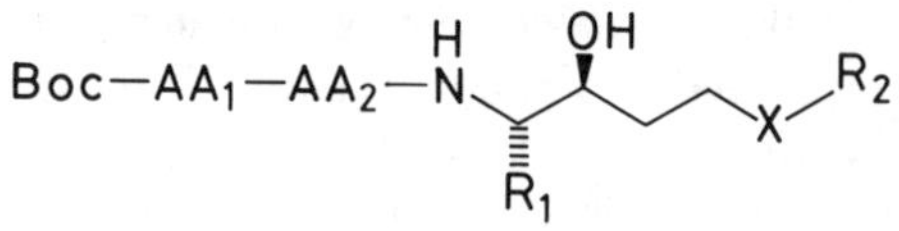

Boc—AA₁—AA₂—NH ... OH ... X—R₂, R₁

(4)

The striving for a size reduction in the case of peptide analog of ANG that are potent inhibitors of human renin also stands out in the work by Bolis et al.[17] They synthetized a series of compounds of type 4, which are derived from the renin substrate ANG by modification of the scissile amide bond with a hydroxyethylene isostere, replacement of the Val amide bond with a biosteric substituent (X = S, SO_2, elimination of the remainder of the protein, and incorporation of a small hydrophobic substituent (R = isobutyl or cyclohexylmethyl; R_2 = i-C_5H_{11} or CH_2CH_2Ph). The more potent congeners of this series show a high specificity toward human renin, and high activity, so they are potential antihypertensive agents.

The compound which contains a fluoroketone as a replacement for the enzymatically cleaved substrate amido bond might act as a "transition state analog inhibitor".[18] Fearon et al.[19] prepared a difluoromethylene ketone group containing a pentapeptide (Boc-Phe-difluorostatone-Leu-Phe-NH_2) which is 7 and 22 times more potent than the analogous statine- and statone-containing peptides, respectively.

Sham et al.[20] synthetized a novel series of conformationally constrained cyclic peptide inhibitors of human renin with ring sizes of 10, 12, and 14 **(5)**. These were based upon a linear hexapeptide inhibitor with a reduced amide replacing the scissile bond at the active site.

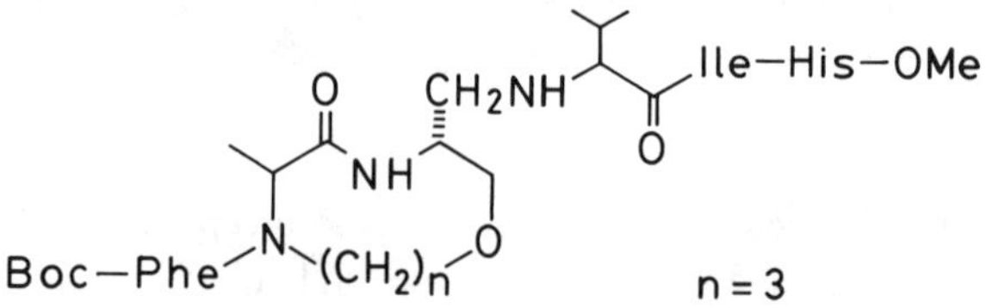

(5)

n = 3
n = 5
n = 7

It was stated that the compound with a 10-membered ring was inactive; it also seemed to fit into the human renin model active site satisfactorily. The 12-membered ring compound exhibited slight activity, but only the compound with a 14-membered ring had a significant effect on renin. Further examination by NMR showed that the Phe P_3-Ala P_2 peptide bond exists in *cis* and *trans* isomeric forms. The *cis* form of the inhibitor cannot fit into the active site of renin, and only the *trans* form can bind to the enzyme. The 10-membered ring is inactive because it is entirely *cis*; the 12-membered ring compound is partially *trans,* and the 14-membered ring derivative is 50% in the *trans* form.

As with the substrate analogue inhibitors, the olefinic peptides Leu Ψ [E–CH=CH] Gly-Val-Phe-OCH_3 and His-Leu Ψ[E–CH=CH] Gly-Val-Phe-OCH_3 were found to be potent.[21]

The results on this relatively young field of rational drug design via molecular modeling are promising even if few of the synthetized compounds are suitable for human application.

REFERENCES

1. **Boger, J.,** *Annu. Rep. Med. Chem.* 20, 257, 1985.
2. **Szelke, M., Leckie, B. J. Hallett, A., Jones, D. M. Sueiras, J., Atrash, B. and Lever, A. F.,** *Nature London,* 299, 555, 1982.
3. **Szelke, M., Leckie, B. J., Tree, M., Brown, A., Grant, J., Hallett, A., Hughes, M., Jones, D. M., Lever, A. F.,,** *Hypertension (Dallas),* 4 (Suppl. 2), 1159, 1982.
4. **Poulsen, K., Burton, J., and Haber, E.,** *Biochemistry,* 12, 3877, 1973.
5. **Burton, J., Poulsen, K., and Haber, E.,** *Biochemistry,* 14, 3892, 1975.
6. **Cody, R., Burton, J., Evin, G., Poulsen, K., Herd, J. A. and Haber, E.,** *Biochem., Biophys. Res. Commun.,* 97, 230, 1980.
7. **Burton, J., Cody, R. J., Jr., Herd, J. A., and Haber, E.,** *Proc. Natl., Acad. Sci. U.S.A.,* 77, 5476, 1980.
8. **Burton, J., Quinn, T., and Poulsen, K.,** in Peptides:Synthesis, Structure-Function, Proc. 7th Am. Peptide Symp., Rich, D. H. and Gross, E., Eds., Pierce Chemical Co., Rockford, 1981, 447.
9. **Thaisrivongs, S., Pals, D. T., Krooll, L. T., Turner, S. T., and Han, F. S.,** *J. Med. Chem.,* 30, 976, 1987.
10. **Boger, J., Lohr, N. S. Ulm, E. H., Poe, M., Blaine, E. H., Fanelli, G. M., Lin, T. Y., Payne, L. S., Schorn, T. W., La Mont, B. I., Vassil, T. C., Stabilito, I. I., Veber, D. F., Rich, D. H., and Boparai, A. S.,** *Nature (London),* 303, 8, 1983.
11. **Rosenberg, S. H., Plattner, J. J., Woods, K. W., Stein, H. A., Marcotte, A. P., Cohen, J., and Perun, T. J.,** *J.Med.Chem.,* 30, 1224, 1987.
12. **Hui, K. Y., Carlson, W. D., Bernatowicz, M. S. and Haber, E.,** *J. Med. Chem.,* 30, 1287, 1987.
13. **Sawyer, T.K., Pals, D. T., Mao, B., Staples, D. J., de Vaux, A. E., Maggiora, L. L., Affholter, J. A., Kati, W., Duchamp, D., Hester, J. B., Smith, C. W., Saneii, H. H., Kinner, J., Handschumacher, M., and Carlson, W.,** *J. Med. Chem.* 31, 1, 1988.
14. **Ten Brink, R. E., Pals, D. T., Harris, D. W., and Johnson, G. A.,** *J. Med. Chem.,* 31, 671, 1988.
15. **Tizuka, K., Kamijo, T., Kubota, T., Akahane, K., Umeyama, H., and Kiso, Y.,** *J. Med. Chem.,* 31, 1, 1988.
16. **Luly, J. R., Yi, N., Soderquist, J., Stein, H., Cohen, J., Perun, T. J., and Plattner, J. J.,** *J. Med. Chem.,* 30, 1609, 1987.
17. **Bolis, G., Fung, A. K. L., Greer, J., Kleinert, H. G., Marcotte P. A., Perun, T. J., Plattner, J. J., and Stein, H. H.,** *J. Med. Chem.,* 30, 1729, 1987.
18. **Wolfenden, R.,** *Annu. Rev. Biophys. Bioeng.,* 5,271,1976.
19. **Fearon, K., Spaltenstein, A., Hopkins, P. B., and Gelb, M. H.,** *J. Med. Chem.,* 30, 1617, 1987.
20. **Sham, H. L., Bolis, G., Stein, H. H., Fesik, S. W. Marcotte, P. A., Plattner, J. J., Rempel, Ch. A., and Greer, J.,** *J. Med. Chem.,* 31, 284, 1988.
21. **Johnson, R. L.,** *J. Med. Chem.,* 27, 1351, 1984.

Chapter 3

ANGIOTENSIN CONVERTING ENZYME (ACE) INHIBITORS

I. MECHANISM OF ACTION

Captopril, enalapril, and lisinopril, the ACE-inhibitory agents commercially available at present, exert their antihypertensive action in qualitatively the same way. They inhibit the cleavage of the decapeptide angiotensin-I to the octapeptide angiotensin-II (A-II) by the competitive inhibition of angiotensin-converting enzyme (ACE). Consequently, the blood level of the vasoconstrictor A-II is reduced, resulting in a lowering of the blood pressure. The pharmacology and mode of action of captopril and ACE inhibitors in general have been reviewed by several authors.[10-15]

A relationship was found between the antihypertensive effect of the ACE-inhibitors and the initial plasma renin activity.[16] Cptopril and the other ACE-inhibitors were shown to potentiate the vasodilatory effect of bradykinin (which is also degraded by ACE). Therefore, the plasma level of bradykinin increases in the presence of ACE-inhibitors, and this effect has been presumed to be a factor in the antihypertensive effect of ACE-inhibitors,[17,18] although this view has not been confirmed by other authors.[19,20]

II. HISTORY AND STRUCTURE-ACTIVITY RELATIONSHIP

The gradual understanding of the physiological role of ACE led to the recognition of a basically new way to elicit a hypotensive effect in the human organism. The possibility that medicinal inhibition of ACE may cause the simultaneous blocking of the clevage of angiotensin-I and bradykinin (see pp. 12 and 28) initiated an intensive search for ACE-blocking agents. As a first milestone, the strong ACE-inhibitory potential of the nonapeptide SQ 20881 **(5)** was identified.[21,22] Although the therapeutic use of this compound was limited by its lacking oral activity, the knowledge that accumulated on the interaction between SQ 20881 and ACE, and especially on the chemical and enzymatic properties of ACE, prompted new possibilities in ACE-inhibitory drug design.

Glu–Trp–Pro–Arg–Pro–Gln–Ile–Pro–Pro

(5)

The theoretical arguments and the practical confirmation of such a new approach were published by the research team at the Squibb Institute in the late seventies.[23] Their designing principle in synthetizing nonpeptidic ACE-inhibitors was based on two experimental findings:

1. It was confirmed that ACE belongs in the group of carboxypeptidases, and in its structure and properties resembles pancreatic carboxypeptidase-A;[24,25] consequently, ACE is a metalloprotein containing one zinc ion/mol as one of the potential binding sites, and a positively charged Arg as a second one; a hydrophobic ''pocket'' which is a selective binding site for carboxypeptidase-A is lacking from ACE, but as a third active site this latter can form a hydrogen bond in which a functional group is located between the previously mentioned two active groups (Figure 1).
2. D-Benzylsuccinic acid was found to be a potent inhibitor of carboxypeptidase-A,[15] and its interaction with the three binding sites of the enzyme has been postulated.[26,27] Assuming that the distance between the zinc ion and the argininium cation in ACE is greater than that in carboxypeptidase-A, Ondetti, Rubin, and Cushman hoped to find

TABLE 1
Drugs Interfering with the Renin-Angiotensin System

Generic name	Chemical structure	Proprietary name®	Therapeutic use	Doses
Captopril	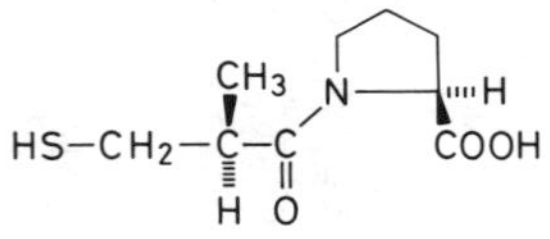	Lopryl, Tensobon, Lopirin, Capoten, Tensiomin	Used in acute and chronic heart failure and in hypertension. Orally administered, Captopril is absorbed very quickly (15 min) and is effective for hours. It has no action in normotensive persons, but depressed the hypertension also in that case, when renin activity of plasma was low. Side effect: hypotension.	Usual doses: 25—450 mg/day orally
Dihydralazine	See under II/6			
Enalapril	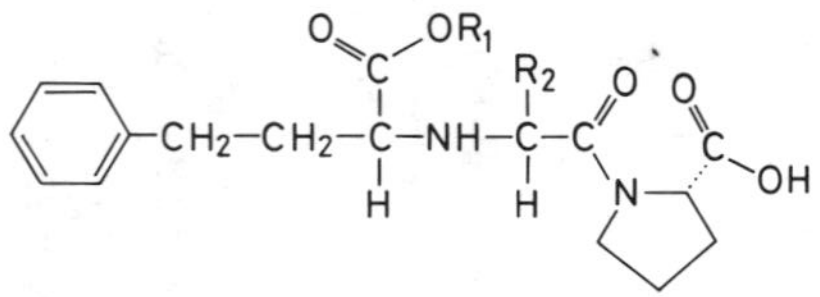 R_1 / R_2 OC_2H_5 / CH_3 — enalapril H / $(CH_2)_4NH_2$ — lisinopril	Vasotec, Xanel	Enalapril acts stronger and is more effective than captopril. Its primary indication is hypertension. Side effects: rash, taste inconvenience.	Usual doses: 10—20 mg once a day
Lisinopril	(Same as above)			
Indoramin	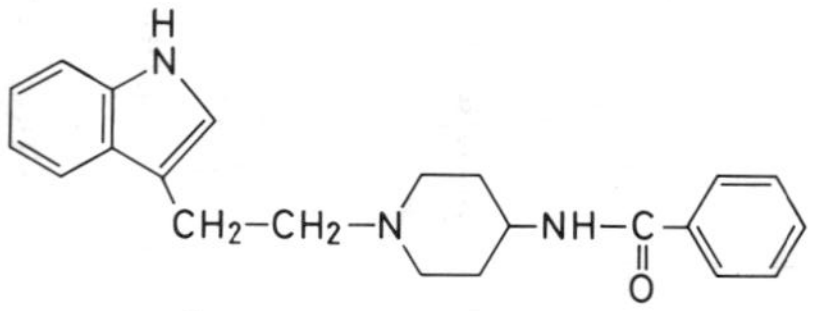	Wydora		Usual dose: 50—150 mg/day
Saralasin	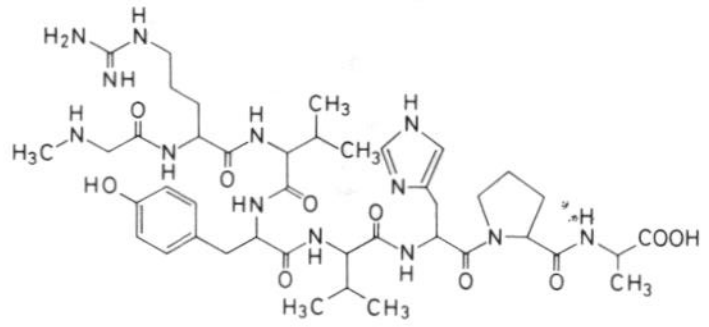	Saralasinacetate, Sarenin	A competitive antagonist of angiotensin II, it has a hypotensive effect on normotensive persons too.	

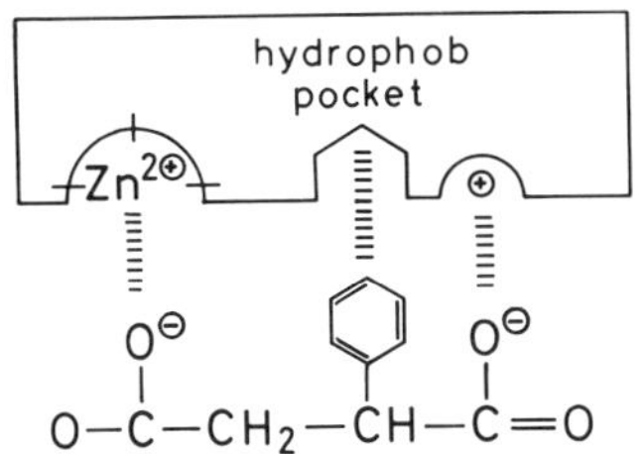

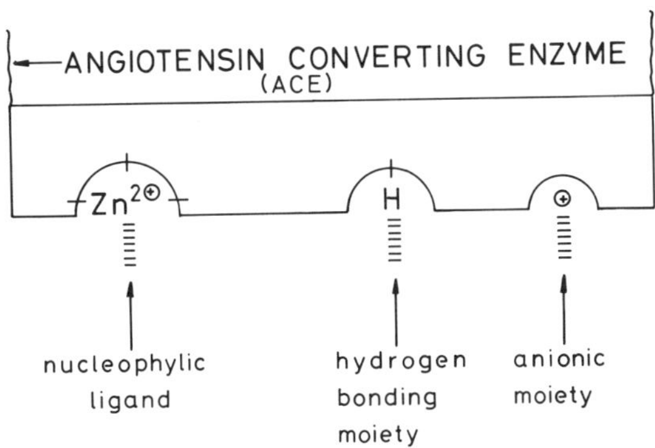

FIGURE 1. Schematic representation of the binding sites of carboxypeptidase-A and of angiotensin-converting enzyme. (From Ondetti, M. A. et al., *Science,* 196, 441, 1977. With permission.)

potent ACE-inhibitors among the succinic acid derivatives lengthened by one amino acid unit. A series of succinylproline derivatives were synthetized,[23] and, of these, compound SQ 14.225 proved to be a highly effective ACE-linhibitor. This compound was given the nonproprietory name ''captopril'', and for a time was the prototype of ACE-inhibitor hyptensive agents. The history and development of captopril are drawn up by early reviews.[28,29]

The three points presumed to bind captopril to the ACE surface are demonstrated in Table 2. The IC_{50} data clearly show the supremacy of the thiol (mercapto) group over the carboxy group in complex formation with the zinc cation. Further, the importance of the conformations in the strength of the ACE-inhibitor interaction are convincingly indicated by the IC_{50} values in Table 2.

Although captopril is a member of the youngest generation of antihypertensive agents, and in some cases its administration as a first-step therapeutic may be reasonable, the search for newer ACE-inhibitors has continued during the eighties.

The general structural requirements for therapeutically useful sulfur (or phosphorus)-containing ACE-inhibitors are to be found in the paper by Kim et al.[30] They synthetized mercaptopropanoyl indolinocarboxylic acids and postulated their binding to the hypothetical active sites of ACE.

A series of 1-[3-(acylthio)-3-aroylpropionyl]-L-proline derivatives was prepared and tested for hypotensive activity by Mc Evoy et. al.[31] As most effective agents were found the acetylthio 3-substituted-benzoylpropionyl derivatives **(6)** proved to be equipotent with captopril.

TABLE 2
ACE-Inhibitory Activities of Different N-Acylproline Derivatives*

Structure (interaction)	ACE-inhibitory activity (IC_{50} μM)	Note
Zn^{2+} ..... H ..... ⊕		
$^{\ominus}OOC-CH_2-CH_2-C(=O)-N$ (proline, $COO^{\ominus}$)	135	Appropriate length, but relatively weak ligand affinity
$^{\ominus}OOC-CH_2-CH(CH_3)-C(=O)-N$ (proline, $COO^{\ominus}$)	12	Same as for 1, but more favourable conformational chances
$^{\ominus}OOC-CH(CH_3)-CH_2-C(=O)-N$ (proline, $COO^{\ominus}$)	340	Same as for 1, but less favourable conformational chances
$^{\ominus}OOC-CH_2-CH_2-CH(CH_3)-C(=O)-N$ (proline, $COO^{\ominus}$)	1.0	Same as for 2, but more favourable conformational chances
$^{\ominus}OOC-CH_2-CH_2-CH(CH_3)-C(=O)-N$ (proline, $COO^{\ominus}$)	230	See the comparison of 2 and 3
$HS-CH_2-CH_2-C(=O)-N$ (proline, $COO^{\ominus}$)	0.04	Same as for 1, but with much stronger ligand affinity
$HS-CH_2-CH(CH_3)-C(=O)-N$ (proline, $COO^{\ominus}$) (Captopril)	0.005	See the comparison of 1, 2, and 3
$HS-CH_2-CH(CH_3)-C(=O)-N$ (proline, $COO^{\ominus}$)	0.50	—

* Modified from Reference 3.

Suh et al.[32,33] reported the design and synthesis of an orally active novel series of substituted (mercaptoalkanoyl) glycines **(7)**.

R-C_6H_4-C(=O)-CH($SCOCH_3$)-CH_2-C(=O)-N (proline, COOH)

(6)

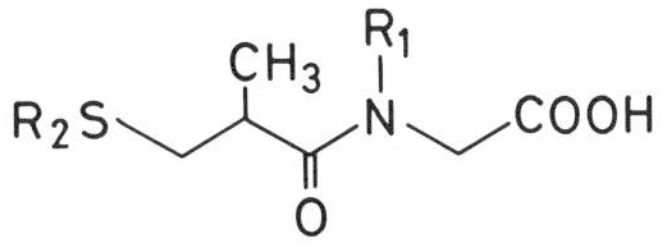

(7)

An interesting and challenging feature of this approach is that this series of compounds contains exclusively the nonchiral glycine as amino acid component, which contradicts the earlier view concerning the essential nature of the proline moiety as a source of maximum ACE-inhibitory (hypotensive) activity.[34-36]

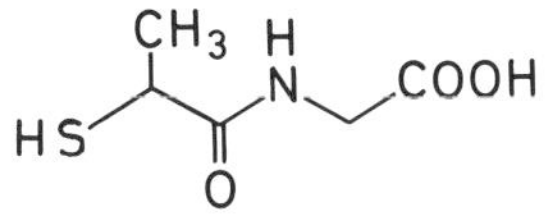

(8)

The simplest member of the tested series mercaptopropanoyl-glycine **(8)** (tiopronin) was found to have a rather large *in vitro* ID_{50} value (1.9 μM) compared with that of captopril (0.005 μM,[23] 0.017 μM[33]), and the moderate activity of tiopronin is diminished further in serum due to the metabolic cleavage of the unsubstituted amide group. The structural modifications aimed at increasing the ACE-inhibitory activity followed two lines:

1. Substitution of the amide-hydrogen by an acyl (mainly acetyl) group;
2. Substitution of the mercapto-hydrogen by an alkyl, cycloalkyl, or aryl group.

The variations within these substitutions resulted in the preparation of a series of active ACE-inhibitors with inhibitory effects similar to or stronger than that of captopril. This is illustrated by the examples in Table 3.

In conclusion, instead of proline, glycine should serve as a basis for the construction of potent ACE-inhibitor compounds. This suitability of glycine may be rooted in the similar positions of the terminal carboxy group and the nitrogen in the two compounds **(9, 10)**:

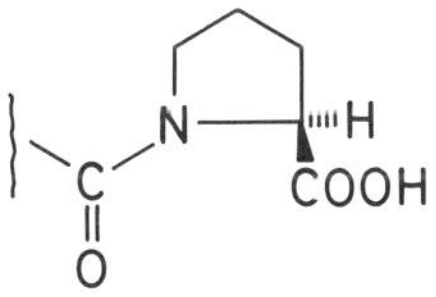

proline based ACE inhibitor

(9)

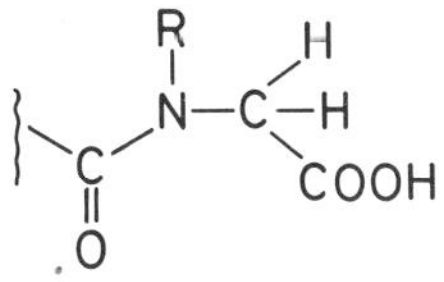

glycine based ACE inhibitor

(10)

The N-substitution of the glycine-based ACE-inhibitor structure by alkyl, cycloalkyl, or aryl (R) groups significantly enhanced the ACE-inhibitory potency in a wide circle of substituents. This experience, together with the rather low value for the nonsubstituted compound (Table

TABLE 3
ACE-Inhibitory Activity of N- and S-Substituted Mercaptopropanoylglycines[33]

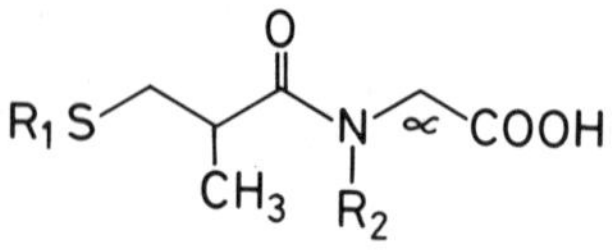

Compound	R_1	R_2	ID_{50}, μM
1	H	H	0.21
2	H	CH_3	0.13
3	H	$CH(CH_3)_2$	0.072
4	CH_3CO	$CH(CH_3)_2$	5.9
5	H	c-C_3H_5	0.03
6	CH_3CO	c-C_3H_5	0.54
7[a]	H	c-C_3H_5	0.079
8	H	c-C_4H_7	0.018
9	CH_3CO	c-C_4H_7	0.22
10	H	c-C_5H_9	0.018
11	CH_3CO	c-C_5H_9	0.082
12	H	c-C_6H_{11}	0.035
13	H	c-C_7H_{13}	0.031
14	H		0.032
15	CH_3CO		0.020
16	H	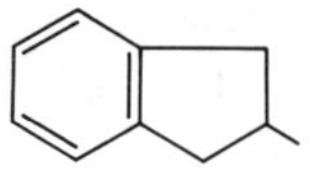	0.031
17	CH_3CO	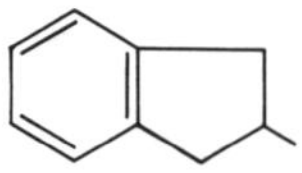	0.34
18	H	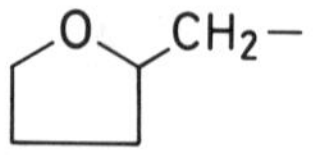 CH_2—	0.13
19	H	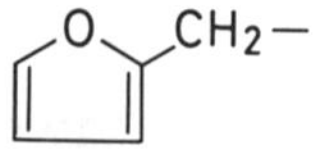 CH_2—	0.17
20	H	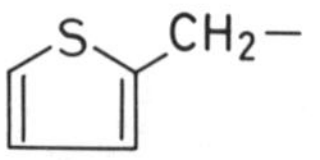 CH_2—	0.055
21	H	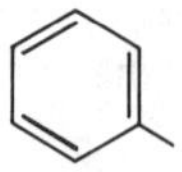	0.30

TABLE 3 (continued)
ACE-Inhibitory Activity of N- and S-Substituted Mercaptopropanoylglycines[33]

Compound	R_1	R_2	ID_{50}, μM
22	H	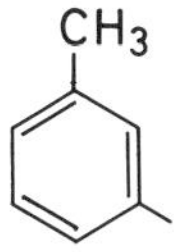 CH_3 (tolyl ring)	0.019
23	H	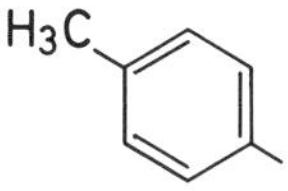 H_3C (tolyl ring)	0.005
24	H	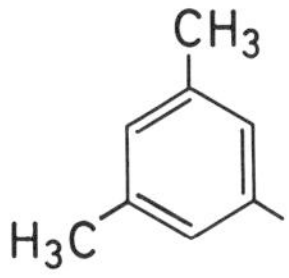 CH_3, H_3C (dimethylphenyl ring)	0.044
25	H	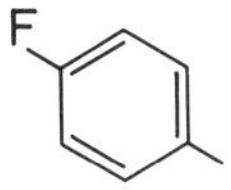 F (fluorophenyl ring)	0.023
26(R)	$(CH_3)_3C{-}C(=O){-}$	c-C_5H_9	100
27(S) (pivopril)	$(CH_3)_3{-}C{-}C(=O)$	c-C_5H_9	3.60

[a] α-methyl derivative.

From Suh, J. T. et al., *J. Med. Chem.*, 28,57,1985. With permission.

3, compound 1) appears to confirm that the amino acid-nitrogen does not bind to an active site of the ACE, and also that the N-substituents take part in the ACE-inhibitor interactions only by influencing the steric structure of the inhibitor. The most active compound prepared, the *p*-tolyl derivative (compound 22), exhibited an IC_{50} value of 0.005 μ*M*, exceeding the value of captopril. The S-substitution of the glycine-based ACE inhibitor structure generally reduced the ACE-inhibitory activity. From the series of *S*-acylated and *S*-alkylated compounds prepared, the pivaloic esters (compounds 26,27) were selected for further development. The moderate inhibitory activity of these compounds was thought to be balanced by the lower side-effects (rashes, loss of taste, etc.), which have been reported to be associated with the presence of the free mercapto group. Moreover, the *in vivo* efficacy (potency and duration) of pivopril was comparable to that of captopril.

In a search of another way to eliminate the side-effects presumed to be associated with the mercaptoalkyl side-chain of sulfur-containing ACE-inhibitors, Menard, Suh et al.[37] synthetized a series of mercaptoaroyl amino acids in which the mercapto group is bound to an aromatic ring and the amino acid is represented by glycine. To influence the chelating ability of the mercapto group, the *S*-aromatic nucleus was substituted by substituents with different electronic and steric properties. Of the compounds tested for ACE-inhibitory activity, those with electronegative aromatic substituents were shown to have fairly strong *in vitro* activity. The IC_{50} of *N*-(3-chloro-2-mercapto-benzoyl)-*N*-cyclopentylglycine **(11)** was 0.28 μM, which was significant, though much larger than that of captopril (IC_{50} = 0.027 μM.

Substitution by the same substituents in positions other than C_3 impaired the ACE activity (IC_{50} of the 4-Cl and 5-Cl derivatives: 100 μM and 80 μM, respectively). To change the chelating ability in a more direct way, the mercapto group was replaced by other polar chelator functionalities, such as a nitro, hydroxy, or carboxyl group. These compounds were inactive, revealing the essential role of the mercapto group.

To eliminate the side effects induced by the free mercapto group, further captopril analogs were synthetized with a cycloalkyl bridge between the proline and mercapto moieties.[38] The strongest ACE-inhibitory activity was shown by the benzoylthiocycloheptyl captopril analog **(12)**. Somewhat earlier, similar efforts seem to be reflected by the papers of Watthey et al.[39] who prepared the mercaptoazepinyl and mercaptoazocinyl captopril analogues **(13, 14)**

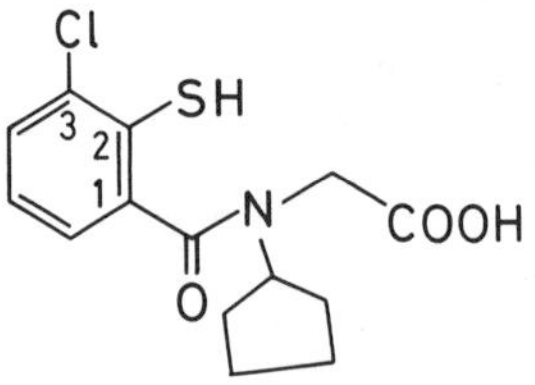

Cl SH 3 2 1 N COOH O

(11)

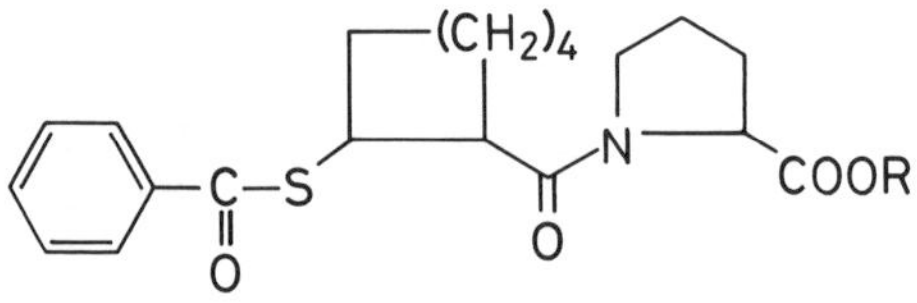

(CH2)4 C S N COOR O O

(12)

(CH2)n SH N O COOH

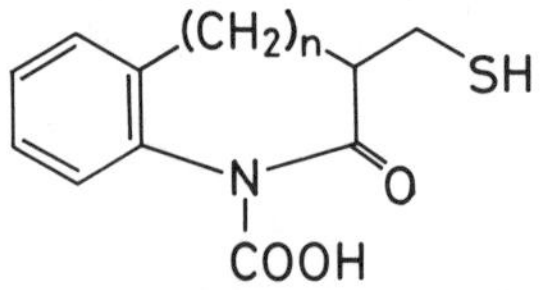

n = 2 (13)
n = 3 (14)

In another approach to improve the ACE-inhibitory selectivity and intensity, Smith et al.[40] recently introduced substituents into the proline C_4 position **(15)**. One of the most potent compounds was that carrying a dithioethylene bridge at this position **(16)**.

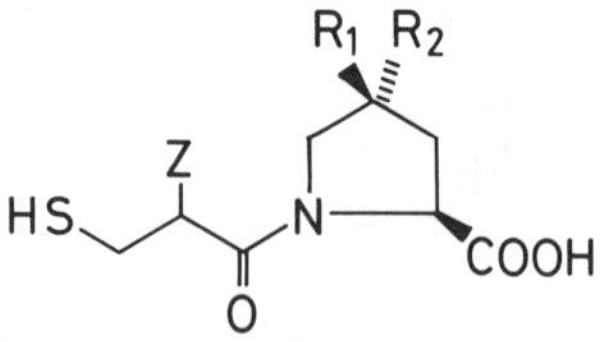

R1 R2 Z HS N COOH O

(15)

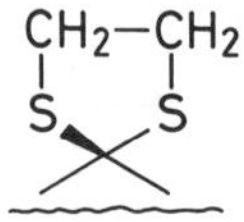

CH2—CH2 S S

(16)

A fundamentally different attempt to eliminate the side-effects due to the mercapto group from the overall effect of ACE-inhibitors was manifested in the effort to synthetize non-sulfurated ACE-inhibitors. In light of the ACE-inhibitory effect of the nonapeptide, SQ 20881, Patchett et al.[41] designed and synthetized the *N*-carboxymethyl dipeptide, enalapril (MK 421). This compound was found to be more potent and longer-acting than captopril and lacked some of the side-effects exhibited by the latter. Another compound with a longer duration of action was the aminobutyl (lysyl) analog of enalapril, lisinopril (MK 521). Enalapril and lisinopril are regarded as pro-drugs, which are activated by esterases in the organism to give the dicarbonic acid derivatives. The therapeutic utility of enalapril is similar to that of captopril, and therefore further search for new ACE inhibitors remained a reasonable task.

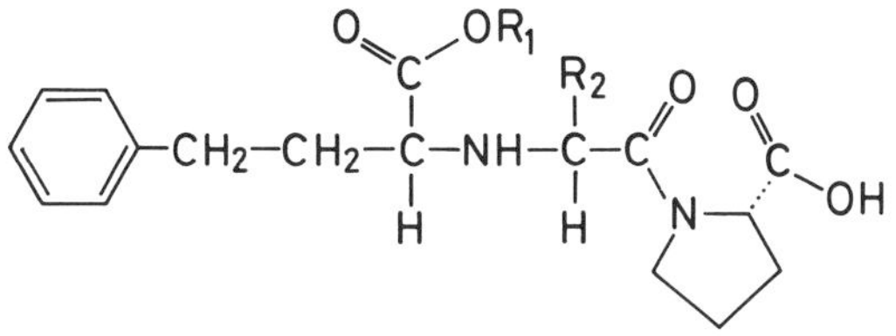

R_1	R_2	
OC_2H_5	CH_3	enalapril
H	$(CH_2)_4NH_2$	lisinopril

Greenlee et al.[42] have prepared substrate analog with an acyltripeptide structure, in which the carbonyl of the scissile peptide bond is replaced by a –CH–COOH group **(18)**. These compounds did not show an ACE-inhibitory effect greater than that of enalapril. This experience was presumed to demonstrate that no additonal interactions take place that involve contributions of the moieties of the extended peptide side-chain. For further support of this view, extended tetra- and pentapeptide analogs were prepared and, indeed, exhibited an *in vitro* inhibitor potency not greater than that of the tripeptide (benzamido) analog **(18)**. The pentapeptide analog synthetized by Almquist et al.[43] was ten times more effective as an ACE inhibitor than the acyltripeptide **(18)**, indicating that cyclobutylcarbonyl L-lysine substitution at the N-terminal provides further binding sites on the ACE surface. The short half-life (24 min) and inactivity in hypertension of renal origin are unfavorable factors for this compound.

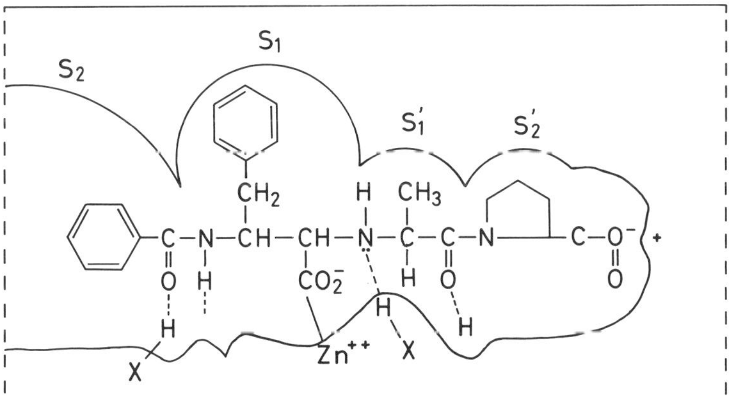

(18)

Binding of inhibitors to the hypothetical active sites of angiotensin-converting enzyme.(From Greenlee, W. J. et al., *J. Med. Chem.*,28,434,1985. With permission.)

On the basis of previous results[44] and a study of the active conformation of enalapril,[45] benzazepin-2-one derivatives together with six- and eight-membered ring analogs were synthetized by Watthey, Stanton et al.[46] The biological profile of a benzazepine bicyclic lactam **(19)** proved to be comparable to that of enalapril (IC_{50}).

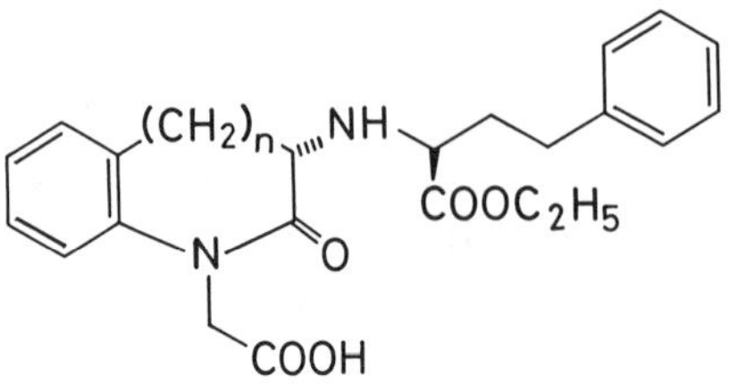

(19)

Analysis of the results allowed further insight into the ACE-S_1-subsite interactive properties. It appears that the S_1-subsite (see **18**) cannot have very specific structural requirements against ACE-inhibitors. This view[32] is based on a comparison of the IC_{50} data of enalapril with those of several other tested compounds.

The synthesis of 1,5-benzthiazepine derivatives provided ACE-inhibitory compounds.[47] The most effective proved to be compound **(20)**.

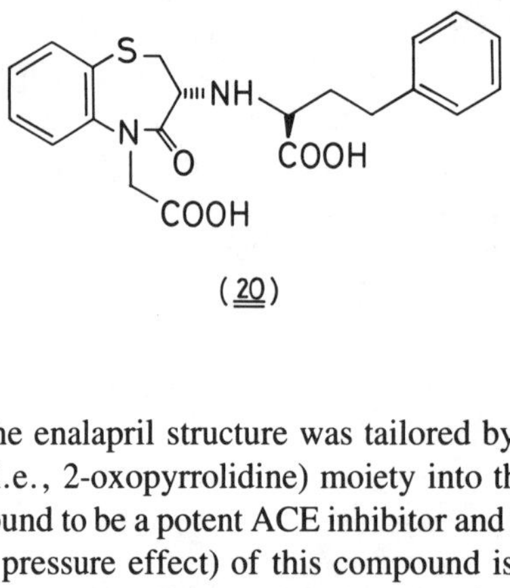

(20)

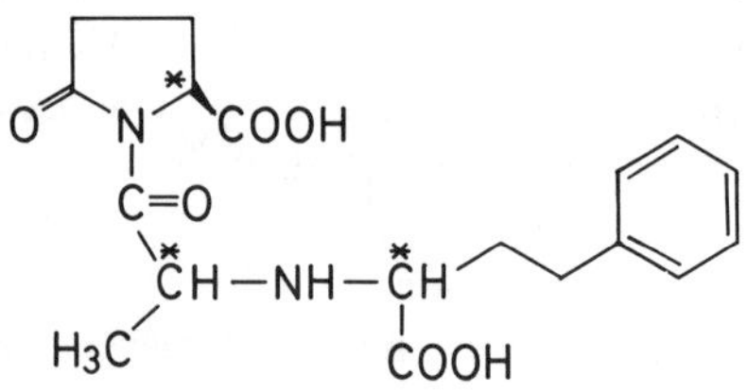

(21)

The enalapril structure was tailored by Johnson et al.,[48,49] who built an L-pyroglutamic acid (i.e., 2-oxopyrrolidine) moiety into the peptide structure; the resulting compound **(21)** was found to be a potent ACE inhibitor and hypertensive agent. The potency (ACE inhibition, blood pressure effect) of this compound is approximately half of that of enalapril, but it is similar to enalapril in its duration of action. To establish the stereochemistry of **(21)** and its congeners, the X-ray crystallography of enalapril was chosen as comparative basis, and the isomer to which the *S,S,S*-configuration was assigned had an ED_{50} value (0.13 mg/kg) one order of magnitude less than that of the *S,R,S*-diastereomer (ED_{50} = 4.0 mg/kg).

A series of nonmercapto dipeptides related to enalapril were prepared by Tinney et al.,[51] who substituted the amino group by isosteric moieties (O, S, SO, SO_2) and replaced the proline residue with different hydrophobic amino acids **(22)**.

An *in vitro* inhibitory activity comparable to that of enalapril was obtained when the NH group was substituted by an etheric oxygen, and the proline by fused ring analogs **(23—25)**. In accordance with earlier findings, the potent compounds have an optical configuration S- or RS at the chirality centers. The poor oral activity of the above compounds is attributed[51] to bioavailability causes.

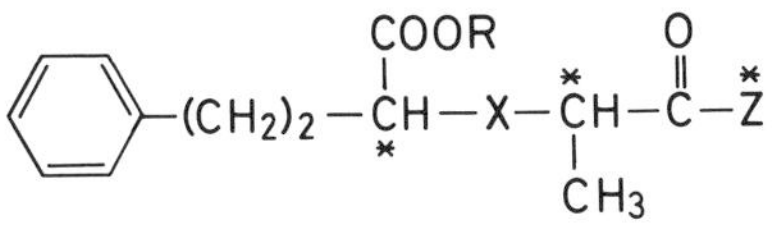

(22)

R	X	Z	IC_{50}
C_2H_5	NH	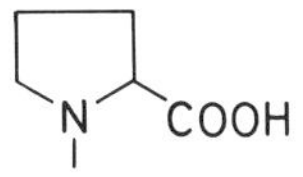 enalapril	1.3×10^{-8}
C_2H_5	O	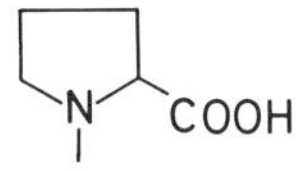 (23)	2.2×10^{-7}
C_2H_5	O	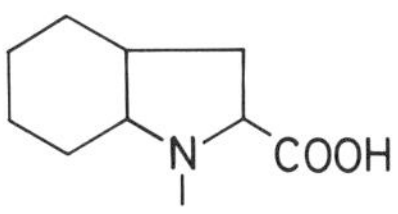 (24)	2.4×10^{-8}
C_2H_5	O	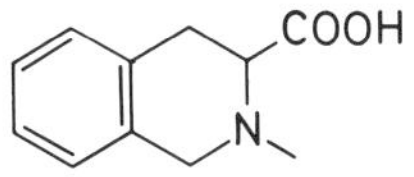 (25)	1.4×10^{-8}

Earlier, the tetrahydroisoquinoline moiety provided significant ACE-inhibitory activity when a ureido carbonyl group was substituted into the molecule **(26)**.[52]

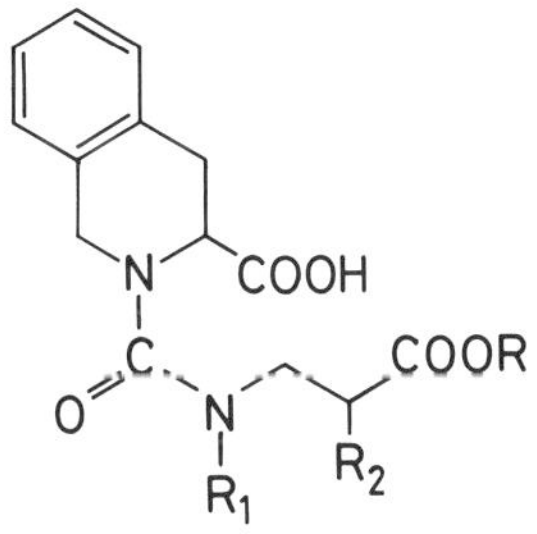

(26)

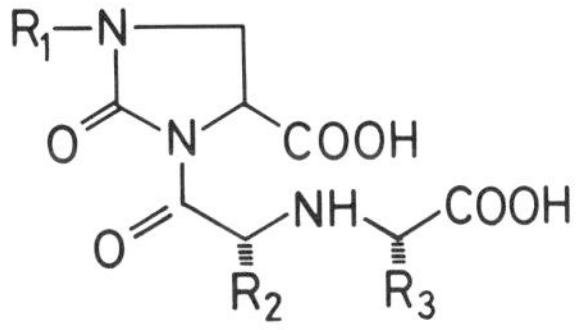

(27)

On this basis, it was expected that if the ureido group was incorporated into a ring, the compound formed would be a potent inhibitor of ACE. A series of 2-imidazolidone carboxylic acids **(27)** was prepared.[53] The compounds with a CH_3 or C_2H_5 substituent as R_2 with (*S*)-configuration; with H, CH_3, or CH_2-C_6H_5 substituents as R_1; and with *n*-octyl, phenylethyl, 5-isobutyl, 5-isopentyl, or benzyl substituents as R_3 had IC_{50} values of 1.1×10^{-8} to 1.5×10^{-9}, i.e. comparable or superior to that of enalaprilate. These results suggest that the coplanarity between the 2-oxoimidazolidine ring and the amido carbonyl in the side-chain provides geometry suitable for effective binding with the active site of ACE and the ureido carbonyl in the ring, and can provide an additional binding site to the enzyme through hydrogen-bonding. The compound with R_1=CH_3, R_2=CH_3 and R_3=$CH_2CH_2C_6H_5$ (S-) was the most active in animal experiments. The ID_{50} value of this compound was 0.24 mg/kg, compared with 0.30 mg/kg for enalapril.[53]

A recent development in the preparation of structurally modified ACE inhibitors is the introduction of phosphinyloxy derivatives.[54] In these compounds, it is proposed that the hydroxyphosphinyl function coordinates to the zinc atom instead of the mercapto group of captopril or the peptide moiety of enalapril. The geometry of the phosphonates **(28)** is planned to approximate to the hydrated amide transition state[53] **(29)** assumed as an activated form of carboxyalkyl dipeptides.

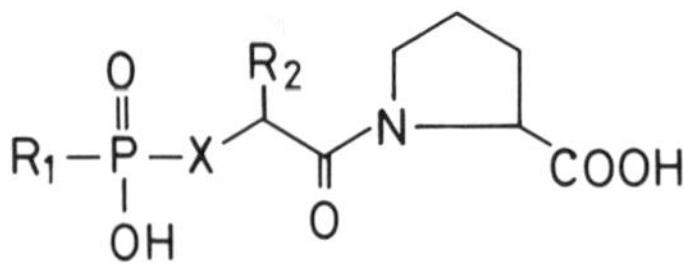

X=O (30)
X=NH (31)
X=CH_2 (32)

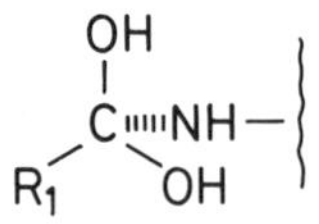

(29)

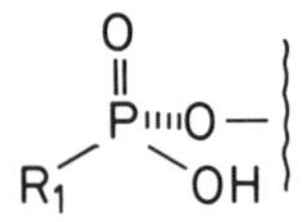

(28)

It has been proved[54] that phosphonamide derivatives **(31)** bind more strongly to the ACE surface than either the phosphonates **(30)** or the phosphinic acid isostere **(32)**. One main cause of this seems to be the formation of a hydrogen bond between the phosphonamide-NH and an active site on ACE.[55] The oral activity exhibits a different sequence. Hydrophobic substitution at position R_2 and a phosphonate structure (X=O) provided the best results (Table 4).

TABLE 4
Comparative Activity Data on Phosphinyloxy Derivatives

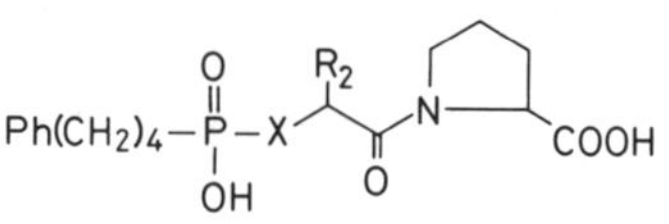

Compound X	Compound R_2	ACE IC_{50} (n*M*)	ED_{50} (μ*M*/kg i.v.)	ED_{50} (μ*M*/kg orally)
NH	CH_3	12	0.050	25
O	CH_3	59	0.750	25
CH_2	CH_3	220	2.20	No data
NH	$(CH_2)_4NH_2$	9.0	0.022	2.6
O	$(CH_2)_4NH_2$	36	0.063	0.53
CH_2	$(CH_2)_4NH_2$	85	0.180	35
Captopril		23	0.092	0.73
Enalaprilate		4.3	0.033	5.13
Enalapril				0.46

As the most promising compound, the derivative with phenylbutyl and aminobutyl side-chains was chosen for clinical study in humans.[54] In a recent publication[56] the same research team reports on the ACE-inhibitory effectiveness of 4-substituted proline derivatives in the mercapto (captopril) and nonsulfurated structural types. Substitution by lipophilic groups at position 4 of proline produces greater potency and longer duration of *in vivo* ACE-inhibitory action than for the parent compound, captopril. Thus, the *S*-benzoyl and 4-*cis*—phenylthio derivative in the form of its Ca salt, as a pharmacon candidate (izofenopril), were subjected to pharmacological[57] and pharmacokinetic[58] evaluation. Of the enalapriltype analogs, spirapril[59] **(33)**, and of the phosphinic acid derivatives, fosinopril[60] **(34)**, an acyloxy pro-drug of **(35)**, proved to be orally active antihypertensive medicine candidates.

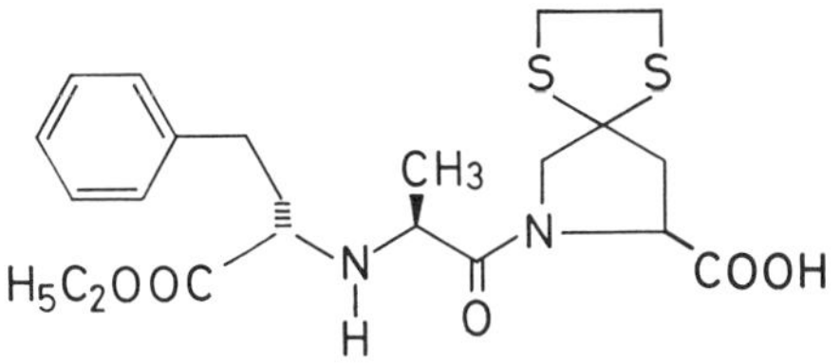

H_5C_2OOC … CH_3 … COOH

(33)

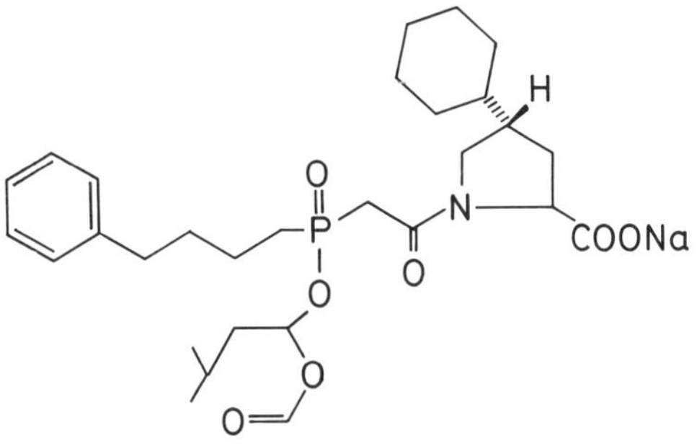

COONa

(34)

COOH

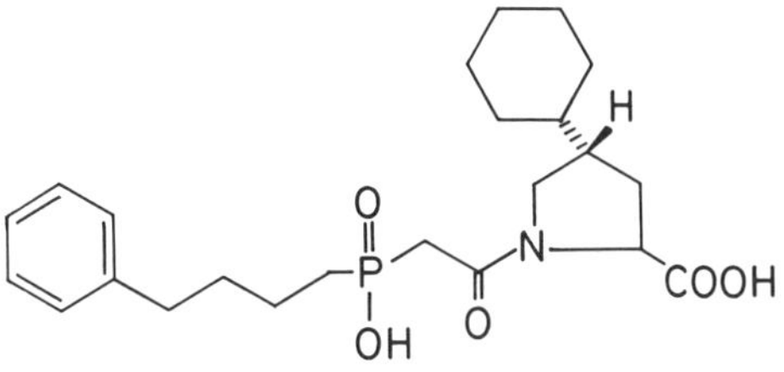

(35)

Mention should be made of the conformational analysis of ACE by Andrews et al.[61] to determine the structural and conformational requirements for ACE inhibition.

Mapping of the ACE active sites suggests that the carboxyl binding group is an arginine-guanidinium ion; the protonated nitrogen is located approximately 2.8 Å from one or both carboxy-oxygen atoms. The binding site of the amide-carbonyl lies on the same side of the proline ring as the carbonyl binding groups. The third and fourth binding groups are the zinc atom and the aromatic binding site. The aromatic groups of the inhibitors seem to be accommodated in a hydrophobic binding pocket.

III. CAPTOPRIL

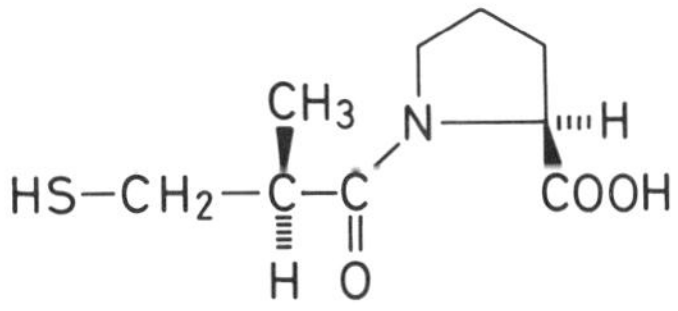

$HS-CH_2-C(CH_3)(H)-C(=O)-N$ … H, COOH

L-proline 1-(3-mercapto-2-methyl-1-oxopropyl)-(*S*)-

$C_9H_{15}NO_3S$ $M_r = 217.28$

A. PROPERTIES

Captopril is a white crystalline powder with two melting points. The labile form melts at 87 to 88°C. Then, on further heating it resolidifies and remelts at 104 to 105°C. It has

TABLE 5
^{13}C Chemical Shifts in Captopril[5]

Carbon atom	Conformation	pH Species	0.59 H_2A	7.43 HA^-	12.30 A^{2-}
C_3	*trans*	—	27.630	27.730	29.689
	cis	—	a	27.560	29.101
C_2	*trans*	—	42.553	42.333	44.832
	cis	—	43.141	42.850	a
CH_3	*trans*	—	16.898	16.901	16.751
	cis	—	a	16.750	16.089
CON	*trans*	—	177.153	176.505	178.476
	cis	—	a	177.373	179.726
C_α	*trans*	—	60.049	62.550	62.475
	cis	—	a	63.210	63.211
C_β	*trans*	—	29.836	30.433	30.497
	cis	—	31.673	32.190	32.261
C_γ	*trans*	—	25.204	25.130	25.131
	cis	—	a	23.290	23.293
C_δ	*trans*	—	48.655	48.800	48.802
	cis	—	47.722	47.921	47.772
COO	*trans*	—	177.006	180.534	178.402
	cis	—	a	a	179.729

Note: a = resonance either too small to be detected, or not resolved from resonance for *trans* isomer.

been found[1] that pure D-captopril melts in the range 105 to 108°C, while a sample containing L-captopril as impurity melted at 103 to 105°C. Captopril has an orthorhombic crystal structure[2] and exhibits an unusual antiplanar information of the carboxy group in the single-crystal due to the presence of a strong intermolecular hydrogen bond. It is moderately soluble in water (~ 50 mg/m1), and soluble in ethanol, ethyl acetate, dichloromethane and benzene.

Several physical-chemical properties and reactions are very similar to those of enalapril, and therefore some instructive information may be found under that head too. As proline-containing peptides were found to exist as an equilibrium mixture of *cis* and *trans* isomers,[3,4] captopril was also expected to be present as a mixture of the *trans* (**1**) and *cis* (**2**) forms in aqueous solution.

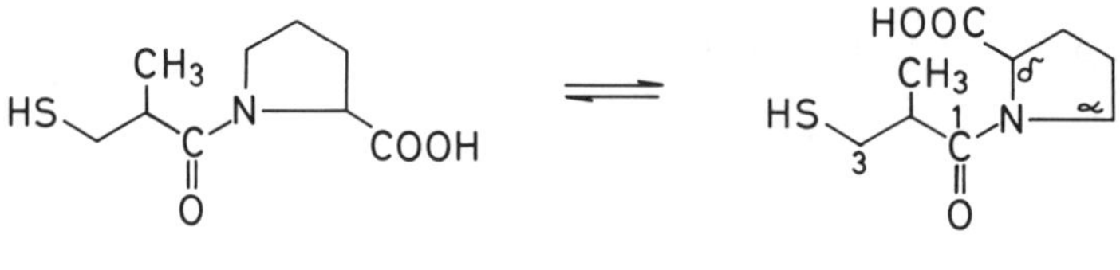

HS, CH3, C, O, N, COOH ⇌ HOOC, CH3, δ, HS, 1, 3, C, O, N, α

(1) (2)

Nothing is known about the ACE-inhibitory activities of (**1**) and (**2**), and therefore the characterization of the above equilibrium was of importance from both practical and theoretical points of view. Rabenstein and Anvarhusein[5] published the results of an ^1H- and ^{13}C-NMR study of captopril in aqueous solution adjusted to pH values between 1 and 13. Due to the slow rate of interchange between the *cis* and *trans* isomer, NMR spectroscopy provided an excellent method for the acquisition of numerical data on the relative concentrations of the two forms. The different chemical shifts for some of the carbon atoms of the two isomers are seen in Table 5, especially at pH 7.43 and 12.30, when the species HA^-

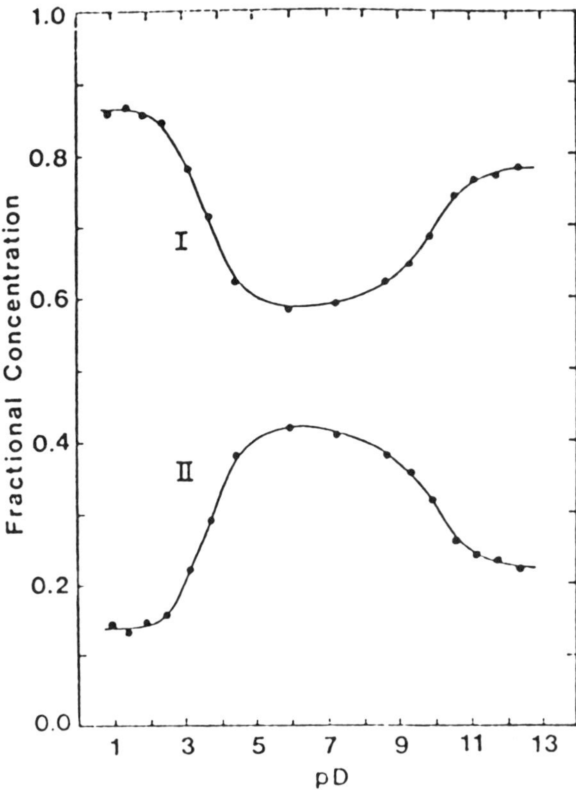

FIGURE 2. Fractional concentrations of the *trans*(I) and *cis*(II) isomers of captopril as a function of pD.(From Rabenstein, D. L. and Anvarhusein, A. I., *Anal. Chem.*, 54, 526, 1982. With permission.)

and A^{2-} are present. The fractional concentrations of the two isomers as a function of the pH (pD) of the solution are shown in Figure 2. For the determinations, the relative intensities of the two multiplet patterns in the region 3.4 to 3.9 ppm were used.

In accordance with earlier observations on proline-containing peptides,[6] the ratio of the *trans* isomer of captopril decreases upon titration of the carboxylic group (i.e., with the increase of pH).

The acid dissociation constants for the carboxylic acid and sulfhydryl groups of the *trans* and *cis* isomers were calculated from the chemical shift titration curves for the α-carbon at the proline residue, and for C-3 of the 3-mercapto-2-methyl-1-oxopropyl group of the *cis* and *trans* conformations of captopril.

	pK_1 (carboxy)	pK_2 (sulfhydryl)
trans	3.52 ± 0.08	9.71 ± 0.05
cis	2.86 ± 0.08	9.99 ± 0.06

The somewhat lower acidity of the *trans* isomer is interpreted in terms of the existence of a hydrogen bond between the carboxylic acid hydrogen and the amide carbonyl oxygen. The same fact explains the higher stability of the *trans* form in the H_2A state.

B. METABOLISM

The most characteristic metabolic pathway of captopril is the formation of a symmetrical disulfide, i.e., oxidative dimerization:

Q—S—S—Q

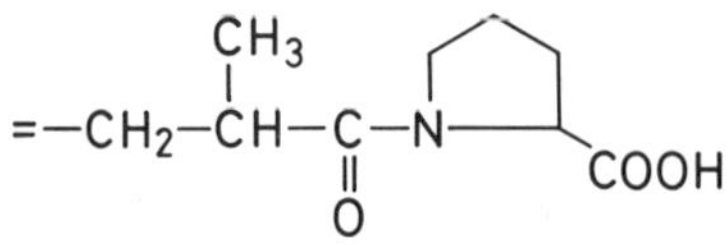

$$Q = -CH_2-CH(CH_3)-C(=O)-N\text{(pyrrolidine-2-COOH)}$$

Due to the strong reactivity of the sulfhydryl group, captopril can function as an alkylating agent readily forming covalent bonds with plasma peptides or proteins.[62] This aspect was studied in a system containing such endogenous sulfhydryl compounds as glutathione (G).[63] The mixed disulfide (G-S-S-R) formed was rapidly hydrolyzed to the mixed disulfide of cysteine and captopril. A study of the *in vitro* biotransformation of captopril disulfide demonstrated the reductive reproduction of captopril (R-S-S-Q $\rightleftharpoons$ 2QSH) with the participation of NADH or NADPH.[64] Following oral and i.v. dosing of the disulfide dimer conjugate of captopril, the metabolism of the dimer to captopril is revealed.[65] The plasma captopril disulfide level was much higher than the plasma levels of captopril after the administration of either the dimer or captopril. In connection with a pharmacokinetic study, the concentrations of captopril, captopril disulfide, and other metabolites were determined in body fluids by Creasey et al.[66]

The 200 MHz NMR spectra of captopril and its metabolic product, the disulfide, were studied.[67] They have similar preferred conformations in aqueous solution and the following ratios of E:Z isomers:

at pD 2.3	1:9
pD 7.5	2:3

C. ANALYSIS

HPLC plays an outstanding role in the separation and assay of captopril. Several HPLC systems were developed for captopril analysis by Perlman and Kirschbaum.[68] The determination of the free and total captopril contents of plasma and urine was most often performed in reversed-phase HPLC systems (Table 6).

7-Fluorobenzofurazan-4-sulfonate (SBD-F) is an agent that is often used for the prelabelling of captopril;[69,70] the disulfide content is reduced with tributylphosphine to give captopril before determination of the total captopril content.[69,70,82,84]

Gas chromatography also provides a sensitive and selective method for the assay of captopril and its sulfur-containing metabolites in biological samples. An antioxidant precolumn reagent that is often used is *N*-ethylmaleimide (see the data in Table 7, and Reference 87).

In the optical analytical methods, fluorimetry has an outstanding role, using different fluorogenic reagents. The main characteristic of the methods are summarized in Table 8.

The colorimetric-spectrophotometric methods for the determination of captopril in pharmaceuticals were compared by Raggi et al.[92] For Ellman's reagent, the Fe(III)-phenanthroline system, and the *N*-ethylmaleimide reagent, the absorbance vs. concentration function was linear in the 1.0 to 7.0, 1.7 to 2.13, and 46.0 to 180.0 μg/ml, respectively. The Fe(III)-phenanthroline method provided the best accuracy. This latter reagent was used by Mohamed et al.[93] for the determination of captopril in tablet formulations. Beer's law was obeyed for 2.5 to 15.0 μg/ml.

Determination of captopril by means of adsorptive cathodic differential pulse stripping voltammetry was applied by Passamonti et al.[94]

TABLE 6
HPLC Methods for the Analysis of Captopril

Compound(s)	Stationary phase	Mobile phase	Detection	Sensitivity, linearity	Ref., Note
Captopril (C) C-Disulfide DL-1-(3 mercapto-1-oxypropyl)-piperidine-2-carboxylic acid (OPCA) (I.St.)	Supelcosil LC-8-DB	0.09 M H_3PO_4-MeOH 4:1, containing 10m*M* Na-heptanesulphonate	F exc 385 nm emiss 515 nm	10 ng/ml 50—1000 ng/ml	69 In plasma and urine. Derivatization: SBD-F
C C-disulfide OPCA (I.St.)	μBondapak C_{18}	MeOH-1% aqu. H_3PO_4 7 13	F exc 385 nm emiss 515 nm	240-270 pg/ml 25-420 ng/ml	70 In plasma. Several drugs do not interfere
C (I) C-mixed disulfides (II)	μBondapak C_{18}	CH_3CN-MeOH-1% acetic acid	UV	10 ng/ml (I) 50 ng/ml (II) 250 ng (I) in plasma 2500 μg (I) in urine	71 Derivatization: N-(4-benzoyl-phenyl) maleimide
C C-disulfide	Partisil ODS-3	CH_3CN-0.1 *M* citric acid buffer (pH 3.1)	UV	10 ng/ml 10-700 ng/ml	72 In plasma Several drugs do not interfere
C			Electrochem. detection (E)	10 ng/ml	73 In plasma Derivatization: N-(4-dimethylaminophenyl)-maleimide
C	ODS	MeOH-phosphate buffer (pH = 2) 35:65	E	1 p*M*	74 In physiological fluids
C	ODS-Hypersil	MeOH-0.1 *M* ammonium acetate buffer (pH = 2) 1:3 containing 1m*M* EDTA	E		75
C C-disulfide Hydrochlorothiazide	RP Phenyl	Gradient elution: MeOH (25—45%)- 0.05% aqu. H_3PO_4 (75—55%)	UV (210 nm)		76 In tablets

TABLE 7
Gas Chromatographic Methods for the Analysis of Captopril

Compound(s)	Stationary phase	Carrier gas	Detection	Sensitivity, linearity	Ref., Note
C (I) *S*-benzoyl-C (II) 4-fluoro-C-N-ethylsuccinimide (III) (I. St.) *S*-benzoyl-4-fluoro-C (IV) (I.St.)	CP Sil 19CB	He 190—285°C	GCMS m/e = 230 (I-II) = 248 (III-IV)	3 ng/ml	76,77 Derivatization: *N*-ethylmaleimide (NEM) + methylester formation
C C-disulfide	10% Doxosil 300 GC column		GCMS	—	78 In biological fluids. Derivatization: NEM + hexafluoroisopropyl ester (HFIE) formation
C 4-F-C S-CH_3-C	Silanized-glass column + 3% OV-101 on Supelcoport	CH_4 190°C	GCMS m/e = 70	—	79,80 In human plasma. Derivatization: methyl ester formation
C (I) C-disulfides			GCMS	20 ng (I)	81 In plasma Derivatization: NEM
C (I) S-CH_3-C (II) *S*-methylcaptopril sulfone (III)	3% OV-101 OH Chromosorb W AW DMCS	He 150—290°C	GCMS	1 ng/ml (I) 25 ng/ml (II) 10 ng/ml (III)	82 In human and rat plasma and urine Derivatization: HFIE formation
C C-disulfides	3% OV-101 on chromosorb W-HP	Ar-CH_4 (19:1) 220°C	ECD	20—200 ng/ml in blood 50—1000 ng/ml in plasma	83 In blood and in plasma Derivatization: NEM + HFIE formation
C *S*-benzoyl-C (I.St.)	2% OV-1 on Gas-Chrom Q	He 230°C	GCMS	100 pg 0.1-5 ng	84 In blood and urine Derivatization: NEM + α-bromo-2,3,4,5,6-pentafluorotoluene
C *S*-benzoyl-C Desmethyl-*S*-benzoyl-C (I.St.)	Capillary column coated with Chrompak CP SH-19 CB	He 220-250°C	ECD	10 ng/ml 25—1000 pg	85 In human urine Derivatization: methyl ester formation

TABLE 8
Fluorimetric Methods for the Analysis of Captopril

Compound(s)	Fluorogenic reagent	λnm	Sensitivity, linearity	Ref., Note
C C-disulfide	1-(7-dimethylamino-4-methyl-2-oxo-2H-1-benzopyran-3-yl)-1H-pyrroline-2,5-dione	exc 380	2 ng/ml	88 In plasma. Fluorimetric determination after HPLC separation. Reduction of C-disulfides with tributylphosphine
C (thiol compounds)	Ammonium 7-fluorobenzo-2-oxo-1,3-diazole-4-sulfonate		43—520 p*M*/ml (for thiols)	89
C	*N*-(7-dimethylamino-4-methylcumarin-3-yl)-maleimide	exc 366 emiss 390	20 ng/ml in plasma; 80 ng/ml in urine; 20—800 ng/ml in plasma, 0.08 μg—10 μg/ml in urine	90 Fluorimetric determination after TLC separation

Various volumetric methods have been described Mohammed et al.[95,96] Argentometric and oxidimetric[95] titrations, based on the reaction with the thiol group, have been described for the analysis of captopril-containing pharmaceuticals. The titration with 0.05 *M* mercuric(II) nitrate, using end-point detection with diphenylcarbazone as indicator, and by differential pulse polarography is suitable for the determination of 2 to 45 ppm of captopril.[96] A radioimmunoassay for plasma captopril was based on the use of antibodies raised in rabbits and ^{125} I-labelled captopril-*N*-ethyl-maleimide as radioligand. Captopril in amounts of 0.05 to 2000 ng could be determined.[97,98] An enzyme immunoassay[99] and competitive inhibitor binding assay[100] for the determination of captopril have also been published.

IV. ENALAPRIL

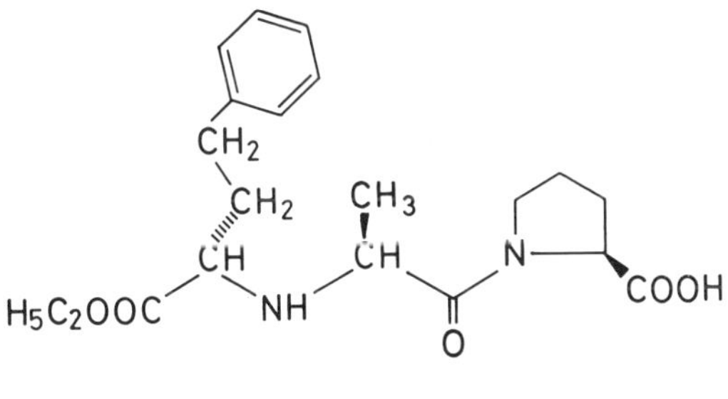

(*S*)-[*N*-[1-(Ethoxycarbonyl)-3-phenylpropyl]-L-alanyl]-L-proline
$C_{17}H_{19}N_2O_5$ $M_r = 271.28$

A. PROPERTIES

Enalapril is commercially available in the form of its maleic acid salt. Enalapril maleate is a white, odorless, crystalline powder. Its melting point is 148 to 151°C. The epimer with S,S,S-configuration (which can be seen in the formula above) is biologically more active than the R,S,S epimer, when the phenylethyl moiety binds in the β-position. As in the case of captopril, in aqueous solution enalapril exists as a mixture of *cis* **(3)** and *trans* **(4)** rotamers.[7]

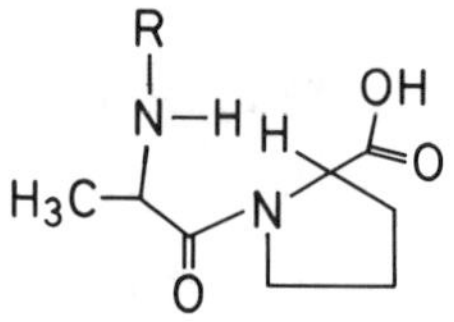

(3) 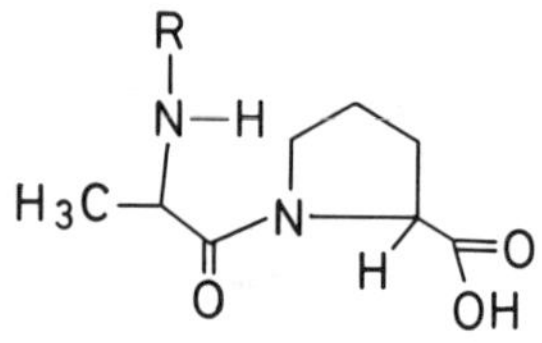 (4)

The presence of such a rotamer equilibrium can be seen in the ^{1}H-NMR spectrum of enalapril maleate recorded in H_2O[(1)]. The adjacent two doublets at 1.52 and 1.56 ppm correspond to the alanine-CH_3 group of a *cis* and *trans* rotamer. From the chemical shifts in the ^{13}C-NMR spectrum[7] all the C-atoms in enalapril could be accounted for. The relaxation times of these rotamers are of the order of minutes, and therefore HPLC proved suitable for the detection of the existence of these rotamers. Under appropriate experimental conditons, splitting of the peak of the pure all-S isomer could be observed.[7]

As expected, enalapril maleate has a rather "simple" UV absoroption spectrum, exhibiting only one maximum near 210 nm. The IR spectrum of enalapril maleate displays strong bands due to the presence of carbonyl groups (at 1753 cm^{-1}: stretch vibration of ethyl ester, at 1731 cm^{-1}: stretch vibration of maleic acid carboxylic groups); the strong band at 1650 cm$^-$1 corresponds to the tertiary amide group. The monosubstituted aromatic nucleus gives rise to intense peaks in the region 710 to 670 cm^{-1}.

The mass spectra of enalapril and enalaprilate were studied and fragmentation schemes were put forward by Oklobdzija et al.[7]

Enalapril has appeared recently in the text of pharmacopoeias. Relatively few reactions and analytical methods can be encountered which are suitable either for the purity control or for the quantitative assay of enalapril. In the HPLC method of Oklobdzija et al.,[7] enalapril maleate splits to give the peaks of the maleic acid and enalapril components; the all-S and the R,S,S isomers can also be separated by this procedure if RP-18 is applied as stationary phase and methanol-water (57:43) as eluent, adjusted to a pH of 3.5 (Figure 3). A prescription has been published for the HPLC assay of enalapril in tablets by the above method.[7]

Enalapril maleate, dissolved in a mixture of glacial acetic acid and acetic anhydride, can be quantitated by nonaqueous acidimetric titration. The flow-injection spectrophotometry of enalapril in pharmaceuticals has been reported by Kato.[8] The method is based on the ion-pair formation with bromothymol blue and the extraction of the ion-pair into dichloromethane. Beer's law is obeyed in the range 1.5 to 60 $\mu g\ ml^{-1}$. Degradation products of enalapril and the common excipients of the tablets do not interfere.[8]

The enalaprilate content of plasma was determined by means of RIA.[9] The standard graph ranged between 2 and 200 pmol ml^{-1}.

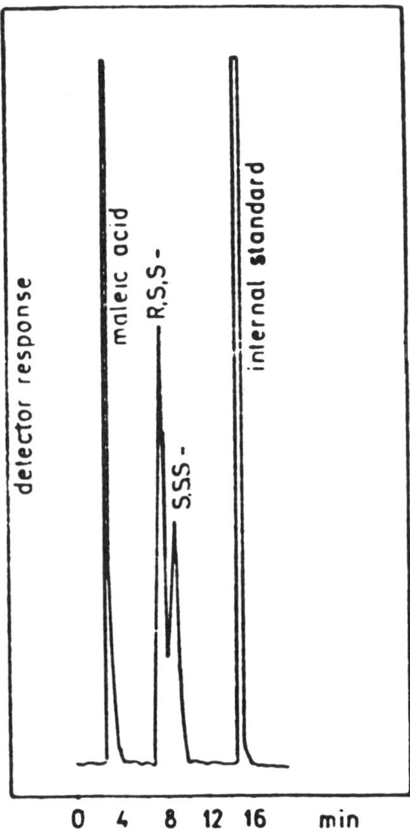

FIGURE 3. HPLC chromatogram of enalapril maleate.(From Oklobdzija, M. et al., *Acta Pharm. Jugosl.*, 38,167,1988. With permission.)

REFERENCES

1. **Xu,Z.,** *Yao Hsueh Ting Pao,* 23, 361, 1988.
2. **Fujinaga, M. and James, M. N. G.,** *Acta Crystallogr. Sect. B,* B36, 3196, 1980.
3. **Madison, V. and Schellman, J.,** *Biopolymers,* 9, 511, 1970.
4. **Gerig, J. T.,** *Biopolymers,* 10, 2435, 1971.
5. **Rabenstein, D. L. and Anvarhusein, A. I.,** *Anal. Chem.,* 54, 526, 1982.
6. **Evans, C. A. and Rabenstein, D. L.,** *J. Am Chem. Soc.,* 96, 7312, 1974.
7. **Oklobdzija, M., Kuftinec, J., Hohnjec, M. and Kajfez, F.,** *Acta Pharm Jugosl.,* 38, 167, 1988
8. **Kato, T.,** *Anal. Chim. Acta,* 175, 339, 1985.
9. **Worland, P. J. and Jarrott, B.,** *J. Pharm. Sci.,* 75, 512, 1986.
10. **Antonaccio, M. J.,** INSERM Symp., 17, 249, 1981.
11. **Johnston, C. I., Jackson, B., Cabelo, R., and Arnoldo, R.,** *Clin. Exp. Hypertens. Part A.,* A6, 551, 1984.
12. **Hopkins, S. J.,** *Med. Actual.,* 17, 233, 1981.
13. **Ferguson, R. K. and Vlasses, P. H.,** *Am. Heart J.,* 101, 650, 1981.
14. **Liedtke, R. K.,** *Therapiewoche,* 31, 5258, 1981.
15. **Borek, M., Charlop, S. and Frishman, W.,** *Pharmacotherapy (Carlisle, Mass.)* 7, 133, 1987.
16. **Bradshaw, D., Franz, P. H., and Sugden, L.,** *Br. J. Pharmacol.,* 76, 163, 1982.
17. **Hosoda, S. and Shimamoto, K.,** *Sapporo Igaku Zasshi,* 57, 395, 1988.

18. **Rubin, B., Laffan, R. J., Kotler, D.G., O'Keefe, E. H., De Maio, D. A., and Goldberg, M. E.,** *J. Pharmacol. Exper. Ther.*, 204, 271, 1978.
19. **Mikami, H., Ogihara, T., Ohde, H., Naka, T., and Kumahara, Y.,** *Jpn. Circ. J.*, 46, 393, 1988.
20. **Staroukine, M., Giot, J. M., Jacobs, V., and Vernieri, A.,** *Bull. Mem. Acad. R. Med. Belg.*, 137, 723, 1982.
21. **Ondetti, M. A., Williams, N. J., Sabo, E. F., Pluscec, J., Weaver, E. R., and Kocy, O.,** *Biochemistry,* 10, 4033, 1971.
22. **Gavras, H., Brunner, H. R., Laragh, J. H., Sealey, J. E, Gavras, I., and Vukovich, V. R.,** *N. Engl. J. Med.*, 291, 817, 1974.
23. **Ondetti, M. A., Rubin, B., and Cushman, D. W.,** *Science,* 196, 441, 1977.
24. **Cushman, D. W., Pluscec, J., Williams, N. J., Weaver, R. E., Sabo, E. F., Kocy, O., Cheung, H. S., and Ondetti, M. A.,** *Experientia,* 29, 1032, 1973.
25. **Das M. and Soffer, R. I.,** *J. Biol. Chem.*, 250, 6762, 1975.
26. **Byers, L. D. and Wolfenden, R.,** *Biochemistry,* 12, 2070, 1973.
27. **Bunting, J. W. and Myers, C. D.,** *Can. J. Chem.*, 52, 2053, 1974.
28. **Andersson, K. V.,** *Sven. Farm. Tidskr.*, 83, 71, 1979.
29. **Horovitz, Z. P.,** *Med. J. Aust.*, 2 (Spec. Suppl.) 2, 1979.
30. **Kim, T. H., Guinosso, C. H. J., Burley, G. C., Jr., Herbst, D. R., Mc Caully R. J., Wicks, T. C., and Wendt, R. L..,** *J. Med. Chem.*, 26, 394, 1983.
31. **Mc Evoy, F. I., Lai, T. M., and Albright, J. D.,** *J. Med. Chem.*, 26, 381, 1983.
32. **Suh, J. T., Skiles, J. W., Williams, B. E., and Schwab, A.,** U.S. Patent 4256761, 1981.
33. **Suh, J. T., Skiles, J. W., Williams, B. E., Youssefyeh, R. D., Jones, H., Loev B., Neiss, E. S., Schwab, A., Mann, W. S., Khandwala, A., Wolf, P. S., and Weinryb, I.,** *J. Med. Chem.*, 28, 57, 1985.
34. **Cushman, D. W., Cheung, H. S., Sabo, E. F., and Ondetti, M. A.,** *Biochemistry,* 16, 5484, 1977.
35. **Cushman, D. W. and Ondetti, M. A.,** *Prog. Med. Chem.*, 17, 41, 1980.
36. **Cushman, D. W., Cheung, S. H., Sabo, E. F., and Ondetti, M. A.,** *Prog. Cardiovasc. Dis.*, 21, 176, 1978.
37. **Menard, P. R., Suh, J. T., Jones, H., Loev, B., Neiss, E. S., Wilde, J., Schwab, A., and Mann, W. S.,** *J. Med. Chem.*, 28, 328, 1985.
38. **Ciabatti, R., Padova, G., Ballasis, E., Tarzio, G., Depaoli, A., Battaglio, F., Cellentavi, M., Barone, D., and Boldoli, E.,** *J. Med. Chem.*, 29, 411, 1986.
39. **Watthey, J. W. H., Gavin, T., and Desai, M.,** *J. Med. Chem.*, 27, 816, 1985.
40. **Smith, E. M., Swiss, G. F., Neustadt, B. R., Gold, E. H., Sommer, J. A., Brown, A. D., Chin, P. J. S., Moran, R., Sybertz, E. J., and Baum, T.,** *J. Med. Chem.*, 31, 875, 1988.
41. **Patchett, A. A., Harris, E., Tristram, E. W., Wyvratt, M. J., Wu, M. T., Taub, D., Peterson, E. R., Ikeler, T. J., ten Broeke, J., Payne, L. G., Ondeyka, D. L., Thorsett, E. D., Greenlee, W. J., Lohr, N. S., Hoffsommer, R. D., Joshua, H., Ruyle, W. V., Rothrock, J. W., Aster, S. D., Maycock, A. L., Robinson, F. M., Hirschmann, R., Sweet, C. S., Ulm, E. H., Gross, D. M., Vassil, T. C., and Stone, C. A.,** *Nature (London),* 288, 280, 1980.
42. **Greenlee, W. J., Allibone, P. L., Perlow, D. S., Patchett, A. A., Ulm, E. H., and Vassil, T. C.,** *J. Med. Chem.*, 28, 434, 1985.
43. **Almquist, R. G., Jennings-White, C., Chao, W., Steeger, T., Wheeler, K., Rogers, J., and Mitome, Ch.,** *J. Med. Chem.*, 28, 1062, 1985.
44. **Manis, P. A. and Rathke, M. W.,** *J. Med. Chem.*, 28, 1062, 1985.
45. **Watthey, J. W. H., Stanton J. L., Desai, M. N., Babiarz, J. E., and Finn, B. M.,** *J.Med. Chem.*, 28, 1511, 1985.
46. **Stanton, J. L., Watthey, W. H., Desai, M. N., Finn, B. M., Babiarz, J. E., and Tomaselli, H. C.,** *J. Med. Chem.* 28, 1603, 1985.
47. **Slade, J., Stanton, I. L., Ben-David, D., and Mazrenga, G. C.,** *J. Med. Chem.*, 28, 1517, 1985.
48. **Johnson, A. L., Price, W. A., Wong, P. C., Vavala, R. F., and Stump, J. M.,** *J. Med. Chem.*, 28, 1596, 1985.
49. **Johnson, A. L.,** U. S. Patent 4,296,110, 1981.
50. **Wyoratt, M. J., Tristram, E. H., Ikeler, T. J., Lahr, N. S., Joshua, H., Springer, J. P., Arison, B. H., and Patchett, A. A.,** *J. Org. Chem.*, 49, 2816, 1984.
51. **Roark, W. H., Tinney, F. J., Cohen, D., Essenburg, A. D., and Kaplan, H. R.,** *J. Med. Chem.*, 28, 1291, 1985.
52. **Hayashi, K., Nunami, K., Sakai, K., Ozaki, Y., Kato, J., Kinashi, K., and Yoneda, N.,** *Chem. Pharm. Bull.*, 31, 3153, 1983.
53. **Hayashi, K., Nunami, K., Kato, J., Yoneda, N., Kubo, M., Ochiai, T., and Ishida, R.,** *J. Med. Chem.*, 32, 289, 1989.
54. **Karanewsky, D. S., Badia, M. C., Cushman, D. W., De Forrest, J. M., Dejneka, T., Loots, M. J., Perri, M. G., Petrillo, E. W., Jr., and Powell, J. R.,** *J. Med. Chem.*, 31, 204, 1988.

55. **Tronrud, D. E., Holden, H. M., and Matthews, B. W.,** *Science (Washington D.C.),* 235, 569, 1987.
56. **Krapcho, J., Turk, Ch., Cushman, D. W., Powell, J. R., De Forrest, J. M., Spitzmiller, E. R., Karanewsky, D. S., Duggen, M., Rowniak, G., Schwartz, J., Natarajan, S., Godfrey, J. D., Ryono, D. E., Neubeck, R., Atwal, K. S., and Petrillo, E. W., Jr.,** *J. Med. Chem.,* 31, 1148, 1988.
57. **Cushman, D. W., Powell, J. R., De Forrest, J. M., Petrillo, E. W., Krapcho, J., Rubin, B., and Ondetti, M. A.,** *Prog. Pharmacol.,* 5, 5, 1984.
58. **Dean, A. V., Kripalani, K. J., and Migdalof, B. H.,** *Fed. Proc. Fed. Am. Soc. Exp.Biol.,* 43, 349, 1984.
59. **Sybertz, E. J., Baum,. T., Ahn, H. S., Desideris, D. M., Pula, K. K., Tedesco, R., Washington, P., Sabin, C., Smith, E., and Becker, F.,** *Arch. Int. Pharmacodyn. Ther.,* 286, 216, 1987.
60. **Powell, J. R., De Forrest, J. M., Cushman, D. W., Rubin, B., and Petrillo, E. W.,** *Fed. Proc. Fed. Am. Soc. Exp. Biol.,* 43, 733, 1984.
61. **Andrews, P. R., Carson, J. M., Caselli, A., Spark, M. J., and Woods, R.,** *J. Med. Chem.,* 28, 393, 1985.
62. **Park, B. K., Grabowski, P. S., Yeung, J. H. K., and Breckenridge, A. M.,** *Biochem. Pharmacol.,* 31, 1755, 1982.
63. **Komai, T., Ikeda, T., Kawai, K., and Kemeyama, E.,** *J. Pharmacobio-Dyn.,* 4, 677, 1981.
64. **Lan, S. J., Weinstein, S. H., and Migdalof, B. H.,** *Drug Metab. Dispos.,* 10, 306, 1982.
65. **Drummer, O. H., and Jarrott, B.,** *Biochem. Pharmacol.,* 33, 3567, 1984.
66. **Creasey, W. A., Morrison, R. A., Singhvi, S. M., Willard, D. A.,** *Eur. J. Clin. Pharmacol.,* 35, 367, 1988.
67. **Aboul-Enein, H. Y., Maat, L., and Peters, J. A.,** *Spectrosrosc. Lett.,* 18, 419, 1985.
68. **Perlman, S., and Kirschbaum, J.,** *J. Chromatogr.,* 206, 311, 1981.
69. **Boekens, H., Foullois, N., and Mueller, R. F.,** *Z. Anal. Chem.,* 330, 431, 1988.
70. **Toyooka, T., Imai, K., and Kawahara, Y.,** *J. Pharm, Biomed. Anal.,* 2, 473, 1984.
71. **Hayashi, K., Miyamoto, M., and Sekine, Y.,** *J. Chromatogr. Biomed. Appl.,* 39, 161, 1985.
72. **Pereira, C. M., Tam, Y. K., Collins-Nakai, R. L., Ng, P.,** J. *Liq. Chromatogr.,* 7, 96, 1984.
73. **Simada, K., Tanaka, M., Nambara, T., Imai, Y., Abe, K., and Yoshinaga, K.,** *J. Chromatogr.,* 227, 445, 1982.
74. **Perrett, D. and Drury, P. L.,** *J. Liq. Chromatogr.,* 5, 97, 1982.
75. **Perrett, D., Rudge, S. R., and Drury, P. L.,** *Biochem. Soc. Trans.,* 12, 1059, 1984.
76. **Kirschbaum, J. and Perlman, S.,** *J. Pharm Sci.,* 73, 686, 1984.
77. **Ivashkiv, E., McKinstry, D. N., and Cohen, A. I.,** *J. Pharm. Sci.,* 73, 1113, 1984.
78. **Jemal, M., Ivashkiv, E., and Cohen, A. I.,** *Biomed. Mass Spectrom.,* 12, 664, 1985.
79. **Matsuki, Y., Ito, T., Fukuhara, K., Nakamura, T., Kimura, M., Ono, H., and Nambara, T.,** *J. Chromatogr.,* 239, 58, 1982.
80. **Cohen, A. I., Ivashkiv, E., McCormick, T., and McKinstry, D. N.,** *J. Pharm. Sci.,* 73, 1493, 1984.
81. **Cohen, A. I., Devlin, R. G., Ivashkiv, E., Funke, P. T., and McCormick, T.,** *J. Pharm Sci.,* 71, 1251, 1982.
82. **Ivashkiv, E., McKinstry, D. N., and Cohen, A.,** *J. Pharm. Sci.,* 73, 1113, 1984.
83. **Drummer, O. H., Jarrott, B., and Louis, W. J.,** *J. Chromatogr. Biomed. Appl.* 30, 83, 1984.
84. **Bathala, M. S., Weinstein, S. H., Meeker, F. S., Jr., Singhvi, S. M., and Migdalof, B. H.,** *J. Pharm. Sci.,* 73, 340, 1984.
85. **Ito, T., Matsuki, Y., Kurihara, H., and Nambara, T.,** *J. Chromatogr. Biomed. Appl.,* 61, 79, 1987.
86. **Jemal, M. and Cohen, A. I.,** *J. Chromatogr. Biomed. Appl.,* 43, 186, 1985.
87. **Jemal, M. and Cohen, A. I.,** *Anal. Chem.,* 57, 2407, 1985.
88. **Ivashkiv, E.,** *J. Pharm. Sci.,* 73, 1427, 1984.
89. **Imai, K., Toyooka, T., and Watanabe, Y.,** *Anal. Biochem.,* 128, 471, 1983.
90. **Toyooka, T. and Imai, K.,** *Analyst (London),* 109, 1003, 1984.
91. **Cawello, W. and Bonn, R.,** *Z. Anal. Chem.,* 327, 29, 1987.
92. **Raggi, M. A., Cavrini, V., DiPietra, A. M., and Lacche, D.,** *Pharm. Acta Helv.,* 63, 19, 1988.
93. **Mohamed, M. E., Tawakkol, M. S., and Aboul-Enein, H. Y.,** *Zbl. Pharm. Pharmakother Laboratoriumsdiagn.,* 122, 1163, 1983.
94. **Passamonti, P., Bartocci, V., and Pucciarelli, F.,** *J. Electroanal. Chem. Interfacial Electrochem.,* 230, 99, 1987.
95. **Mohamed, M. E., Aboul-Enein, H. Y., and Gad-Kariem, E. A.,** *Anal. Lett.,* 16, 45, 1983.
96. **Mohamed, M. E., Gad-Kariem, R. A. and Aboul-Enein, H. Y.,** *Anal. Instrum. (N. Y.),* 13, 79, 1984.
97. **Duncan, F. M., Martin, V. I., Williams, B. C., Al-Dujaili, E. A. S., and Edwards, C. R. W.,** *Clin. Chim. Acta,* 131, 295, 1983.
98. **Tu, J., Liu, E., and Nickoloff, E. L.,** *Ther. Drug Monit.,* 6, 59, 1984.
99. **Kinoshita, H., Nakamaru, R., Tanaka, S., Tohira, Y., and Sawada, M.,** *J. Pharm. Sci.,* 75, 711, 1986.
100. **Gronhagen-Riska, C., Tikkanen, I., and Fyhrquist, F.,** *Clin. Chim. Acta,* 162, 53, 1987.

Chapter 4

CENTRALLY ACTING AND ADRENERGIC α-BLOCKING ANTIHYPERTENSIVE AGENTS

I. RESERPINE

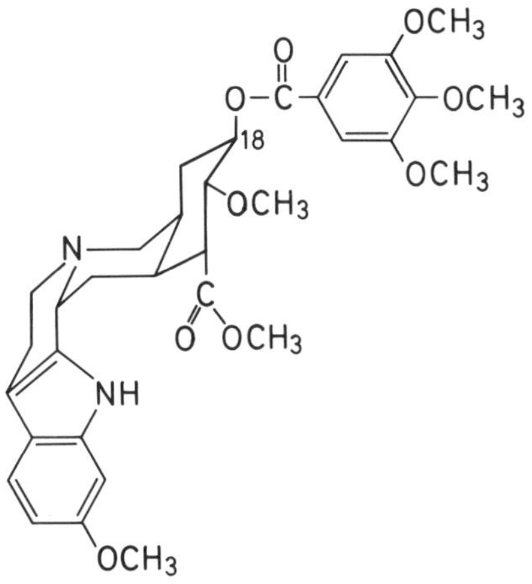

(−)-Yohimbane-16-carboxylic acid, 11,17-dimethoxy-18-[(3,4,5-trimethoxybenzoyl)oxy], methyl ester, 3β, (*R*)16β, (*S*)17β, (*R*)18β, (*R*)20α (*S*)-
(−)-Methyl 18β-hydroxy-11,17α-dimethoxy-3β,20α-yohimbane-16β-carboxylate 3,4,5-trimethoxybenzoate
$C_{33}H_{40}N_2O_9$ $M_r = 608.69$

A. PROPERTIES

Reserpine is a white crystalline powder. Due to its light-sensitivity it is generally discolored yellowish. It is odorless and almost tasteless (very poor solubility in the saliva). Mp: about 270°C (with degradation). Reserpine is practically insoluble in water and in ether; slightly soluble in benzene and alcohol (1:2000); only somewhat better soluble in acetone; freely soluble in chloroform (1:6) and in acetic acid. The solubilization of reserpine by different surfactants has been studied[1] the solubilizing effectiveness decreased in the following sequence: Na laurylsulfate > benzalkonium chloride > polysorbate 60 > polysorbate 20.

Reserpine is a moderately weak base; the published pK_a values lie in the range 6 to 7.5.[2] The center of basicity is expected to be on the nitrogen at position 4.

The UV absorption spectrum of reserpine may be regarded as the sum of the UV absorption increments of the two main structural moieties, i.e., the indole and the trimethoxybenzoic acid elements. For analytical purposes, it would be advantageous to work at a wavelength close to 300 nm, but the flatness of the spectral line and the relatively low sensitivity do not favor UV spectrophotometry for the analysis of reserpine products.

The specific rotation of reserpine may range between −113° and 123°, determined in a solution containing 100 mg reserpine per 10 ml. Although the reserpine molecule contains six centers of chirality (at positions 3, 15, 16, 17, 18, and 20) and reserpine is only one of the possible 64 stereoisomers, none of the other stereoisomers known so far display significant pharmacological activity. The NMR spectra, the chemical shifts, and most of the main coupling constants of reserpine (and three other basic alkaloids) have been recorded.[3]

TABLE 1
Centrally Acting and Adrenergic α-Blocking Antihypertensive Agents

Generic name	Chemical structure	Proprietary name®	Therapeutic use	Doses
Centrally acting agents				
Methyldopa (USP) R = H	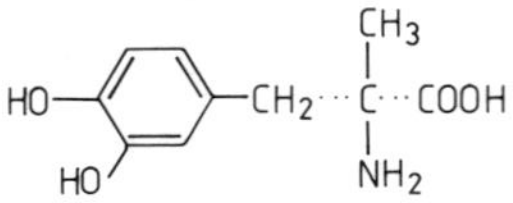	Baypresol, Dopamet, Dopegyt, Equibar, Hyperten, Hydromet, Midopren, Methyldopa, Mopatil, Novomedopa, Presolisin	Its primary site of action has been found to be within the CNS. Exhibits gradual and prolonged effect. Not always effective; delayed onset of action. Side effect: sedative.	Initial dose: 2 × 0.25 g/day. Increased every 3 days up to 3 g/day.
Methyldopate hydrochloride (USP) R = C_2H_5		Aldomet, Presinol		
Clonidine Clonidine hydrochloride (USP)	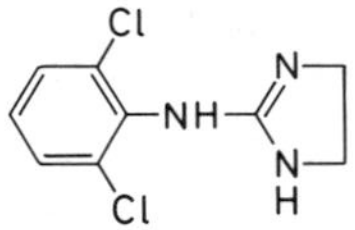	Atensina, Baypresan, Catapres, Catapressan, Clophazolin, Clonidine hydrochloride, Clonisin, Hyposin, Namestin, Prescatan	A potent central antihypertensive α-adrenoreceptor-stimulating drug. Side effects: sedation and dry mouth.	Total daily dose: from 200 μg to 2 mg.
Guanfacine	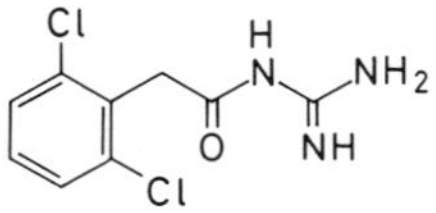	BS 100—141, Guanfacine hydrochloride, LON-798, Estulic, Tenex	Main indication: essential arterial hypertension. It is absorbed quickly, the maximum plasma concentration being attained within 2 or 3 h. Patients tolerate it well; somnolence and dizziness occur rarely.	Initial dose: 1 mg/day. This can be increased up to 2 mg/day.
Tolonidine	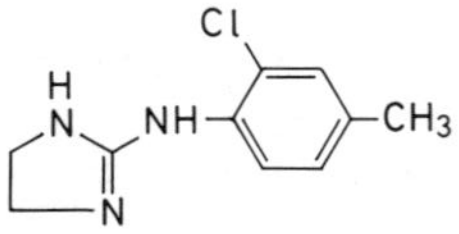	CERM 10137, St 375, Euctan	Used in all kinds of arterial hypertension. The oral doses are absorbed well and the maximum effect is achieved within 2—4 h. Side effects: seldom somnolence, bradycardia, Na-retention.	Initial dose: 20—40 drops, which can be increased if necessary.

Guanoxabenz	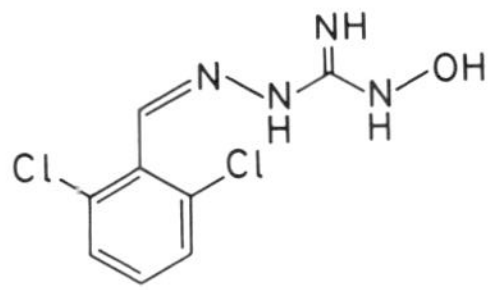	Benzerial, Benzeriel, Guanoxabenz hydrochloride	Diminishes the arterial systolic and diastolic pressure without causing any hypotension. Side effects: sedation, somnolence, rarely orthostatic hypotension.	Usual dose: 1 tablet/day (= 28 mg). Duration of effect: 12 h.
Guanabenz Guanabenz acetate (USP)	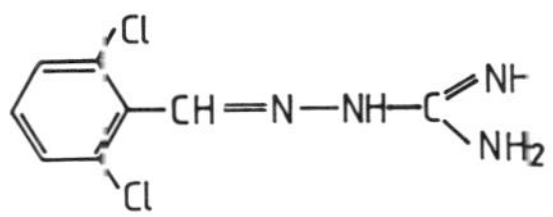	Wytensin, Br-750, Wy-8678	Used alone in the treatment of hypertension, or in combination with thiazide diuretics. It is an orally active central α_2-adrenergic agonist. Side effects: dry mouth, sedation, weakness, headache.	Initial dose: 4 mg twice a day. This may be increased up to 8 mg/day.
Lofexidine vasodilators Arteriolar vasodilators	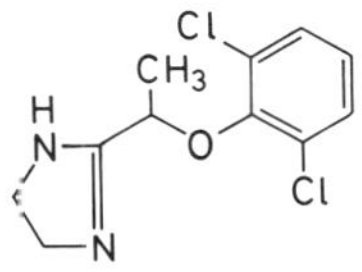			
Hydralazine Hydralazine hydrochloride (USP)	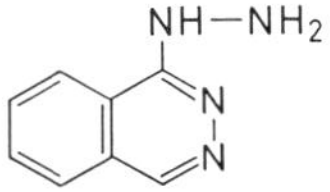	Aprelazine, Apresoline, Eralazin, Hydralazine, Hydrapress, Unipress	Used for the treatment of hypertensive crisis. The effect develops gradually over 15—20 min.	The usual oral dose is 100—200 mg daily.
Dihydralazine	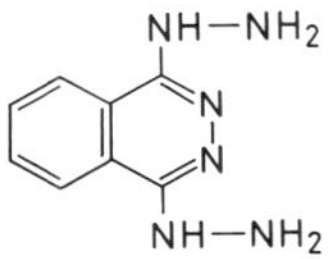	Nepresol, Depressan, Trasipressol	A peripheral vasodilator. Used primarily in cardiac therapy and arterial hypertension. Side effects: headache, nausea, vomiting, tachycardia, orthostatic hypotension.	Usual dose: 30—40 mg daily, orally, 5—10 mg/h in infusion. Doses are slowly increasable up to 200 mg, but not more.
Diazoxyde (USP)	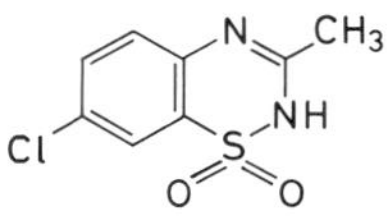	Hyperstat, Hypothiazide, Proglycem	Effective in hypertensive crisis, lowering the blood pressure rapidly and consistently. In a fixed dosage tachycardia, hypotension, and nausea may occur.	Usual dose: 150—300 mg daily, orally or parenterally. Duration of action: 4—12 h.

TABLE 1 (continued)
Centrally Acting and Adrenergic α-Blocking Antihypertensive Agents

Generic name	Chemical structure	Proprietary name®	Therapeutic use	Doses
Minoxidil	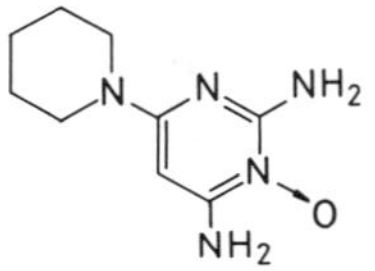	Loniten, Lonoten, U-10858	Used in severe hypertension. Side effects are similar to those of dihydralazine and diazoxide.	Initial dose: 5 mg. Effective range is usually 10—40 mg daily. Maximum dose: 100 mg.
Arteriolar and venular vasodilators Sodium nitroprusside (USP)	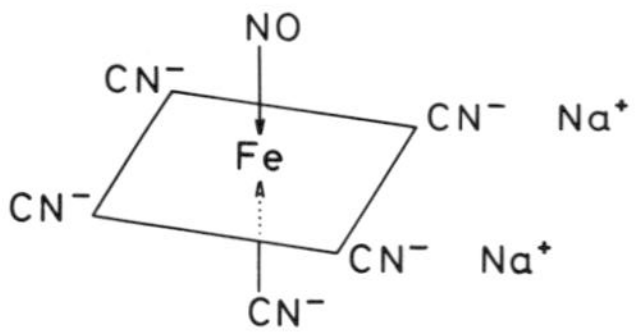	Acetest, Hypoten, Na-nitroferricyanidum, Natrium nitroprussicum, Nipride, Nipruss, Nitropress	Given in hypotensive emergencies, it acts instantly (within 1 min.), continuously and consistently. Also used in surgery when hypotension is required to minimize the side effects (nausea, hypoxia, etc.).	In 5% solution for intravenous infusion. Average adult dose: 3 mg/kg/min.
Prazosine Prazosine hydrochloride (USP)	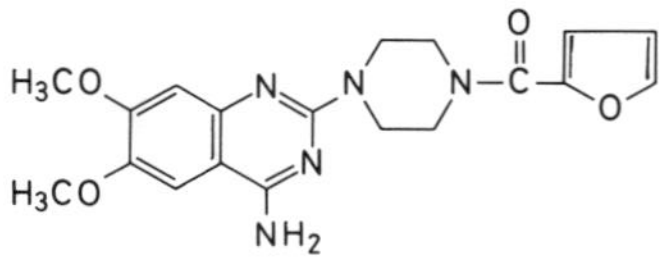	CP-12-299, prazosine hydrochloride	Acts predominantly as a α-adrenergic blocking agent on vascular smooth muscle. It is used in mild or moderate hypertension. Generally combined with diuretics and/or adrenergic blocking agents.	Initial dose: 0.5 mg 2 or 3 times daily, gradually increased up to 3 × 1 mg/day, and finally to 20 mg daily.
Urapidil	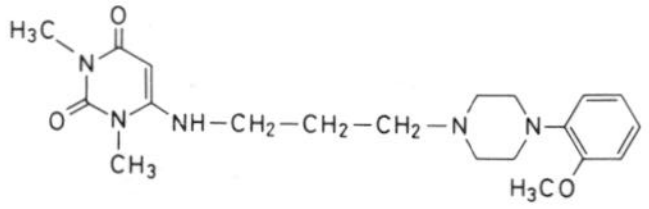	B 66256, Ebrantil	Its vasodilator activity is related to the stimulation of central receptors. It depresses the systolic and diastolic pressure, and reduces the peripheral resistance too. Side effects: nausea, headache, weakness.	30—90 mg/day orally (acts after 2—4 h); 2—4 mg/h i.v. (acts immediately).

Nifedipine (USP)	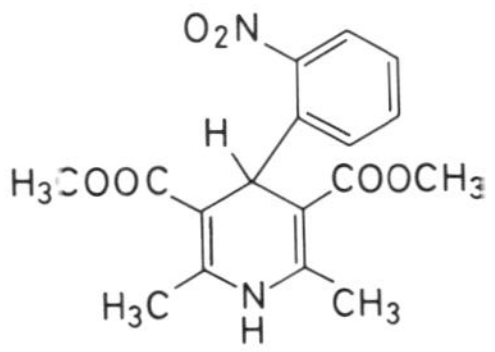	Adalat, Adarat, Corinfar, Nifelat, Procardia, Nifecor	A calcium antagonist, it causes a significant coronary and peripheral vasodilatation. Efficacious agent for the therapy of vasospastic angina, arrhythmia, and hypertension.	Usual dose: 3 × 10 mg daily, increasable up to 180 mg/day.
Adrenergic receptor blocking drugs α-Adrenergic blocking agents Pre- and postsynaptic acting drugs Phenoxybenzamine Phenoxybenzamine hydrochloride (USP)	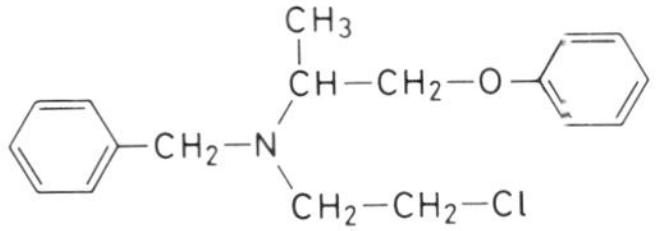	Benzylatum, Blocadren "Benzon", Dibenzyline, Dibenzyron	Used in the treatment of peripheral vascular diseases characterized by a high α-adrenergic tone, such as frostbite, Rayneaud's disease, thrombophlebitis. Side effects: postural hypertension, tachycardia.	Oral dose: 20—60 mg daily. 1 mg/kg in a solution of 5% glucose for infusion, given very slowly (during 1 h).
Phentolamine	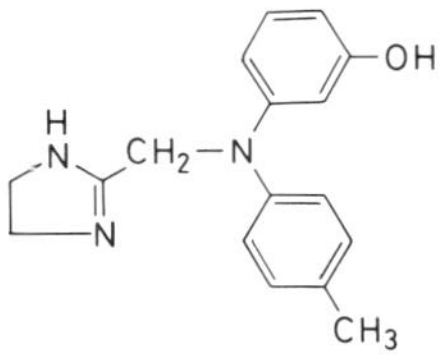	Dibasin, Regitine, Rogitine	In pheochromocytoma-induced hypertension and in adrenal tumour, which produce excessive amounts of norepinephrine. Side effects: acute hypertension, tachycardia, arrhythmia, nausea.	50 mg 4—6 times daily orally, or 5 mg i.v.
Tolazoline Tolazoline hydrochloride (USP)	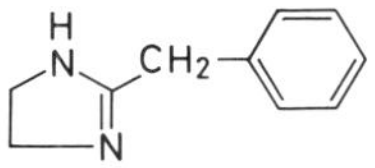	Antitens, Arterodyl, Bendarin, Benizol, Benzazolium hydrochloride, Benzimidon, Diladon, Divescol, Pervasol, Priscol, Toline, Vasodil, Vasopan, Zoline, Priscoline	Given intravenously, it produces cardiac stimulation and vasodilation by direct action on vascular smooth muscle. Used in different disturbances of the blood vessels in angiospasm to prevent dermal necrosis in rheumatic diseases. Side effects: tachycardia, gastrointestinal symptoms.	Orally: 25 mg 4—6 times/day, or 80 mg every 12 h (as long-acting dose). I.v., i.m. or s.c.: 50—200 mg in one dose.

TABLE 1 (continued)
Centrally Acting and Adrenergic α-Blocking Antihypertensive Agents

Generic name	Chemical structure	Proprietary name®	Therapeutic use	Doses
Drugs acting on postganglionic sympathetic nerve endings				
Guanethidine Guanethidine monosulfate (USP) Guanethidine sulfate (USP)	$N{-}CH_2{-}CH_2{-}NH{-}C({=}NH){-}NH_2$	Abapressine, Azetidine, Eutensol, Guanelin, Fuanexil, Ismelin, Isobarin, Octotenzin, Sanotensin, Visutensil	Used in moderate to severe hypertension and in mild hypertension too when other antihypertensives are not efficacious. Also given in hyperthyroidism. Side effects: orthostatic hypotension, collapse.	Initial dose for adults: 10—20 mg daily 2 or 3 times in equal doses. After 7—14 days the average dose can reach 25—150 mg daily.
Bethanidine Bethanidine sulfate	$CH_2{-}NH{-}C({=}N{-}CH_3){-}NH{-}CH_3$	Batel, Benzanidin, Bethanid, Regulin, Medica, Nyco, Esbatal	Indicated in all grades of hypertension. Because of its speed of action, it is of special value in malignant hypertension and hypertensive crises. Side effects: faintness, sweating, muscle fatigue, headache	Usual starting dose: 10 mg 3 times daily.
Bretylium tosylate	$H_3C{-}H_2C{-}\overset{+}{N}(CH_3)_2{-}CH_2$, Br	Bretylan, Bretylin, Bretylium-tosilat, Darethnin, Hypotyl, Isodaren, Ornid, Bretylol	Used primarily for the treatment of severe or life-threatening ventricular arrhythmias or fibrillation and tachycardia. Side effects: hypotension, occasionally nausea, vomiting.	Initial dose: 5—10 mg/kg 6—8 h for up to 5 days.

Debrisoquine Debrisoquine sulfate	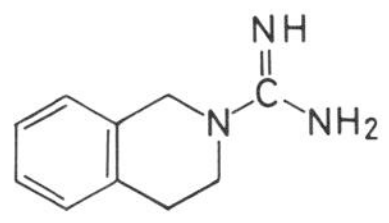	Debrisochin sulfate, Declinax, Equitensor, Equitonil, Tendor, Roche 10	Lowers blood pressure markedly and promptly in the hypertensive patient. Effect usually lasts 9—24 h. There is no accumulation. Side effects: weakness, giddiness, and fatigue.	Usual dose: 10 mg once or twice daily. Dosage range: 20—60 mg/day.
Guanoxan	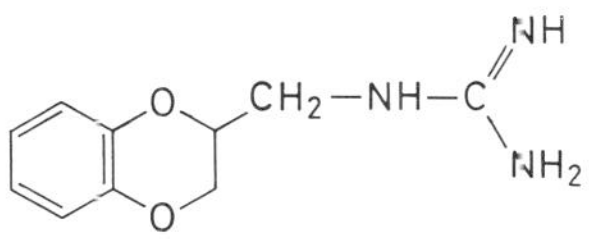	Enacar, Envacar		
Reserpine (USP)	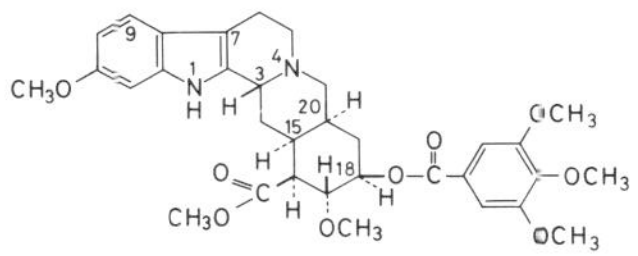	Alserin, Bioserpine, Cardioserpin, Eserpine, Hiposerpil, Neo-Serp, Rausedyl, Rauvopur "Byk", Resedrex, Reserp Reserpil, Reserpin, Serpasil, Tenserpine	Given in labile hypertension with concomitant mild or moderately severe tachycardia. The delayed onset of action and the sedative, cumulative effect may cause hypotension. Side effects: central nervous, systemic, and gastrointestinal disturbances.	Usual dose: 0.1—1 mg/day. Maintenance dose: 0.25 mg/day.
Monoaminooxidase inhibitors Pargyline Pargyline hydrochloride (USP)	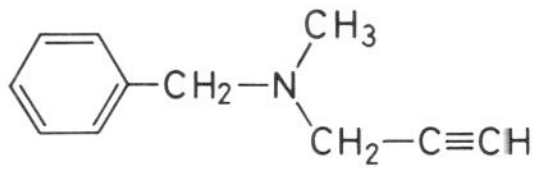	Pargyline hydrochloride, Eutonyl "Abbott", Supirdyl	In moderate to severe hypertension alone or concurrently with other antihypertensives. Side effects: changes in blood pressure, occurrence of hypotensive crises precipitated by its interaction with numerous drug and food substances. The latter is avoidable if recommended dosage is used.	Usual dosage: 25 mg once daily. May be increased up to 200 mg.

TABLE 2
UV Absorption Characteristics of Reserpine

Solvent/symbol	Methanol	Acetic acid 5 *M*
λ_{max}	294	289
	266	269
$A^{1\%}_{1cm}$	168	190
	274	280
ϵ	10,230	11,570
	16,680	17,040

Reserpine is one of the most unstable alkaloids. Depending on the circumstances, it is subject to different forms of degradation, i.e., isomerization (epimerization), oxidation, and hydrolysis.

The simple fact that no other Rauwolfia alkaloid containing the β-carboline moiety **(1)** exhibits therapeutically significant properties indicates the importance of steric factors in the reserpine-receptor interaction.

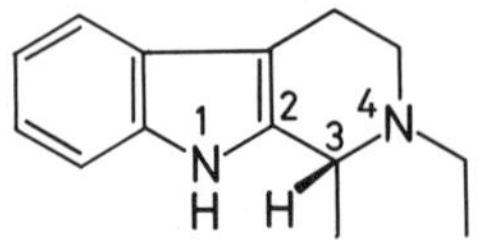

(1)

Consequently, the C-3-isomer (epimer) of reserpine **(2)**, 3-isoreserpine **(3)**, is inactive as expected, i.e., it has no hypertensive effect. This finding clearly shows the importance of C-3 in the reserpine-receptor interactions.

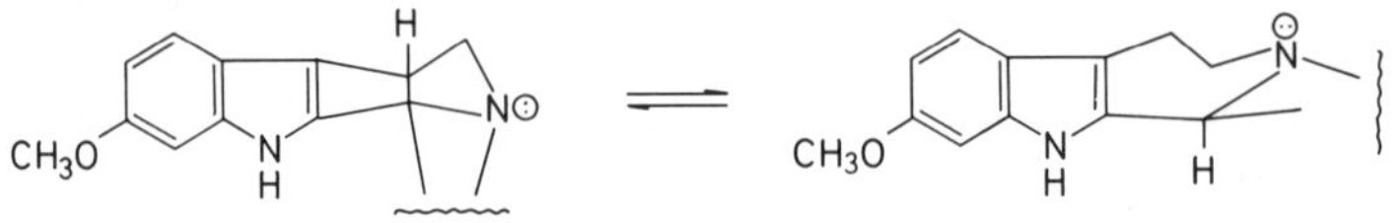

(2) (3)

The transisomerization of reserpine to isoreserpine is catalyzed by UV light. These isomers can readily be distingusihed through their different melting points or their specific rotations:

	Mp(°C)	Specific rotation
Reserpine	270 (degrad)	−120°
Isoreserpine	152—156	−164°

Chromatographic methods have been reported for the separation of the two substances (see ''Analysis''). Schiffl and Pindur[4] studied the mechanism of the epimerization and, on the basis of the differences in the intramolecular interactions of the substituents, postulated that reserpine is the kinetically and isoreserpine is the thermodynamically controlled isomer.

The hydrolytic cleavage of reserpine, which may take place in either acidic or alkaline solution, has been investigated by several authors.[5-8] The first step of hydrolysis preferentially involves position 18, when methyl reserpate **(4)** and trimethoxybenzoic acid **(5)**) are formed. Hydrolysis of the 16-carbomethoxy group proceeds relatively slowly, resulting in the formation of reserpic acid **(6)**.

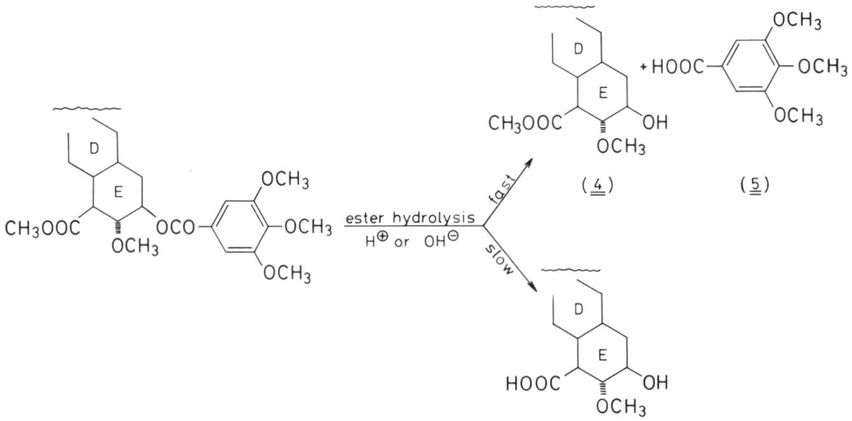

CH_3OOC D E OCH_3 OCO OCH_3 OCH_3 OCH_3 —ester hydrolysis, $H^{\oplus}$ or $OH^{\ominus}$→ fast: (4) CH_3OOC D E OH OCH_3 + (5) HOOC OCH_3 OCH_3 OCH_3; slow: (6) HOOC D E OH OCH_3

The optimum relative stability was found in acidic medium in the pH range 3 to 4. The stability of reserpine injections (and tablets) stored in hospital pharmacies across the U.S. was recently studied by Coffmann et al.[9] The great majority of the 93 injection samples examined appeared to be stable under the actual marketing conditions. The amounts of oxidation products found in some samples were under the limits of USP.

Light-catalyzed autooxidative degradation is of more importance in the protective storage and purity control of reserpine.

The main primary product of oxidative degradation of reserpine is 3-dehydroreserpine. This is a yellowish-green substance, and its formation is therefore indicated by the color of the solution. The autooxidation of reserpine is catalyzed by light and heat. The 3-dehydroreserpine impurity causes a very characteristic change in the UV-VIS spectrum of reserpine by the appearance of a new maximum at 380 nm in acidic solution. Its specific absorption is approximately 400; the limit of detection in reserpine is about 1%. After alkalinization, this peak at 380 nm disappears, and two new maxima become visible at 318 and 321 nm.

An equilibrium between protonated **(7)** and zwitterion **(8)** forms of dehydroreserpine may be assumed:

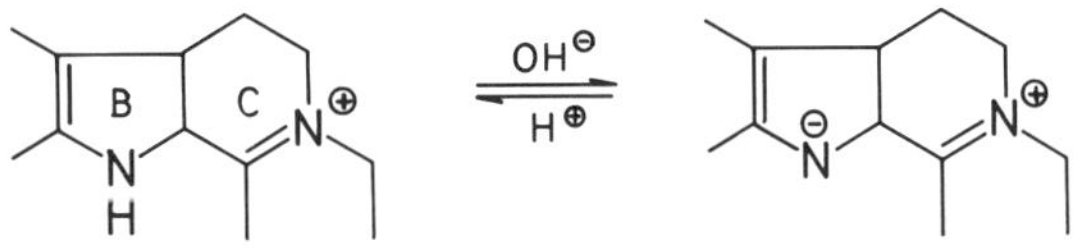

(7) B C N H $N^{\oplus}$ ⇌ ($OH^{\ominus}$ / $H^{\oplus}$) (8) $N^{\ominus}$ $N^{\oplus}$

The conversion of reserpine to 3-dehydroreserpine by molecular oxygen in acetic acid medium was thoroughly studied by Savory and Turnbull,[10] and a similar stability examination was carried out by Pfeifer et al.[11] The final product of the photodegradation of reserpine was identified as 3,4,5,6-tetradehydroreserpine (lumireserpine, **9**), a substance with intense blue fluorescence.[12-14] This compound was found to be abundant in a chloroformic solution of reserpine irradiated with UV light.[15] When reserpine tablets were examined, a small amount of 3-dehydroreserpine, but no **(9)** was found.

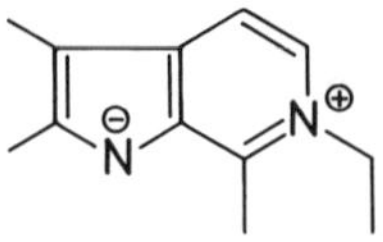

(9)

The kinetic pattern of reserpine oxidation indicated first-order kinetics in two consecutive steps when the oxidation was performed in acidic medium with hydrogen peroxide or peroxidisulfate.[16,17]

Besides protection from light, reserpine solutions can be stabilized by ensuring the optimum pH range (3 to 4) and an oxygen-poor atmosphere by the addition of antioxidant agents. Ascorbic acid and phenolic compounds (Bu hydroxyanisole, gallic acid esters, nor-dihydroguajaretic acid, dl-α-tocopherol, etc.) are mostly used for this purpose. The protective strength of propyl gallate was convincingly demonstrated by Sato et al.[18] and the same effect of a thiocarbamide-methionine combination was reported by Pötter and Voigt.[7]

The thermal degradation of reserpine has been the subject of several investigations.[19-21] The kinetics of degradation in reserpine injections stored under different conditions was investigated by Asker et al.[22] In the pH range 1.5 to 3.0, the overall thermal degradation of reserpine proved to be acid-catalyzed and followed first-order kinetics. The metabolism of reserpine has been studied by a few authors.[23-25] Methyl reserpate was found to be the main metabolite of reserpine.[23]

B. MECHANISM OF ACTION

The pharmacology of reserpine has been reviewed by several authors.[26-31] The hypotensive effect of reserpine proved to be due to the depletion of norepinephrine and serotonin stores in peripheral and central sympathetic neurons. The decrease in adrenergic neurotransmitter concentration causes a definite lowering of the blood pressure without a significant change in the cardiac output. This therapeutic effect is based on inhibiton of the active transport of norepinephrine into tissue storage sites. The basis of this competition or of other features of inhibition has not been elucidated so far. The free (unbound) amines are rapidly inactivated by monoamine oxidase, inducing further periods of the depletion process. The prolonged dilation of the blood vessels, a lasting therapeutic effect of reserpine, is based on its long-term residence on the surface of the storage proteins.

In spite of the fact that reserpine was introduced into the armament of hypotensive therapy more than 35 years ago, it has retained its position, and even today it is one of the most effective blood pressure-lowering medicines. It is currently mostly given in combination with another hypotensive agent (dihydralazine, etc.). Reserpine has the advantage that only one dose (~0.05 mg) a day is enough to reach and maintain a significant and lasting hypotensive effect: the average decrease in diastolic pressure was found to be about 17 mm Hg.[32] Serious side effects have been met only rarely.

The debate about the carcinogenicity (in the breast) of reserpine during chronic administration does not appear to have been resolved. Ambiguous views are to be found in the literature as regards the results and interpretation of retrospective studies[33,34] in human therapy and of renewed animal experiment, and the problem seems to be coming into prominence again.[35]

C. STRUCTURE-ACTIVITY RELATIONSHIP

In spite of the great number (~60) of alkaloids isolated and characterized from *Rauwolfia serpentina*,[36] reserpine is obviously the major active constituent of the plant. Of the natural alkaloids, rescinnamine **(10)**, which was also isolated from *R. serpentina*[37] and from several other R. species,[38] has similar pharmacological and therapeutic properties to those of reserpine itself. Its blood pressure-lowering activity is only somewhat less than that of reserpine, but it has milder adrenolytic and spasmolytic side effects.[39] Ranoxydin **(11)**, reserpine *N*-oxide, also exhibits reserpine-like hypotensive and sedative effects.[40] Deserpidine **(12)**, a desmethoxy derivative of reserpine, and raunescine **(12)**, des-*O*-methyldeserpidine, have pharmacological properties similar to those of reserpine. The despyrrolo derivative mediodespidine **(14)** is an interesting structural variant indicating that the primary role of the ring system in the reserpine molecule must be the optimization of partitioning behavior. Mediodespidine is equipotent with reserpine and causes no sedation.[41,42]

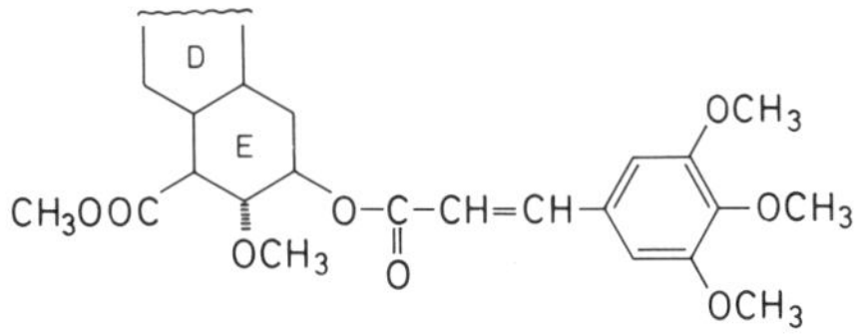

(10)

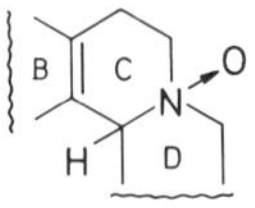

(11)

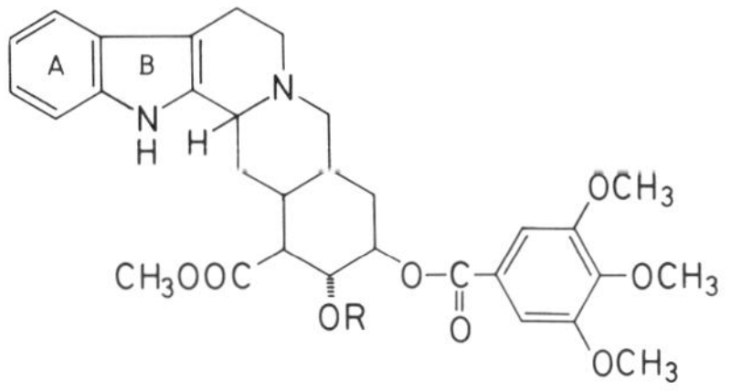

R = CH_3 deserpidine (12)

R = H raunescine (13)

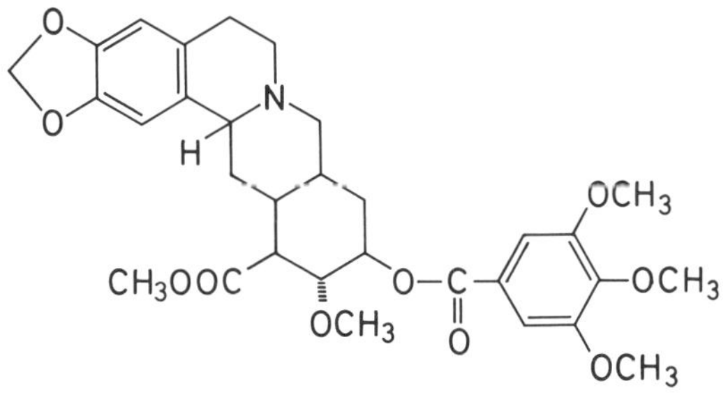

(14)

From the understanding of certain details about the structure-activity relationship for naturally occurring, synthetic, and semisynthetic R. alkaloids, it has become an accepted principle that the aromatic ester moiety is an essential part for the hypotensive activity. In this respect, it is worth mentioning that the Rauwolfia bases with anhydronium structure proved to display a selective hypotensive effect without exhibiting sedative activity. Serpentine (**15**) and ajmalicine (**16**), representatives of this type, were isolated[43] from *R. serpentina*. The structure of serpentine was elucidated in two steps.[44,45]

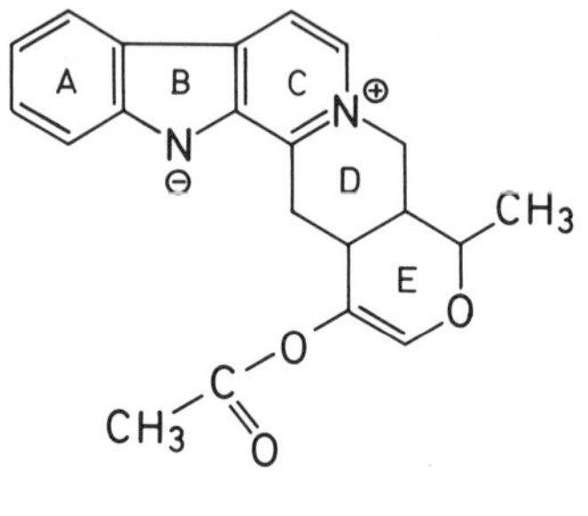

(15)

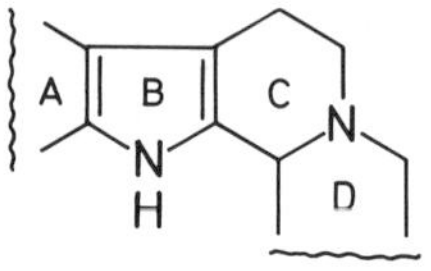

(16)

Ajmalicine was identified as tetrahydroserpentine, and could be prepared from serpentine by catalytic hydrogenation.[46,47] All these findings seem to indicate that the 18-esters moiety plays merely a carrier role in the biological activity of reserpine. In spite of this, when the trimethoxybenzoyl group was replaced with other 18-*O*-acyl groups, very few of the resulting products had a strong hypotensive activity similar to that of reserpine;[48] β-cyclopentylpropionyl and propionyl may be mentioned as effective groups in the latter category.[49,50] However, reserpine homologue syrosingopine (3′,5′-dimethoxy-4′-ethoxybenzoylmethyl reserpate) has an almost selective hypotensive effect.[51] The fact that a methoxy group at position 11 is not essential for the hypotensive effect is demonstrated by the effectiveness of deserpidine (**12**) and 10-methoxydeserpidines.

Transformation of the 16-methoxycarbonyl group to carboxamide abolishes the biological activity of reserpine;[52] this may indicate the importance of enzymatic hydrolysis of the 16-ester group before the drug-receptor interaction. β-Diethylaminoethyl substitution of reserpine[53] did not cause a weakening of the hypertensive activity (bietaserpine, **17**).

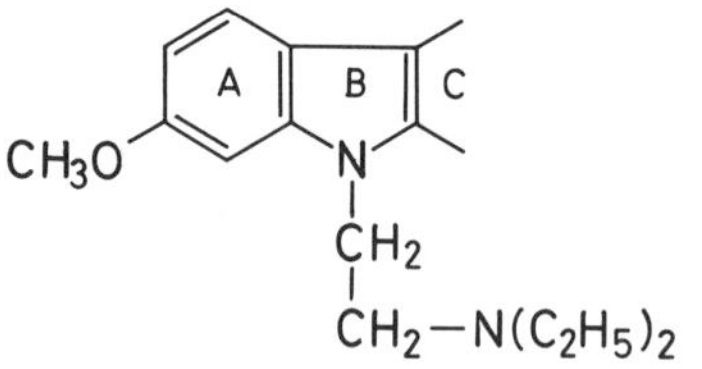

(17)

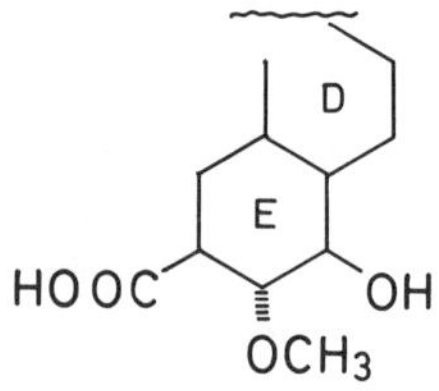

(18)

The highly polar reserpic acid (**18**), formed by double hydrolysis of reserpine, has no activity. The N^1-alkyl derivatives of reserpine are inactive, and unsaturation in rings C-D-E abolishes the hypotensive activity.

D. ISOLATION AND SYNTHESIS

Although extracts of the roots of *Rauwolfia serpentina* and of other Rauwolfia species have been used for medical purposes for centuries, a new era of medication with Rauwolfia

TABLE 3
Thin Layer Chromatographic System for the Separation of Reserpine and Other Rauwolfia Alkaloids

From Ref. 61:			
1.	Silica gel G	(a)	Chloroform-acetone-diethylamine 50:40:10
		(b)	Chloroform-diethylamine 90:10
		(c)	Cyclohexane-chloroform-diethylamine 50:40:10
2.	Silica gel G +0.1*M* NaOH		Methanol
3.	Aluminum oxide G		Cyclohexane-chloroform 30:70 + 0.05% diethylamine
4.	Aluminum oxide	(a)	Chloroform-acetone 85:15
		(b)	Abs ethanol
		(c)	Chloroform-ethanol-acetone 90:5:5
5.	Cellulose + Formamide		Heptane-methylethylketone 50:50
From Ref. 64:			
	Silica gel G	(a)	Xylene-isooctane-ethylacetate-diethylether 15:45:5:40
		(b)	*n*-Butanol-ethylacetate-ethylenedichloride 10:30:60
		(c)	Acetone-petroleumether-carbon tetrachloride-isooctane 35:30:20:15
		(d)	Methanol-methylethylketone-*n*-heptane 8.4:33.6:58
		(e)	Acetone-*n*-butanol-isooctane 33.6:8.4:58
		(f)	Acetone-petroleumether-glacial acetic acid 45:45:10
		(g)	Acetone-Methanol-glacial acetic acid 70:25:5
		(h)	*n*-Butanol-glacial acetic acid-water 4:1:1
		(i)	Acetone-petroleumether-diethylamine 20:70:10
		(j)	Acetone-methanol-diethylamine 70:20:10

began when papers on the isolation of the active principles of the plants were published in the early thirties.[54] The hypotensive effect in man of purified extracts of Rauwolfia roots containing a mixture of reserpine and other alkaloids with related structures was reported by Vakil[55] in 1949. Reserpine was isolated by Mueller, Schlitter, and Bein[56] in 1952, and its determining role in the hypotensive effect of Rauwolfia extracts was also proved.[57] The total synthesis was solved first by Woodward et al.[58] but was approached in different ways by other authors too.[59,60]

Throughout the past 30 years, the efforts of drug research in this field have been concentrated on learning the structure activity relationship, and on the preparation of semi-synthetic derivatives of reserpine equipotent with reserpine in hypotensive efficacy but devoid of the undesirable central effects of the parent compound. Irrespective of certain developments (e.g., syrosingopine), from the latter respect no newer compound that is more effective and/or more selective than reserpine has yet been prepared.

E. ANALYSIS

Since reserpine is only one of the numerous related natural alkaloids in plants, it is an important task to separate each of them from the others and to identify and/or determine all of them. For this purpose, the most often used general method is TLC. Reserpine (Rauwolfia alkaloids) has been chromatographed on different layers with different solvents or solvent systems.[61] Table 3 details some chromatographic systems, of which 1/a,b,c; 3; and 4/a,c are the most suitable for the separation of reserpine from other Rauwolfia alkaloids.

For the chromatographic purification of reserpine and its separation from rescinnamine, reserpiline, and raubasine, a TLC method was developed by Devred et al.[62] The quantitative determination was effected by UV spectrophotometry or colorimetry. As a result of a very thorough study, Le Xuan et al.[63] recommend the two-dimensional separation of Rauwolfia alkaloids on unmodified silica gel under mild conditions. It is important that the chromatographic process is performed in the dark because of the light-unstable behavior of reserpine. Acetic acid, acetone, 1,2-dichloroethane, diisopropyl ether, diethylamine, ethanol, ethyl

TABLE 4
Fluorescence Colors Produced by Rauwolfia Alkaloids Viewed in Screened UV Light[a]

Alkaloid	Proportion in test mixture (mg/10ml)	Sorbent Kieselgel G Wavelength 350 nm	Sorbent Kieselgel GF_{254} Wavelength 254 nm
Ajmalicine	1	apple-green	blue
Ajmaline	20	violet	brown
Aricine	10	orange	brown
Deserpidine	10	blue-green	blue
Rescinnamine	1	blue-green	pale blue
Reserpiline	20	yellow	brown
Reserpine	10	apple-green	pale blue
Reserpinine	10	apple-green	pale blue
Serpentine	20	intense blue-violet	pale blue
Tetraphyllicine	10	violet	brown
Yohimbine	10	blue	blue-violet
α-Yohimbine	10	blue	blue-violet

[a] From Court, W. E. and Habib, M. S., *J. Chromatogr.*, 80, 101, 1973. With permission.

acetate, methanol, methyl ethyl ketone, and propanol were used as mobile phase in the first direction. Either neutral or acidic solvents were administered for the next development: diisopropyl ether-methanol (85:15), 1,2-dichloroethane-methanol (90:1), acetone-acetic acid (90:10), or methyl ethyl ketone-acetic acid (90:10).

Court and co-workers[64-69] have performed an extensive study on the TLC of Rauwolfia alkaloids. In an attempt to obtain an efficient separation for quantitative work, they investigated the systems for the separation of 12 Rauwolfia alkaloids.[64] The systems used are outlined in Table 3.

They found that, of the 300 examined solvent systems, there was no one which could adequately separate the available Rauwolfia alkaloids for quantitative determination. The same solvents were used in later research for the study of an extended range of alkaloids.[65] In both papers, the relationship between the chromatographic properties and the chemical structure was studied. R_F values were tabulated. Amato et al.[70] separated Rauwolfia alkaloids on a silica gel layer containing a fluorescent indicator by using the MeOH-NH_4OH (100:15), $CHCl_3$-MeOH (9:1), Me_2CO-NH_4OH (58:2), and EtOAc-isoPrOH-NH_4OH (45:35:5) solvent systems.

Common methods employed for the detection of these alkaloids are irradiation with UV light at 254 and 350 nm, providing a fluorescent spot, or spraying with color reagents. Tables 4 and 5 list the fluorescence colors[64] and chromogenic reactions,[64,66] respectively, of some Rauwolfia alkaloids. Amato et al.[70] and Cavazzutti et al.[71] use HIO_3 in 90% H_2SO_4 as a generally suitable spray reagent which yields a colored spot with all molecules containing an aromatic ring activated by electrophilic substitution.

The TLC separation, detection, and quantitative determination of reserpine in plants,[69] in the presence of thiazide diuretics and antihypertensives,[72] in pharmaceutical preparations,[73,74] and in biological fluids[75,76] is an everyday task. As solutions of these tasks, extraction of reserpine from plants with alkaline chloroform,[69] from equine plasma with alkaline benzene,[75] and from human plasma with mildly alkaline dichloromethane[76] have been used. The leading method for the quantitation of reserpine is *in situ* fluorometry; this is a potent method for the assay of μg amounts in plant extracts,[77] pharmaceutical preparations,[78] and biological fluids.[76] In the last case,[76] linearity in the range 0.05 to 4 ng/ml plasma has been achieved with a recovery of 87 to 104%. From the methods suitable for

TABLE 5
Chromogenic Reactions of Some Rauwolfia Alkaloids

Alkaloid	1	2	3	4	5	6	7	8	9	10	11	12	13
Ajmalicine	gr	g	gr	p	y	g	gb	b	p	v-b	b	y-g	y-g
Ajmaline	c	r	deep r	p	deep r	deep r	c	c	g	p	r	p	p
Aricine	b	deep g	o-b	p	pale g	pale g	b	b	pu	r-v	y-g	g	y-b
Deserpidine	g-b	g	y-b	p	w	pale g	g	g	p	bl-v	g	g	y
Rescinnamine	g-b	y-g	y-b	p	y-g	y-g	g-b	g-b	b	r-v	y-g	y-g	y-g
Reserpiline	v	p	b	b	y-g	—	bl-b	v	b	g	g	b	b
Reserpine	g-b	y-g	y-g	p	y-g	y-g	g-b	g-b	b	m	y-g	y-g	y-g
Reserpinine	y-b	g	y	p	y-g	y-g	—	—	—	—	—	—	—
Serpentine	—	—	—	b	bu	bu	bu	bu	—	y	bu	b	bu
Tetraphyllicine	c-r	deep r	p	p	deep r	deep r	r	c-r	g	p	r	p	p
Yohimbine	g	y-g	y	p	y-g	g	g	g-b	p	v-bl	g	g	y
α-Yohimbine	g	y	y	p	y-g	g	g	g-b	p	v-bl	g	g	y

Note: 1 = 1% Ceric sulfate in 10% sulfuric acid; 2 = Sulfomolybdic acid (Frohde's reagent); 3 = 5% Ferric chloride in 50% nitric acid; 4 = Iodoplatinate reagent; 5 = 0.5% Phosphomolybdic acid in 50% nitric acid; 6 = 1% Ammonium vanadate in 50% nitric acid; 7 = Ferric chloride in perchloric acid; 8 = Ceric sulfate; 9 = Van Urk; 10 = Vanillin; 11 = Van Urk Salkowski; 12 = Ninhydrin; 13 = Hydroxylamine-ferric chloride.

b = brown; bl = blue; bu = buff; c = crimson; g = green; gr = grey; m = mauve; o = orange; p = pink; pu = purple; r = red; y = yellow; v = violet; w = white.

From Court, W. E. and Habib, M. S., *J. Chromatogr.*, 80, 101, 1973; Court, W. E. and Timmins, P., *Planta Med.*, 27, 319, 1975. With permission.

the development of fluorescence (exposure to light,[14,68] to acid vapor[69] or to oxidizing agents,[107-109]) the procedure of treating the plate with acetic acid vapor was selected. It should be noted that the potential metabolites, such as reserpic acid, methyl reserpate, and syringomethyl reserpate, are more polar and move with much smaller R_F values than that of reserpine; hence, they do not interfere.

HPLC is the most convenient method for the detection and quantitative determination of reserpine in the presence of other Rauwolfia alkaloids or in drug formulations. Table 6 lists some data on the methods described.

Because of the high molecular mass and low volatility, only a few attempts have been made at gas chromatographic determination.

Settimj et al.[94] described a GC procedure based on esterification with diazomethane, after the alkaline hydrolysis of reserpine and rescinnamine; Chromosorb W + 15% Apiezol L was used as stationary phase, with nitrogen as carrier gas, at 230°C. The calibration curve was linear in the concentration interval 25 to 150 and 100 to 600 μg/5 ml, respectively. Other authors offer a GC determination of reserpine after methylation with 0.2 *M* trimethylanilinium hydroxide in methanol.[95,96] The GC procedure was carried out at 180°C on a column of 3% OV-101, with N_2 as carrier gas. The response to reserpine is linear over the range 10 to 200 μg. The method can be adapted for the assay of pharmaceutical formulations too. A highly sensitive and specific combination of a GC procedure with mass fragmentographic detection was utilized to determine nanogram amounts or reserpine-like material in rat brain.[97] The possibility of the use of computer based GC-UV for the analysis of complex mixtures of biological compounds (among others reserpine) was discussed by Horning et al.[98]

Fission fragment-induced desorption[99] and laser desorption chemical ionization mass spectrometry[103] as well as ^{13}C NMR[100,101] and two-dimensional NMR technique[102] can be found among the newer methods of reserpine investigation. One of the methods frequently used for the quantitative determination of reserpine is spectrofluorometry.[104-109] The fluorimetry can be based on the native fluorescence (or phosphorescence) of the alkaloids, which has been studied by several authors under different conditions. The fluorescence characteristics of reserpine, rescinnamine, etc. have been measured in concentrated sulfuric acid solution[110] and in concentrated solutions.[111] In the latter case, a new fluorescence emission band was found and was assigned to the anion, in which the NH group of the indole ring is deprotonated. The lifetime of natural fluorescence has been calculated by Hidalgo et al. The fluorescence and phosphorescence spectra of reserpine and related alkaloids were recorded under extreme thermal conditions (298° and 77°K); the data obtained provide a basis for the analytical determination of reserpine in low concentrations. In a recent paper, Hidalgo et al.[114] made a detailed study of the fluorometric characteristics of seven Rauwolfia alkaloids. The main characteristics of the fluorescence spectra are presented in Table 7.[114]

The method described was found to be suitable for the determination of reserpine in combination with phenobarbital, methylphenobarbital, and diprophylline,[115] and for determination of the content uniformity of tablets containing reserpine, dihydralazine, and hydrochlorothiazide.[116]

Fluorimetry based on the oxidation of reserpine with N_2O_5 or vanadium pentoxide was applied for its automated determination in tablets containing reserpine alone,[117] or in combination with hydralazine and hydrochlorothiazide[118,119] or syrosingopine.[120]

Another means of fluorometric assay involves the classical nitrite procedure:[121] nitrous acid oxidizes reserpine to a yellowish-green fluorescent product. The fluorogen is measured at an excitation wavelength max of 390 nm and a fluorescent wavelength max of 510 nm. The fluorescence intensity markedly increases in the sequence: natural alkaloids < 3,4-dehydro compounds < 3,4,5,6-dehydro compounds. A method based on the dehydrogenation by NO_2^- or *p*-toluenesulfonic acid was developed by Jung and Jungouá.[122]

TABLE 6
HPLC Systems for the Separation of Reserpine and Rauwolfia Alkaloids

Compound(s)	Stationary phase	Mobile phase	Detection	Sensitivity, linearity	Ref., Note
Reserpine, rescinnamine	BAS Biophase ODS (5 μm)	MeOH-0.05 *M* KH_2PO_4 (pH = 4.5)-0.005 *M* 1-heptane sulfonic acid	Amperometry	0.9 ng reserpine 0.8 ng rescinnamine 5—40 ng/ml reserpine 10—50 ng/ml rescinnamine	79
Yohimbine, reserpine	RP Novapak C_{18} Spherisorb C_{18} μBondapak C_{18} NP Partisil 5	Different compositions of MeOH-water	Fluorimetry		80
Reserpine, chlorthalidone (acetazolamide; I.St.)	μBondapak NH_2	Chloroform-MeOH 75:25	Spectrophotometry 254 nm	4.8—288 μg/ml	81 In single and combined dosage forms
Reserpine, rescinnamine	NP Silica	MeOH	Fluorimetry exc: 280 330 nm emiss: 360 435 nm		82 In tablets and in Rauwolfia serpentina powder (extraction: from acidic-aq soln with chloroform)
Reserpine	YW G CDS	MeOH-acetic acid-water 300:70:5	Spectrophotometry 254 nm	0—1 ng/100 ml	83 In tablets; extracted by chloroform
Reserpine	LiChrosorb RP-8	MeOH-0.05 *M* $NaHPO_4$	Fluorimetry 330 nm/470 nm	0.025—0.4 ng/ml	84 In tablets; extraction: with EtOAc saturated water
Reserpine	LiChrosorb RP C_{18}	5 g/l NH_4Cl-MeOH 4:6	Spectrophotometry 267 nm		85 In injection (containing benzylalcohol, polysorbate 80, glycerol acetic acid ingredients)
Reserpine, chlorothiazide	RP C_8	Abs MeCH-aq 0.5% NH_4Cl (pH = 5.6) 55:45			86 In multicomponent antihypertensive diuretic mixtures

TABLE 6 (continued)
HPLC Systems for the Separation of Reserpine and Rauwolfia Alkaloids

Compound(s)	Stationary phase	Mobile phase	Detection	Sensitivity, linearity	Ref., Note
Reserpine, hydrochlorothiazide, hydralazine	Octadecytrichlorosilane (C_{18})	Acetonitrile-0.1% ammonium acetate, or abs MeOH-0.5% ammonium chloride	Differential spectrophotometry		87 In multicomponent antihypertensive-diuretic mixtures
Reserpine	RP	Acetonitrile-0.005 *M* NaH_2PO_4 buffer (70:3) pH = 6.0	Fluorimetry exc: 395 nm emiss: 470 nm	200 pg 10—100 ng/ml	88—90
Rauwolfia serpentina extr.	μPorasil	MeOH	Fluorimetry 330 nm (rescinnamine) 280 nm (reserpine)		91 Extraction: with ethyl acetate-warm MeOH; rubascine, ajmalicine, yohimbine were also detected
Reserpine (+ methy-18-triethoxy benzoyl reserpate; I.St.)	RP C_1-trimethylsilyl	0.1 *M* acetate buffer (pH 4.2)-acetonitrile containing 5 m*M* heptanesulfonate and 0.01 *M* triethylamine	Fluorimetry exc: 460 nm emiss: 570 nm (after oxidation with vanadium pentoxide)	0.3 ng/ml	92 In plasma
Reserpine, clopamide, dihydroargocristine	RP-18 (5 μm)	Acetonitrile-0.4 *M* phosphoric acid (425:75)	Spectrophotometry 260 nm		93 In tablets (content uniformity testing)

TABLE 7
Maximum Excitation and Fluorescence Wavelength for Some Rauwolfia Alkaloids in Various Solvents and Their Limits of Detection in Ethanol[a]

Compound	Solvent	Excitation (λ_{max})	Fluorescence (λ_{max})	Limit of detection (ng/ml)	Group
Reserpine	Ethanol	271	356	2	R_1 = OCH_3
	Methanol	271	360		R_2 = OOC–CH_3
	Acetonitrile	270	350		R_3 = OCH_3
	5 *N* Acetic acid	271	366		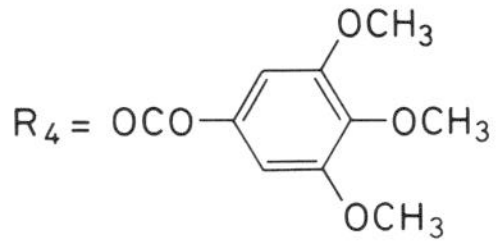 R_4 = OCO–(3,4,5-tri-OCH_3-phenyl)
Rescinnamine	Ethanol	303	425	6	R_1 = OCH_3
	Methanol	303	436		R_2 = OOC–CH_3
	Acetonitrile	303	423		R_3 = OCH_3
	5 *N* Acetic acid	303	460		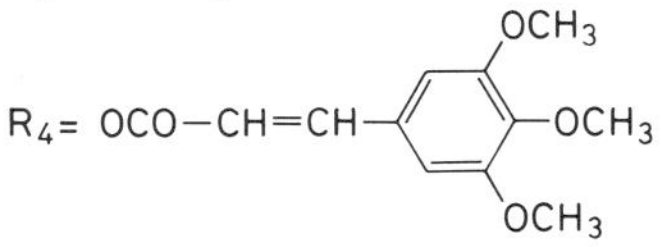 R_4 = OCO–CH=CH–(3,4,5-tri-OCH_3-phenyl)
Corynanthine	Ethanol	282	350	0.2	*cis*-isomer
	Methanol	282	350		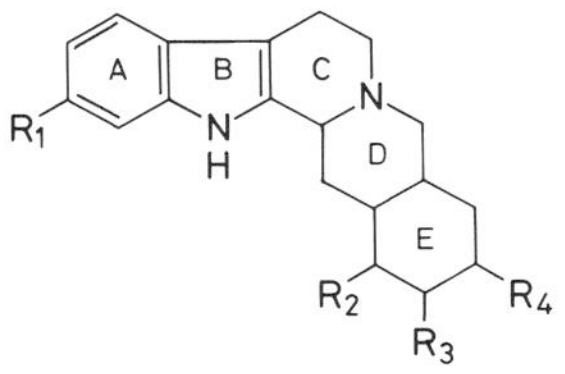 of yohimbine
	Acetonitrile	280	345		
	5 *N* Acetic acid	280	346		

R_1, A, B, N, H, C, N, D, E, R_2, R_3, R_4

TABLE 7 (continued)
Maximum Excitation and Fluorescence Wavelength for Some Rauwolfia Alkaloids in Various Solvents and Their Limits of Detection in Ethanol[a]

Compound	Solvent	Excitation (λ_{max})	Fluorescence (λ_{max})	Limit of detection (ng/ml)	Group
Yohimbine hydrochloride	Ethanol	282	352	0.2	R_1 = H
	Methanol	282	352		R_2 = OCO–CH_3
	Acetonitrile	280	338		R_3 = OH
	5 *N* Acetic acid	280	348		R_4 = H
	Water	280	351		
Ajmalicine	Ethanol	282	352	3	R_1 = R_2 = H
	Methanol	282	352		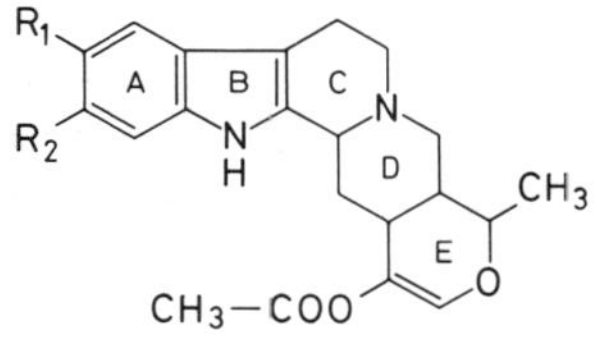
	Acetonitrile	282	348		
	5 *N* Acetic acid	282	350		
Reserpiline hydrochloride	Ethanol	303	335	100	R_1 = R_2 = OCH_3
	Methanol	303	335		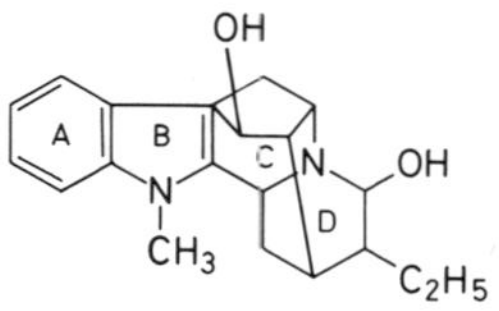
	Acetonitrile	303	330		
	5 *N* Acetic acid	300	331		
	Water	301	336		
Ajmaline	Ethanol	251	358	4	
	Methanol	250	353		
	Acetonitrile	251	358		
	5 *N* Acetic acid	252	358		

[a] From Hidalgo, J. et al., *Pharm. Acta Helv.*, 61, 89, 1986. With permission.

The reaction with arylaldehydes is often used in the analysis of reserpine. USP prescribes a reaction for identification with vanilline; this results in a pink colour which becomes a deep violet-red within a few minutes. With 4-dimethylaminobenzaldehyde, a green product is formed.[123] Pindur states that the reaction involves the formation of a diindolylmethane derivative **(19)** which is oxidized to the green **(20)** product.

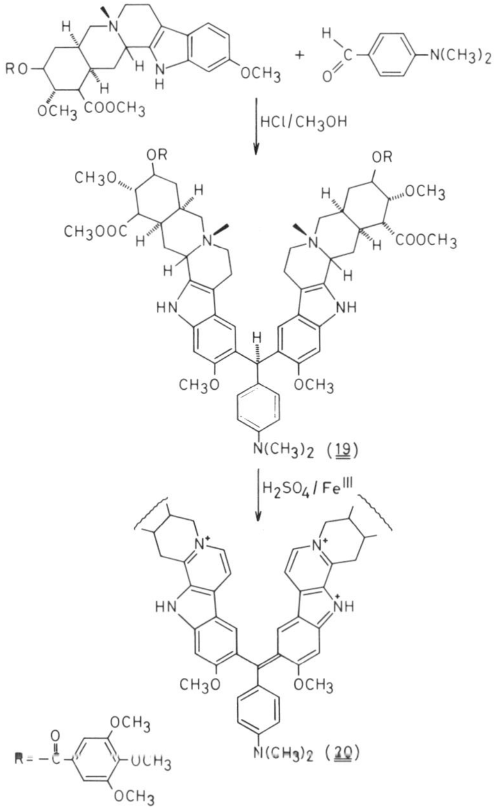

The general reaction path of the color reactions of reserpine with arylaldehydes such as vanillin, 4-methoxybenzaldehyde, and 1-napthylaldehyde was investigated by Pindur and Schiffl.[124] It is worth mentioning that no interference was caused by the acid-catalyzed epimerization of reserpine in the course of the color reaction.

Colorimetric assays based on oxidation with sodium nitrite,[125] sulfovanadic acid,[126] and potassium iodate[127] have been described.

Reactions with xanthydrol[128] and a hydroxyquinolphthalein[129] have been used for the colorimetric quantitation of reserpine in pharmaceutical preparations.

Ferric hydroxamate complex formation through the 16-ester group was used for the colorimetric assay of reserpine in formulations and biological fluids by Karawaya et al.[130] and Dyanowska and Iovchev.[131]

The red-violet complex formed with citric acid in acetic anhydride medium was suitable for the colorimetric determination of reserpine in biological fluids in the range 5 to 400 μg/10 ml.[130] The salt formation that takes place in the reaction product is measured spectrophotometrically.[132] The determination of 10 to 160 μg/ml reserpine in dosage forms was possible.

Sensitive adsorptive stripping voltammetric measurement of reserpine from urine was developed on carbon electrodes.[133] An amount of 2.10^{-9} *M* of reserpine could be detected and measured.

II. α-METHYLDOPA

L-Tyrosine, 3-hydroxy-α-methyl sesquihydrate
L-3-(3,4-dihydroxyphenyl)-2-methylalanine sesquihydrate
$C_{10}H_{13}NO_4 \cdot 3/2H_2O$ $M_r = 238.24$

A. PROPERTIES

Methyldopa is a white to yellowish-white, odorless powder. Moderately soluble (1:200) in water; slightly soluble in alcohol; practically insoluble in ether. It is freely soluble in mineral acids such as hydrochloric acid (salt formation). Melting point: approximately 300°C (with decomposition). The UV absorption spectrum of methyldopa exhibits a maximum at 283 nm in methanolic solution ($\epsilon = 2510$); acidification with hydrochloric acid causes a hypsochromic effect, and the maximum is shifted to 280 nm ($\epsilon = 2790$). The IR spectrum of methyldopa reveals an intense peak of the hydroxy groups, at 3000 to 3500 cm^{-1}, which is broadened by intra- and/or intermolecular interactions. In therapy, the lavorotatory isomer, (L)-(−)-methyldopa is used; $[\alpha]_D^{25°}$: −25° to −28° (USP).

A method for the resolution of the optical isomers (e.g., control of optical purity) of methyldopa was published by Martens et al.[134] The simple and fast method (Chiralplate, Macherey-Nagel; methanol-water-acetonitrile 5:5:3 as mobile phase) allows the detection (ninhydrin) of D-methyldopa in the L-isomer in an amount of ≤3%.

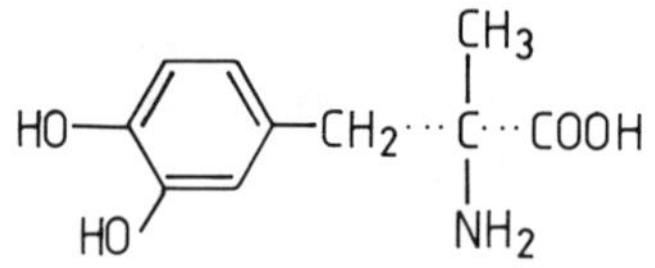

B. MECHANISM OF ACTION AND STRUCTURE-ACTIVITY RELATIONSHIP

When it was clearly demonstrated that the antihypertensive effect of methyldopa can be abolished[135-140] by central inhibition of the enzyme aromatic amino acid decarboxylase, the earlier concept of the "false transmitter" theory[141-144] could no longer be maintained. The experimental findings proved the mainly central feature of the mechanism of the hypertensive action: methyldopa is transported to the brain, and thereafter is converted on the analogy of noradrenaline biosynthesis to α-methylnoradrenaline (Figure 1 a,b). This latter compound is the one which interacts with the central α-adrenergic receptors, decreasing the sympathetic outflow[145] and reducing the arterial pressure and the heart rate.[146] The high affinity of the methyldopa metabolite (−)-erythro-α-methylnorepinephrine for α-receptors supports the suggestion that the hypotensive effects of methyldopa are mediated by α-receptors.[147] Although it seems to be proved that the major part of the antihypertensive effect of methyldopa is due to its CNS action, the contribution of peripheral mechanism cannot be excluded.[148-149] For decades, methyldopa was one of the most important and most frequently used antihypertensives. Following the discovery of new types of antihypertensive agents, the use of methyldopa seems to have decreased somewhat in recent years. A review has been published on the use of methyldopa in the treatment of hypertension.[150]

Studies of the structure-activity relationship[151-167] indicated that relatively very few methyldopa derivatives posses a hypotensive effect at a therapeutically useful level. Elongation of the side chain by even one carbon atom completely abolishes the hypertensive activity.[151]

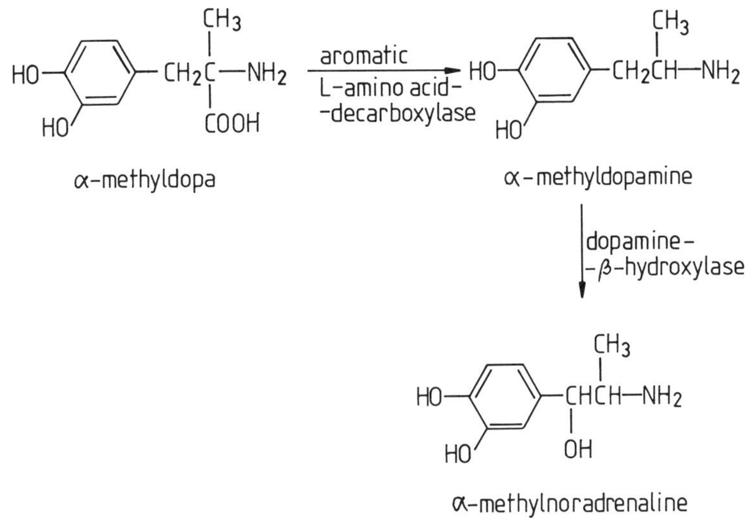

FIGURE 1a. The transformation of α-methyldopa to α-methylnoradrenaline.

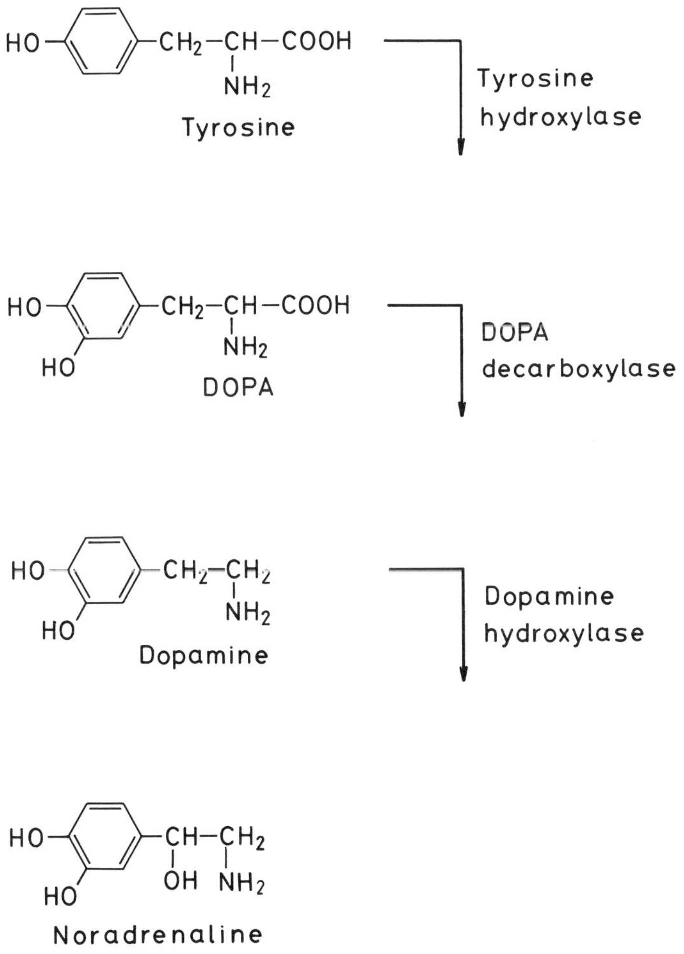

FIGURE 1b. The biosynthesis of noradrenaline.

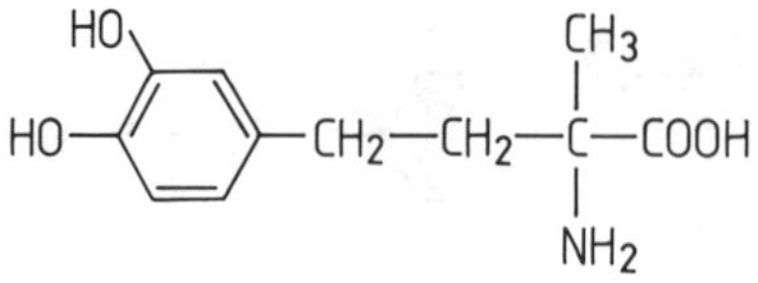

(1) 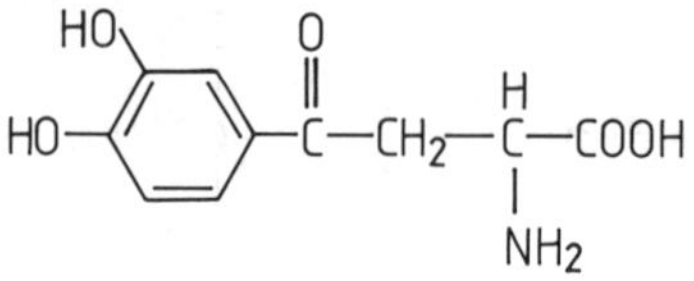 (2) 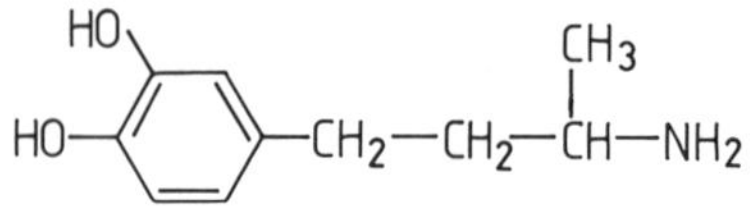 (3)

α-Methylhomodopa **(1)** cannot cross the blood-brain barrier, which is assumed to be the main cause of its inactivity.

Similarly, keto-homodopa **(2)** did not give rise to any decrease in blood pressure.[151] None of the tested dopamine analogs **(3)** caused an increased renal blood flow.[151] The significance of the phenolic group in the receptor binding is reflected by the findings that etherification **(4)** or substitution **(5)** causes a reduction of the hypertensive activity.[152,153] Accordingly, saligenin **(6)**, and analog of methyldopa **(7)**, was synthetized[154] on the basis of earlier observations with catecholamines.[155]

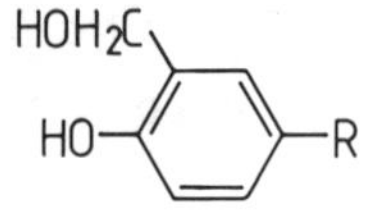

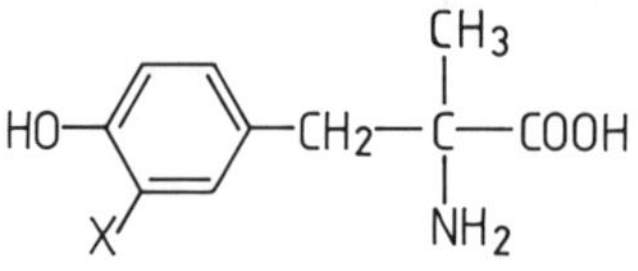

R = H **(6)**

X = OCH_3 **(4)**

= C_2H_5 **(5)**

R = $-CH_2-C(CH_3)(NH_2)-COOH$ **(7)**

α-Hydroxymethyldopa **(10)** was prepared[156] from dopa **(8)** through an azlactone **(9)**[157]; hydroxylation of the α-methyl group destroyed the antihypertensive activity:

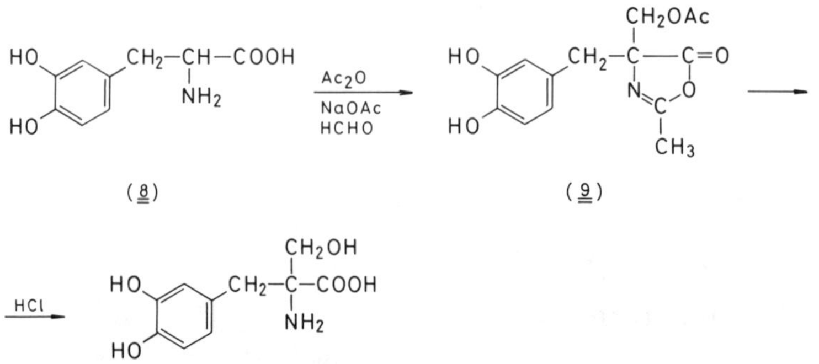

(8) (9) (10)

Displacement of the alcoholic hydroxy group with a fluoro atom **(11)** led to retention of the antihypertensive effect in rats,[158,159] but this compound was highly toxic, due to its very strong inhibition of aromatic L-amino acid decarboxylase. Similar results were obtained

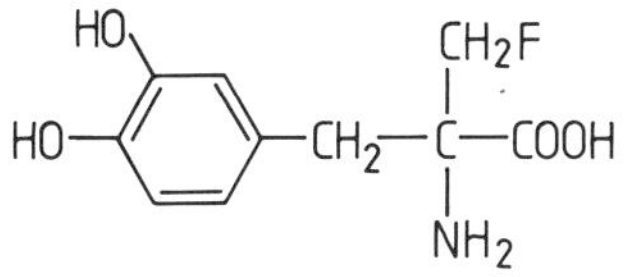

(11) 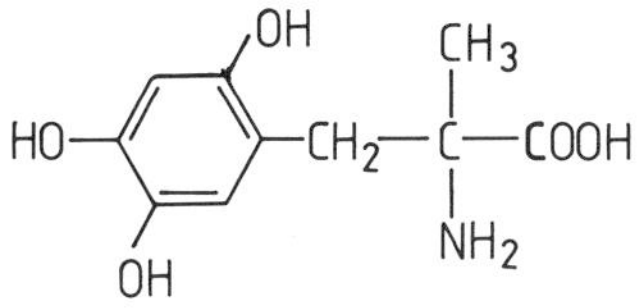 (12) 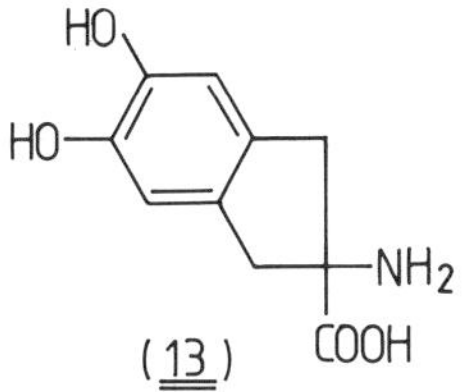 (13)

with the 6-hydroxy derivative of methyldopa **(12)**. Only the (R)-enantiomer caused a reduction of blood pressure, but it also proved to be toxic.[160]

In one interesting experiment, a cyclic variant of dopa, an indane derivative **(13)**, was prepared.[161] The inactivity found was thought to be a consequence of the steric rigidity and structural symmetry of this compound.

C. ANALYSIS

Besides the IR and UV spectra, USP prescribes a color reaction for the identification of methyldopa. With triketohydrindene hydrate (ninhydrin) in sulfuric acid medium, a dark-purple color is produced. On the addition of water to the mixture, this color changes to a pale brownish-yellow. The reaction is typical of amino acids.

Methyldopa is a catecholamine, and also has an amino acid function. All of the reactions of both of these functions are suitable for its identification and assay. There are three main aspects of methyldopa analysis: the assay of methyldopa in biological fluids, in dosage forms, and in the presence of its metabolites. For certain selected tasks, some simple, fast, but not selective color reactions can be used (Table 8).

Titrimetric methods — USP prescribes nonaqueous titration with 0.1 *N* perchloric acid in the presence of crystal violet as indicator. Indian authors have published a series of oxidimetric methods for the determination of methyldopa in the presence of propranolol in tablets: titration was carried out with *N*-bromosuccinimide, with methyl red as indicator.[187] As new oxidimetric agents, bromamine-T, dibromohydantoin, and *N*-bromophthalimide were introduced; in these cases, methyldopa acted as a self indicator (red — colorless).[179] The electroanalytical titrimetry of methyldopa was carried out by using an enzyme electrode.[188] On the basis of its α-amino acid function, methyldopa can be determined via the reaction with an excess of Cu(II) phosphate, using a Cu^{2+} ion-selective electrode. Methyldopa has also been determined in pharmaceuticals without interference by the ingredients.[189]

The behavior of methyldopa (and other catecholamines) has been investigated on rotating disc electrodes of Pt and Au. Rapid voltammetric determination was possible. Methyldopa displayed a single 4-electron wave at the Pt electrode, but there was evidence of a 2-electron wave at the Au electrode.[190]

The stability constants of the copper(II) iron (III) and zirconium(IV) chelates of methyldopa and dopamine have been determined. The Fe(III) complexes are the most stable.[191]

As regards the metabolism and especially the active metabolites of methyldopa, rather divergent opinions are to be found in the literature in recent years. Goldberg et al.[147] investigated the differential competition by methyldopa metabolites for the adrenergic receptors in the rat forebrain. In spite of the fact that α-methylnorepinephrine appears as a main metabolite, α-methylepinephrine seems to be the most potent active metabolite of methyldopa as far as its influence on the blood pressure is concerned.[192,193]

A compilation of the recent HPLC methods is given in Table 9.

TABLE 8
Spectrophotometric Assay of Methyldopa Through Color Reactions

Reagent	Product	Wavelength nm	Range of linearity	Note	Ref
0.1% phloroglucinol + 0.5% NaOH	—	413	0.8—4.0 μg/ml	In tablets	168
Ninhydrin + buffer, pH 7.0	Violet color	570	10—80 μg/ml	In tablets (in presence of bethanidine)	169
0.004 *M* $NH_4Fe(SO_4)_2$ + 0.004 *M* *m*-aminophenol	Green fluorescence	—	0.06—1.6 μg/ml	For methyldopa in bulk and other physiologically active catecholamines too	170
Folin-Ciocalteau phenol reagent + 20% Na_2CO_3 ad soln	—	680	1—7 μg/ml	In tablets	171
$(NH_4)_6Mo_7O_{24}{\cdot}4H_2O$ + 5% H_2SO_4	Yellow color	360	25—250 μg/ml	In tablets	172
0.3 *M* NH_4 metavanadate + 0.1 *N* H_2SO_4 soln	—	680	120—250 μg/ml	In tablets; common excipients do not interfere	173
0.002 *M* $FeCl_3$ (or $FeSO_4$, $MnSO_4$, $K_3Fe(CN)_6$) soln + buffer, pH = 10.5	Pink color (adrenochrom)	485	0.04—1.5 mg	In tablets, injections, solutions, syrups, and for other sympathomimetic amines	174
p-Methoxybenzaldehyde or *p*-methylbenzaldehyde at pH = 4.7	Purple color	540	0.24—1.68 mg	Also for other sympathomimetic amines	175
4-Dimethylaminocinnamaldehyde in alkaline soln	Orange color	440	—	Also for noradrenaline	176
Phenylfluorene + Fe(III) at pH = 8.9—9.9	A complex is formed	—	—	Sensitivity is highest with 5% Brij 35.	177
p-Benzoquinone in buffered aq ethanolic soln	—	495	8—48 μg/ml	—	178
Bromamine-T or dibromohydantoin or *N*-bromophthalimide or *N*-bromosuccinimide	Red (adrenochrom)	485	40—650 μg/ml	—	179
Orcine in NaOH soln	Yellow color	415	1—7 μg/ml	In tablets; no interference from excipients or coloring matter	194

III. GUANABENZ

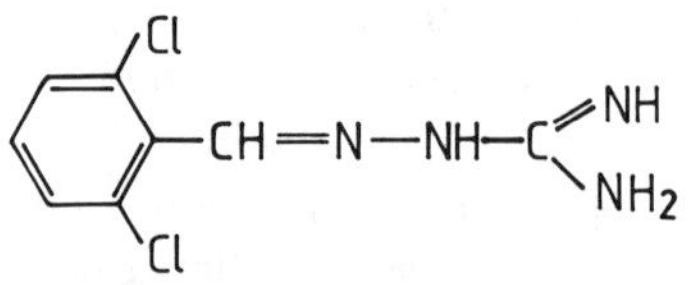

Hydrazine carboximidamide, 2-[2,6-dichlorophenylene]
2-[(2,6-dichlorobenzylidene)-amino]-guanidine acetate
$C_{10}H_{12}N_4O_2Cl_2$ $M_r = 231.08$
$C_{10}H_{12}N_4O_2Cl_2 \cdot CH_3COOH$ $M_r = 291.14$

A. PROPERTIES

The physical and chemical properties of guanabenz were reviewed by Shearer.[195] Guanabenz acetate is a white to off-white crystalline odorless substance. Its melting point is 188 to 190°C.

Certain solubility data, in mg/ml: water 11, alcohol 50, chloroform 0.6, ethylacetate 1, propylene glycol 100. The pK_a (guanidinium, in 40% ethanol-water mixture): 8.1. UV absorption characteristics:

Solvent	Max (nm)	Absorptivity
0.1 *N* HC1	270	44.5
0.1 *N* NaOH	295	43.4
Ethanol	312	45.2

The UV spectrum differs considerably when nonaqueous solvents are used. This is a consequence of the appearance of a more conjugated tautomeric form of guanabenz. Guanabenz exists in two isomers: E and Z. The compound used in therapy is the E isomer. On irradiation with UV light, a partial transformation to the Z isomer may be observed in the presence of light. In mildly acidic, neutral or mildly basic solution there is also a reversible reaction, with transformation of the E isomer to the Z isomer. The two isomers can easily be distinguished by means of spectrometry of TLC.

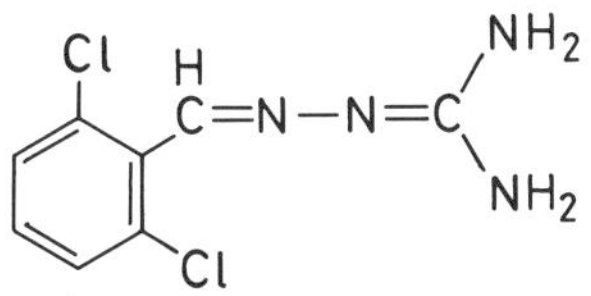

These methods[196] are suitable for stability-indicating assay too.

Guanabenz was the first herbicidal hydrazine compound to be synthetized in 1966.[197] As a substituted benzylidene-hydrazine, it was considered for the treatment of hyperglycemia, obesity, and inflammation.[198]

For this purpose, 2-dichlorobenzaldehyde was refluxed with aminoguanidine (or a salt) in ethanol. Among the *o*-halobenzylidene-aminoguanidines applied as central nervous system (CNS) depressants, cardiovascular depressants, and/or antidepressants, guanabenz was synthetized[199] from aminoguanidine bicarbonate with hydrochloric acid in butanol. In a recent method,[200] guanabenz was prepared from 2,6-dichlorobenzylidene hydrochloride **(1)** and the HBr salt of *S*-methylthiourea **(2)**:

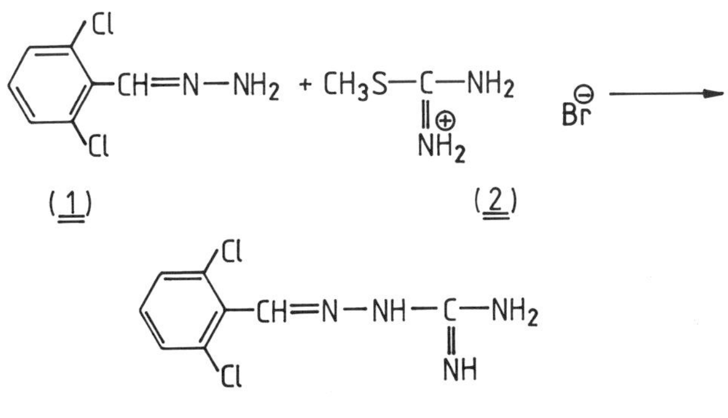

(1) (2)

Another method includes the condensation of 2,6-dichlorobenzaldehyde with aminoguanidine in the presence of acetic acid.[201]

TABLE 9
Methyldopa Assay Through HPLC Methods

Compound(s)	Stationary phase	Mobile phase	Detection nm	Range of linearity	Note	Ref
Methyldopa + hydrochlorothiazide	μBondapak C_{18}	Acetonitrile-water-acetic acid 15:84.4:0.6			In tablets	180
Methyldopa + hydrochlorothiazide (theobromine as int standard)	μBondapak C_{18}	3% Acetic acid-methanol 96.4	Spectrophotometric 280	0.03—0.39 μg/ml	In tablets, oral suspensions	181 182
Methyldopa	Partisil-10 SCX (strong cation-exchanger)	33.8 n*M* $HClO_4$-50 m*M* $NaClO_4$-0.2 m*M* EDTA in 20% aq. MeOH	Electrochemical (glassy carbon electrode)		In blood and breast milk (before HPLC the sample was defatted, deproteinized, and deconjugated	183
Methyldopa	μBondapak alkyl-phenyl	0.02 *M* citric acid-0.02 *M* Na_2HPO_4 2:1 + Na-octanesulfonate (0.1%) + Na-EDTA (0.05 *M*) pH = 3.05	Thin layer flow-through electrochemical cell with glassy carbon working and silver-silver chloride reference electrodes	0.05—1 μg/ml	Assay and stability in plasma (blood samples were acidified with $HClO_4$ at 0° C and centrifuged)	184
Methyldopa + levodopa (I.St.)	Hypersil ODS	Methanol (160 ml) + H_3PO_4 (31 g) + KH_2PO_4 (20 g) + PIC B-8 (4.4 g) in H_2O (1.3 l)	Electrochemical at 0.75 V	50—2000 μg/ml	In plasma (+ $HClO_4$, cooling to −20° C and centrifugation before injection)	185
Methyldopa + catecholamines, metabolites	Hypersil u ODS, M.O.S., S.A.S.	0.5 *M* aluminum acetate buffer (pH = 5.1), 0.05 m*M*/ EDTA, 10 m*M* heptanesulfonic acid	Amperometric			186

B. ACTION AND STRUCTURE-ACTIVITY RELATIONSHIP

It was found by Baum et al.[202,203] that 2,6-dichlorobenzylidene aminoguanidine acetate exhibits good hypotensive activity in experimental hypertensive rats and dogs, producing its maximum blood pressure reduction 1.5 h after administration. It proved more potent than reserpine, guanethidine, and α-methyldopa, but less potent than hydralazine and dihydralazine; in its antihypertensive activity guanabenz resembles certain relatived congeners such as guanfacine, guanoxabenz, and clonidine. Clonidine is said to be the most potent of these. As concerns the mechanism of action, guanabenz is an effective central α_2-adrenergic agonist.

As regards the importance of the length of the intermediate chain between the benzene ring and the aminoguanidino group, the introduction of one methylene group was found to give optimum activity.[204] Diamant et al.[205] established that guanabenz also has surprisingly high β-adrenergic activity. Therefore, the three-dimensional structure of crystalline guanabenz was examined. Substantial structural differences were observed between guanabenz and clonidine, which could explain the difference in specificity exhibited by the two compounds. For example, the angle of rotation of guanabenz is Ø = 39.7°, as compared with 76° for clonidine; the rotational restriction in clonidine seems to be much greater than that in guanabenz.

C. ABSORPTION AND METABOLISM

The absorption, distribution and excretion of guanabenz after its oral administration to rats have been investigated by different research groups. It was verified that about 90% of the dose was excreted in the feces and urine within 24 h. The concentration was high in the liver and digestive tract, followed by the pancreas, kidneys, lungs, spleen, plasma, and CNS. The level in the brain was less than 25% of that in the blood.[206]

Guanabenz can decompose by hydrolysis with the formation of 2,6-dichlorobenzaldehyde **(2)**, aminoguanidine **(3)**, and 2,6-dichlorobenzaldehyde semicarbazone **(4)** but the corresponding reactions proceed very slowly.[196]

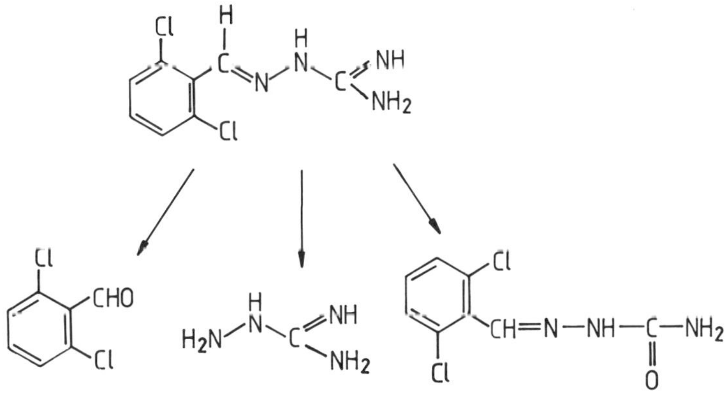

Following guanabenz administration to Rhesus monkeys and rats, the following metabolites have been identified: guanabenz Z isomer, 2,6-dichlorobenzylalcohol, 2,6 dichlorobenzaldehyde, *p*-hydroxyguanabenz, i.e. (4-hydroxy-2,6-dichloro-benzylideneamino)-guanidine,[207,208] and the sulfate and/or glucuronide conjugates of (2,6-dichloro-4-hydroxybenzylidene amino)-guanidine and 3-hydroxyguanabenz. All these metabolites could be dctcctcd by TLC.[209]

D. ANALYSIS

Its IR spectra (USP) and chromatographic behavior serve for the identification of guanabenz acetate.

A microcrystallographic identification involving use of the Reinecke reagent, chloroauric acid, chloroplatinic acid, and 0.1 *N* iodine solution is also known.[210]

In the TLC procedures, silica-gel is generally employed as the stationary phase, and polar mobile phases containing ammonium hydroxide or acetic acid are applied. The TLC method of USP (Suppl.1.) allows the detection of aminoguanidine and other possible impurities too in guanabenz. The developing solvent used is similar to that applied by Shearer and De Angelis (silica gel/chloroform-methanol-ammonium hydroxide).[209]

The gas chromatographic procedure is suitable for the assay of guanabenz in the presence of its degradation products in the bulk substance (USP) or in biological samples.[211,212] One of the methods is based upon the acid hydrolysis of guanabenz to 2,6-dichlorobenzaldehyde, which has strong electron-capturing properties and is volatile enough to be eluted from a gas chromatographic column.[211] It is also possible to determine the monosubstituted guanidines as their hexafluoroacetyl-acetone derivatives by multiple-ion detection.[212]

Guanabenz and mefruside in pharmaceuticals have been assayed by reversed-phase HPLC[213] in a system of C_{18}-acetonitrile-phosphate buffer (pH 3.25) 35:65. UV detection at 265 nm was used. Two possible contaminants, 2,6-dichlorobenzaldehyde and the corresponding semicarbazone, were eluted with retention times different from those of the tested drugs. For the quantitative determination of guanabenz in bulk, USP describes nonaqueous titration with 0.1 *N* perchloric acid in glacial acetic acid medium, using potentiometric end point detection.

IV. GUANETHIDINE

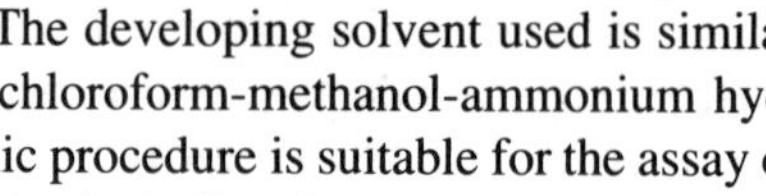

$$\text{(azocan-1-yl)}N{-}CH_2{-}CH_2{-}NH{-}C(=NH){-}NH_2$$

[2-(Hexahydro-1(2H)-azocinyl)ethyl]-guanidine
$C_{10}H_{22}N_4$ $M_r = 198.30$

Guanethidine is a product of the extensive research that began in the late fifties to find potent antihypertensives among amidoxime derivatives. Since then, guanethidine has maintained its position and belongs among the standard antihypertensive agents applied in therapy. In accordance with its polyvalent base character, two salts of guanethidine are available commercially, and are dealt with by monographs in USP.

Guanethidine monosulfate
$C_{10}H_{22}N_4 \cdot H_2SO_4$ $M_r = 296.38$
Guanethidine sulfate
$(C_{10}H_{22}N_4)_2 \cdot H_2SO_4$ $M_r = 494.69$

A. PROPERTIES

The two salts are white or off-white crystalline substances with similar physical and chemical properties. However, guanethidine monosulfate is very soluble in water, whereas guanethidine sulfate, which may be regarded as a basic salt of guanethidine, is only sparingly water-soluble. Each of them is slightly soluble in alcohol, and insoluble in chloroform.

Because of the strong basic character of guanethidine, the solution of the monosulfate exhibits a pH close to neutral (USP: pH 5 to 6).

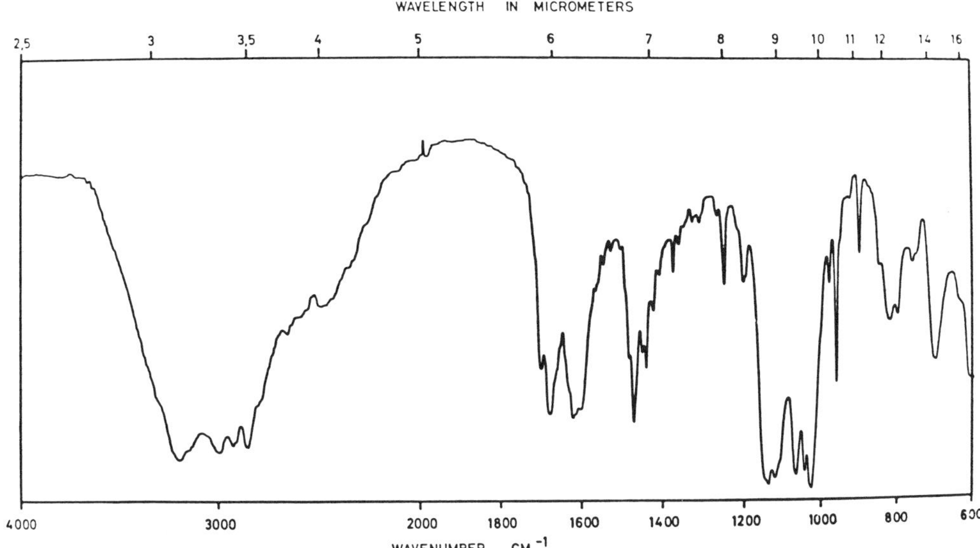

FIGURE 2. IR spectrum of guanethidine.

Tablets of both salts are also official in USP, but most of the commercial products (Ismelin®, etc.) contain the monosulfate as active ingredient.

Due to the lack of a suitable chromophoric group, guanethidine has no UV absorbance above 230 nm. The IR spectrum is used for identification purposes in both cases. The maxima due to the presence of the stretches of the guanidinium $-C{=}N^+$ and $-NH_3^+$ groups can be seen at approximately 1700 cm^{-1} and 3000 cm^{-1}, respectively (Figure 2).

B. HISTORY AND SYNTHESIS

The history of guanethidine dates back to the fifties when Plummer[214] observed the antihypertensive effect of a propionylamidoxime derivative **(1)** that was expected to have a trypanocidal effect. Although this compound (SU 4029) had serious side effects and was therefore withdrawn from clinical trials, amidoximes attracted attention in the search for new antihypertensives. The systematic work following this observation led to the discovery of the structurally related guanethidine **(2)** and to the confirmation of its clinical efficacy in the treatment of hypertension.[215,216] Guanethidine was first synthetized by Mull et al.[217,219]

$N-CH_2-CH_2-C(=NOH)NH_2$ (1) SU 4029

$N-CH_2-CH_2-NH-C(=NH)NH_2$ (2)

starting from octahydroazocinone **(3)**, which was reduced to octahydroazocine **(4)**. The latter was transformed with chloroacetonitrile to octahydro-1-azocinylacetonitrile **(5)**. Hydrogenation with $LiAlH_4$ resulted in the formation of β-(octahydro-1-azocinyl)-ethylamine **(6)** which was reacted with *S*-methylisothiocarbamide **(7)** to yield the guanethidin (G) salts **(8,9)**:

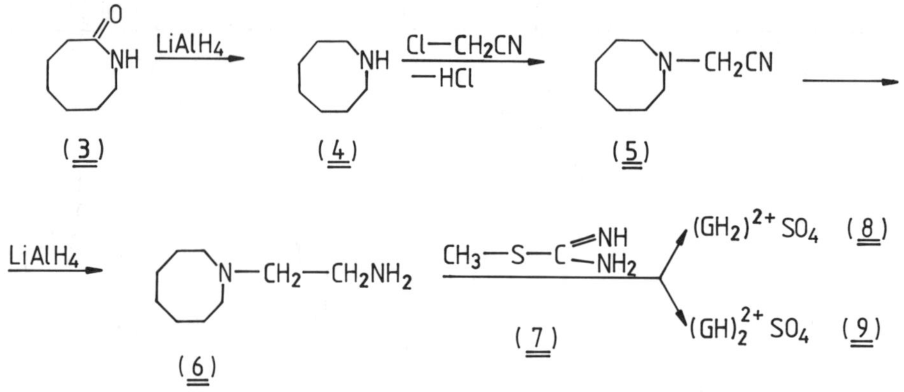

C. STRUCTURE-ACTIVITY RELATIONSHIP

The route to the discovery of guanethidine led through the synthesis and pharmacological trials of hexahydro-1-azepinyl-propionamide oxime. Due to its side effects this compound was withdrawn from further clinical investigations. Continued research led to systematic study of guanidine derivatives[220] and to the selection of guanethidine.[221]

Within the following 10 years, more than 2000 papers were published on guanethidine.[221] In spite of the very intensive preparative activity involving the synthesis of thousands of guanidine (guanethidine) related substances, none of the compounds prepared could be claimed to be as good an antihypertensive agent as guanethidine.

The structural changes made to the guanethidine skeleton concerned the following three main units of the molecule:

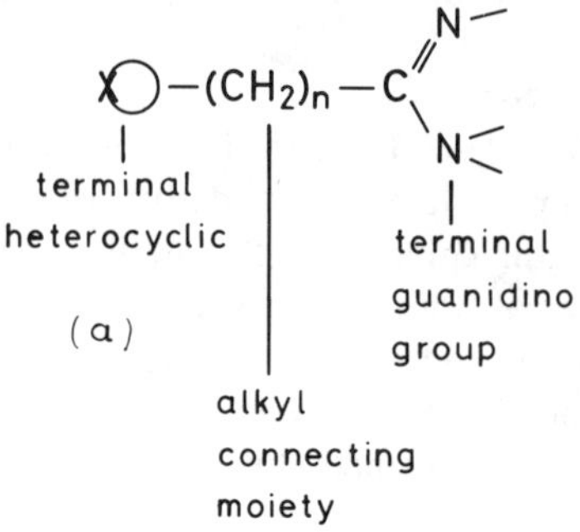

The early changes aiming at decyclization at point (a) were detrimental, resulting in compounds such as (**9** and **10**) without hypotensive activity. Even *t*-octylguanidine, the most promising open-chain variant of guanethidine, did not exhibit an effectiveness corresponding to that of guanethidine.[222] Similarly, phenylethylguanidines, substituted benzylguanidines (**11**)[223] and their analog cinnamylguanidine (**12**)[224] were found to be less active than guanethidine itself.[223]

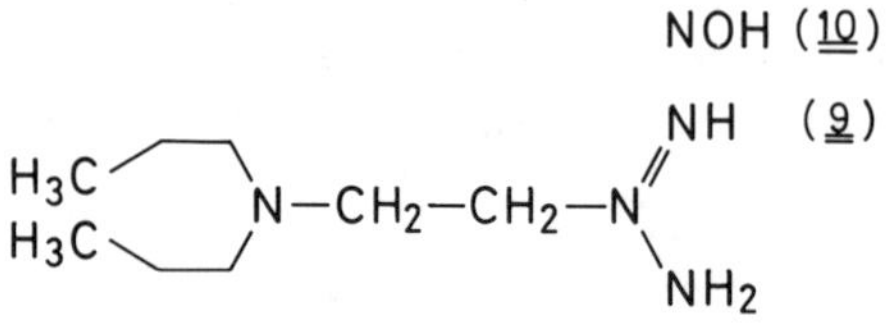

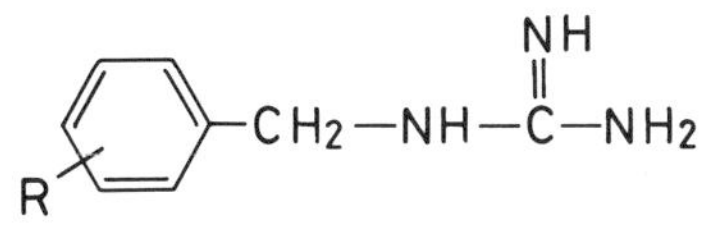

(11) 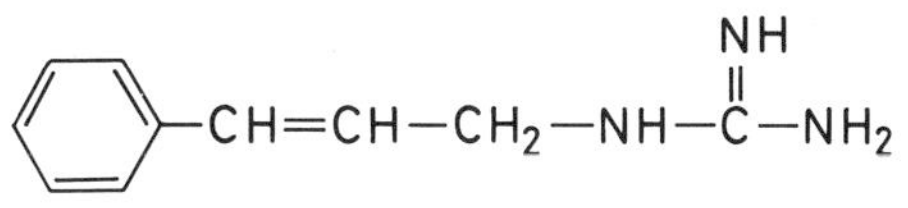 (12)

The above mentioned and numerous other unsuccessful attempts proved the essentiality of the heterocycle in the structure of guanidine derivative hypotensives. As far as the role of the heterocycle in the receptor-pharmacon interaction is concerned, an appropriate basicity, lipophylicity and fitting provided by the heteroatom- and the hetero ring are to be assumed. The optimum distance between the two active poles of the guanethidine molecules is ensured by the connective ethylene bridge of the side chain. Decrease of the ring size in the nitrogen-containing heterocycles (azepine, piperidine) significantly reduced or completely eliminated the effectiveness, while enlargement to a nine or ten-membered ring resulted in compounds with milder activity. Change of the heteroatom to oxygen resulted in the synthesis of such guanethidine congeners as guanoxan **(13)** or certain other active **(14)** but also inactive **(15)** derivatives.

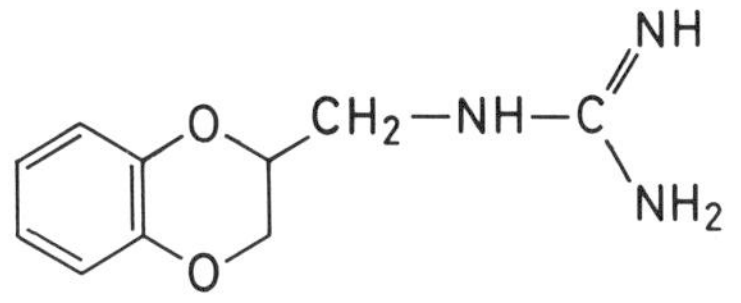

(13) 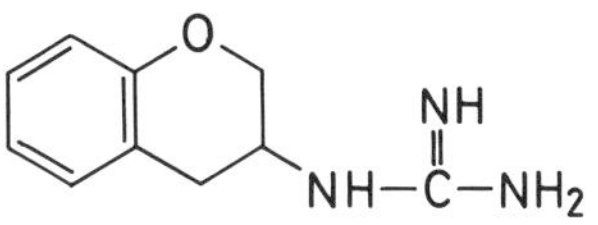 (14)

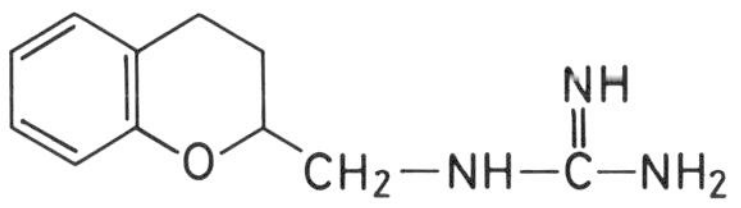

(15)

Replacement of the guanidine group by other nitrogen-containing moieties, such as an amino group, a semicarbazide[226] or carbazide group, etc. yielded less active hypotensives. The distance between the molecular poles seems to play a decisive role, and this is reflected by the inactivity of the homologous benzopyran derivatives.[227]

Substitution of the guanidino-nitrogens with alkyl or aryl groups gave only mildly effective or inactive compounds, due to the decreased reactivity of the nitrogen(s).

D. ANALYSIS

The amines, including guanethidine, give characteristic reactions with different color reagents. The color produced with a modified Dragendorff reagent has been found to be the most sensitive, with a detection limit of 0.65 μg.[228] A combination of various detection methods permitted the distinction of guanethidine from other drugs. USP also applies a color reaction for the identification of the compound. After the addition of α-naphthol, sodium hydroxide, sodium carbonate, water, and 2,3-butanedione, a pinkish-red color develops.

Besides the color reactions, the use of TLC systems and the UV spectrum for the analysis of guanethidine was discussed by Wessinger and Auterhoff,[229] with monitoring with multiple ion detector.

Because of the presence of the guanidino group, guanethidine gives a positive Sakaguchi reaction: the red color observed on the addition of α-naphthol is due to the formation of an indophenol-like structure. In their study of the scope of the Sakaguchi reaction, Casadebaig et al.[230] found bromocaprolactam to be the most suitable reagents for guanethidine.

HO– + H_2N–C(=NH)–NHR $\xrightarrow{+OBr^{\ominus}}$ [HO– N⊕(C(=NH)NHR)= =O] $Br^{\ominus}$

or

O= =N–C(=NH)–NHR (Br, OH)

Guanethidine and its derivatives were determined in biological samples via a modified Sakaguchi reaction, using sulfosalicylic acid and 8-hydroxyquinoline in 25% NaOH solution plus hypobromite as reagents.[231]

A GC-MS method has been described for the determination of monosubstituted guanidines such as guanethidine in blood plasma and urine.[212] Guanidines were derivatized with hexafluoroacetone, forming nonpolar trifluoromethylpyrimidines, which were chromatographed on a 3% OV-1 stationary phase with He as carrier gas and monitored with a multiple ion detector. The Voges-Proshauer reaction has been used for the determination of guanethidine in tablets;[232] diacetyl and α-naphthol were added to an aqueous suspension of guanethidine-containing tablets, followed by heating, cooling, and extraction with isoamyl alcohol. Absorbance was measured at 545 nm. The formation of a volatile nitrosamine via the drug-nitrite reaction is due to the secondary amine moiety of guanethidine. This reaction was tested under physiological conditions.[233] Guanethidine and the other drugs studied were incubated with NO_2^- at pH 3.0 at 37°C for 1 or 4 hours. The volatile nitrosamines formed were determined by means of GC-thermal energy analysis. The important observation was made that ascorbic acid is effective in decreasing the formation of nitrosamine from drugs.

Complexation has also been utilized for the determination of guanethidine. Bromothymol blue, bromophenyl blue, methyl orange, and tropeolin 00 were tested in microreactions with guanethidine. The most sensitive reaction took place with bromothymol blue, and this can be used for a spectrophotometric assay.[235] On the basis of this reaction, an extraction of the drug from tablets was described. The guanethidine-bromothymol blue complex was extracted with chloroform, and the absorbance was measured at 400 nm. Beer's law was obeyed in the concentration range 5 to 100 μg ml^{-1}.[235] The complex extracted with chloroform at pH 7.6 was analyzed by means of IR spectroscopy, which indicated the presence of the quinoidal form of the bromothymol blue moiety in the reaction product.[236]

Guanethidine forms complexes with tetrathiocyanatodiaminechromium(III) reagents, such as $H[Cr(NCS)_4(aniline)_2]$. The red precipitate is collected and dissolved in acetone, and the color is measured at 535 nm. Beer's law is obeyed in the concentration range 0.05 to 0.6 μg/ml^{-1}. In other procedures, the red-violet complex is treated with an excess of EDTA, and the unreacted EDTA is titrated with $Zn(OAc)_2$;[237] or the complex is destroyed by treatment with sodium hydroxide, and the $Cr(OH)_3$ formed is titrated oxidimetrically with $KMnO_4$,KIO_3, or $KBrO_3$.[238]

Among the different amines, guanethidine was determined complexometrically on an ion-exchange resin, Amberlite IR-120 (Zn^{2+} form), from pharmaceutical preparations. This method was found to be highly accurate.[239]

A flow injection HPLC method for guanethidine, based on electrochemical oxidation at a glassy carbon electrode, has been described.[240] The calibration curve was linear in the range 0.5 to 16 $\mu g\ ml^{-1}$; the detection limit was 5 ng (E = +100 mV). Preliminary separation from hydrochlorothiazide, methyldopa, hydralazine, chlorpromazine, and imipramine is required.

Stewart and Clark[241] separated and determined guanethidine salts with an acetonitrile-0.05 *M* aqueous NaH_2PO_4 3:70 mixture containing 0.02 *M* Na pentane sulfonate (pH 2.5) as mobile phase. Amperometric detection utilizing oxidation at glassy carbon electrode was applied. Guanethidine showed linearity in the range 0.5 to 10 $\mu g\ ml^{-1}$.

Two-dimensional GLC utilizing electron capture detection for the analysis of guanethidine in blood plasma has been described.[242] After derivatization with heptafluorobutyric anhydride, dual-GC performed (first column: 1% DEGS on Chromosorb: second column: 5% OV-225 + 0.5 Gas-Qual-L). The detection limit was 0.7 μg guanethidine.

With temperature programming (120 to 270°C), gas chromatography of the silylated derivative of guanethidine gave one peak, whereas under isothermal conditions there was no peak. This observation can be utilized under exact circumstances for the identification of guanethidine.[243]

A radioimmunassay has been developed for guanethidine at the nanogram level in plasma in the presence of its major human metabolites, the N-oxide and ring-opened derivatives.[244] McManus et al.[245] studied guanethidine N-oxidation by human liver microsomes, which was optimum at pH 8.5; the rate of oxidation was reduced to one sixth at pH 7.4. The N-oxidation of guanethidine is mediated by flavine-containing monooxygluase.[245]

V. PRAZOSIN

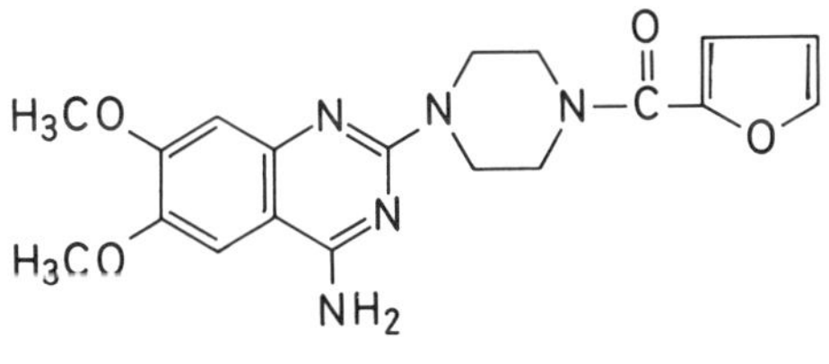

Piperazine, 1-4-Amino-6,7-dimethoxy-2-quinazolinyl)-4-(2-furanyl-carbonyl)-
1-(4-Amino-6,7-dimethoxy-2-quinazolinyl)-4-(2-furoyl)piperazine

$C_{19}H_{21}N_5O_4$ — $M_r = 382.4$

$C_{19}H_{21}N_5O_4 \cdot HCl$ — $M_r = 419.87$

A. PROPERTIES

Prazosin is used in the form of its monohydrochloride (USP). The pure salt is a white powder; due to its light-sensitivity, it most often appears as a yellowish-brown substance with a high melting point at about 280°C. It is only slightly soluble in water or alcohol, and is almost insoluble in the less polar organic solvents such as chloroform or acetone, etc. Thc slight water solubility may be related to the mainly nonelectrolyte nature of its hydrochloride. Prazosin is a weak organic base, with pK_a 6.5.[246] The assignment of the basicity

TABLE 10
Light Absorption Characteristics of Prazosin

Solvent	Absorption maxima nm	$A^{1\%}_{cm}$	ϵ
Methanol	250	1,430	60,100
	331	233	9,760
	342	222	930
0.1 *N* HCl	247	1,343	58,400
	331	257	10,800
0.1 *N* NaOH	252	1,500	62,900
	345	137	5,760

centre does not seem to be a simple task because of the approximate equivalence of its four structural nitrogens with amine character. One possible part of the maximum electron density might be localized in the heteroaromatic (pyrimidine) ring. The electron-attracting effect of this rings allows the existence of several mesomeric structures.

Two crystal forms of prazosin were identified by Bianco[247] and were named the α- and polyhydrate forms. Both were stated to exert hypotensive activity. The results of Lehmann et al.[248] indicated that the α-form of prazosin. HCl is better suited as an antihypertensive agent than the other crystal forms. The α-form was gained by dissolving prazosin in dimethylsulfoxide or dimethylformamide at 40 to 95°C, followed by the addition of hydrochloric acid and crystallization. Recently, a new crystal form of prazosin hydrochloride was gained from aqueous acidic methylene glycol medium and designated the γ-form.[249] The first product (hydrochloride hydrate) was recrystallized from boiling methanol.

The mass spectrum of prazosin exhibits a prominent molecular peak at m/e 383. Other intense peaks indicate cleavage in the piperazine ring and charge retention at the quinazoline pole.[250]

As a consequence of the expanded conjugation of the *n* and π electron pairs, prazosin exhibits an absorption maximum near the visible region (Table 10).

Although purity tests for synthesis intermediates and for degradation products are generally prescribed (USP), a comprehensive study on the stability of prazosin has not been published so far.

B. MECHANISM OF ACTION AND THERAPEUTIC USE

When the early reports confirmed the good efficiency of prazosin against hypertension,[251,252] it became clear that prazosin represents a new stage in antihypertensive therapy. It proved to be a highly selective antagonist of the postsynaptic α_1-adrenoreceptors, abolishing the postsynaptic effect of the neurotransmitter norepinephrine, thereby causing decreases in peripheral resistance and dilatation of the blood vessels. The lack of presynaptic α-adrenoreceptor-blocking activity in prazosin allows the feedback control mechanism of norepinephrine release to function. The pharmacology of prazosin has been reviewed by several authors.[253-264]

The relatively few and usually well-tolerated side effects make prazosin suitable for the chronic treatment of mild to moderate hypertension, alone or in combination.

Orally administered prazosin is readily absorbed from the gastrointestinal tract. It is bound in a very high ratio to proteins. The bound part has been found to be nearly 100%.[266,267] With increasing pH, the free fractions of prazosin in normal human serum decrease.[268]

C. STRUCTURE-ACTIVITY RELATIONSHIP AND SYNTHESIS

A great number of quinazoline derivatives have been found to exert biological activity with different profiles. From the compounds with antihypertensive properties,[269] prazosin was selected for clinical trials.

The career of prazosin began with a study of the hypertensive effect of systematic substitution into the 4-(3H)-quinazolone (**1**) skeleton.[270] The importance of amino substitution also emerged in the early stage of the research.

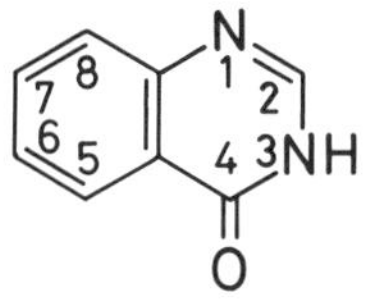

(1)

The average blood pressure decrease measurements (in dogs) indicated the enhancing effect of amino substitution at the 6,7 and 2-positions, whereas 6- or 7-chloro, hydroxy, or methyl substitution did not cause a significant blood pressure-lowering activity. However, the 6,7-dimethoxy-2-amino derivatives proved to be surprisingly highly active. When the blood pressure-lowering abilities of the 6,7-dimethoxy-2-amino derivatives (**2**) were scored, the tertiary amines showed outstanding lowering properties, whereas the primary or secondary amine derivatives exerted very weak or practically no activity (Table 11).

It is also worth noting that the 2-benzyl-6,7-dimethoxy derivative was also ineffective and that the 6,7-diethoxy substituents together with the ''valuable'' 2-diethylamino substituent showed very reduced activity. All these experimental facts are indicative of the lipophilicity and/or basicity-strengthening role of the 2-amino substituent. It would be interesting to compare the effects of the 6,7-dimethoxy and diethoxy 2-methylpiperazino derivatives. The lack of further data makes it unreasonable to draw any conclusions.

All these and some other experiences make it plausible that the furoyl moiety is not in the interacting part of prazosin. The very short half-life and relatively fast metabolism of prazosin with cleavage of the furoyl group allow the presumption that the piperazine-N^1 functions as a basicity centre in the receptor binding of prazosin. If so, the furoyl moiety would belong to the carrier part of the molecule.

In connection with this idea, Honkanen et al.[271] began to study the possibility of replacing the labile furoylpiperazine moiety in prazosin by a piperidino group. A series of new 2-piperidino-4-amino-6,7-dimethoxyquinazoline derivatives (**2**) have been prepared.[271] The results show that the substituted piperidine group can replace furoylpiperazine without loss of the blood presure-lowering effect. This is illustrated by the data in Table 12.[271] The 4-cyclopentylcarbonyl derivatives in higher doses of 10 to 100 μmol/kg appeared to be even more efficacious than prazosin itself.

It can be seen that the activity of some of the 2-piperidino derivatives approximates to that of prazosin. This is especially true for the tested cyclic amides. It should be mentioned that certain ketones were about equally as active as cyclic amides, and some of the compounds exhibited an oral activity comparable with that of prazosin. The question of whether the duration of the hypertensive effect would be prolonged by this mode of quinazoline substitution remained to be answered.

Carbamoylpiperidine derivatives of dimethoxyaminoquinazoline were prepared by Alabaster et al.[272] The 21 derivatives (**3**) almost without exception displayed a high binding affinity for α_1-adrenoreceptors. Several compounds exhibited a potency similar to that of prazosin. Some of the compounds exerted antihypertensive activity in rats.

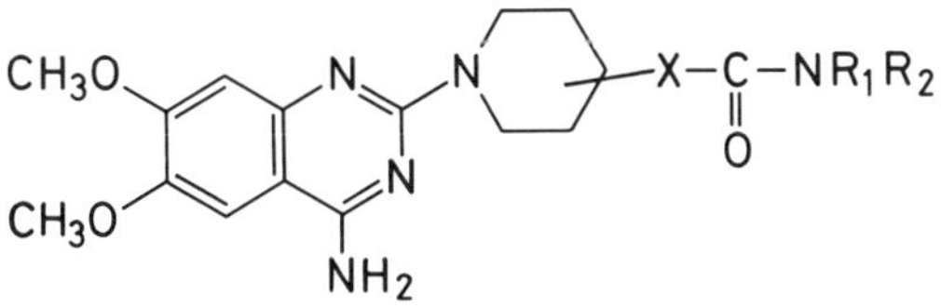

(3)

x = CH_2 or CH_2CH_2, at position 3 or 4

NR_1R_2 = NH_2, NHC_2H_5, $NHCH_2CH_2CH_2CH_3$, $NHCH_2CH_2OCH_3$, $N(C_2H_5)_2$ –N(morpholino)O

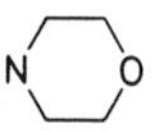

In the interpretation of the high binding affinity of **(3)**, the enhanced basicity of the quinazoline nucleus, and the hydrophobic interaction of the carbamoyl portion are assumed.

For the synthesis of prazosin **(4)**, several modified methods have been reported.[273-275] In the method of Schickaneder and Ahrens,[273] 3,4-dimethoxyanilines were condensed with diphenoxycyanonitrile **(5)** to yield **(6)**; this was treated with piperazine to give piperazino-carboxamide **(7)** (90%). The latter was cyclized by refluxing in 10% HCl to give the quinazoline derivative **(8)** 75%, which was acylated with furoyl chloride to give prazosin (94%).

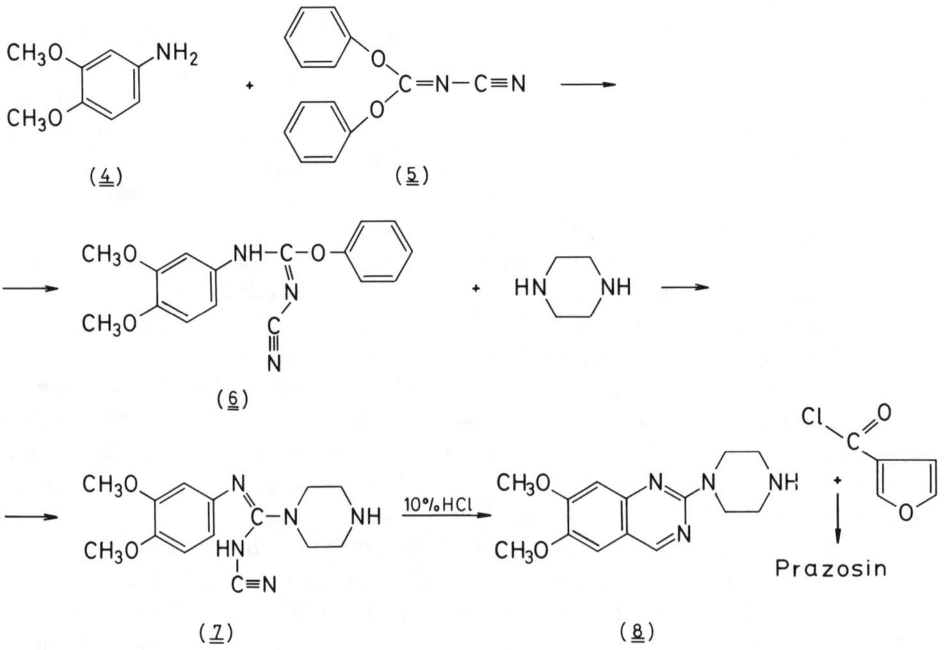

(4) (5) (6) (7) 10% HCl (8) Prazosin

For the preparation of the new quinazoline derivatives, the route which was used in the synthesis of prazosin was again applied.[276] 3,4-Dimethoxy-6-isothiocyanatobenzonitrile **(9)** was condensed with the appropriate piperidine derivatives **(10)**, when thioureas **(11)** were formed. After methylation, the *S*-Methylthiourea derivatives **(12)** formed were cyclized to the quinazolines **(13)** with an excess of ammonium chloride.

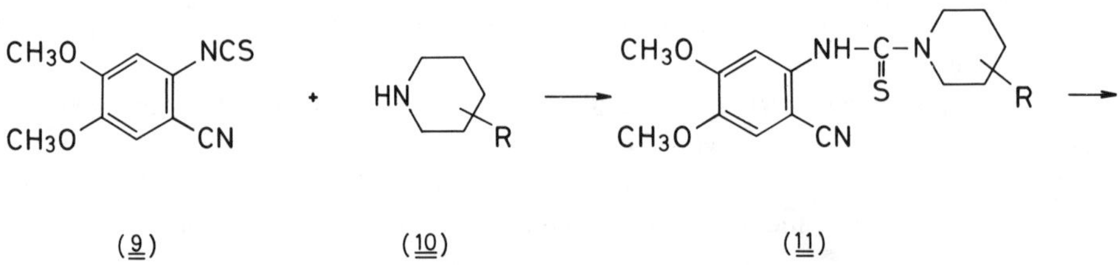

(9) (10) (11)

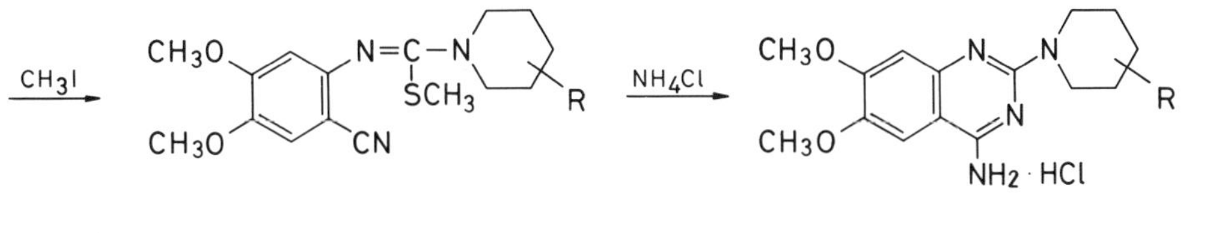

(12)

(13)

Interesting experiments were carried out by Giardina et al.,[277] who synthetized eight open-chain analogs of prazosin and the corresponding benzdioxane derivatives **(15-18)**. Almost all the components exhibited a marked selectivity towards α-adrenoreceptors, but to a much lesser extent than prazosin or WB 4101 **(14)**. There is only one exception. Compound **(15)** displays an outstanding selectivity toward α_1-receptors. Compounds **(15)** and **(16)** are homologs, but there is a difference of two orders of magnitude in their selectivities. Another difference is shown by compound **(17)**, which exhibits no selectivity.

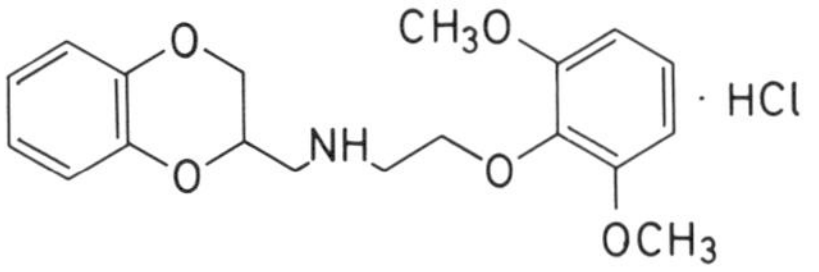

(14)

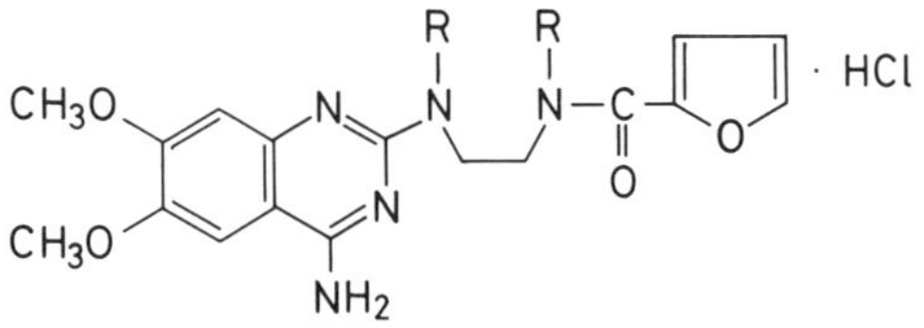

R = H = (15)

R = CH_3 = (16)

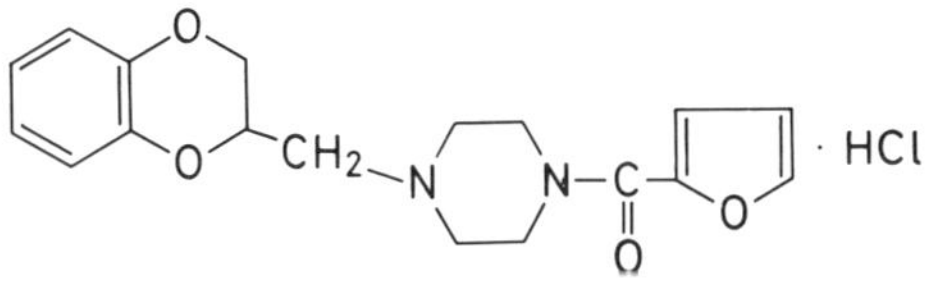

(17)

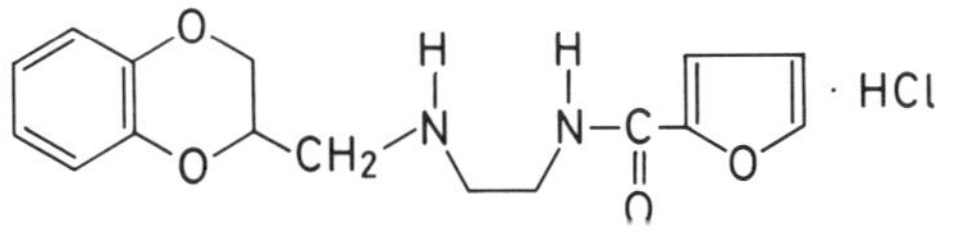

(18)

pA_2 is defined as the negative logarithm of that dose of antagonist which requires a doubling of the agonist dose to compensate for the action of the antagonist.[277]

It can be seen from the data in Table 13 that structural changes which markedly effect the affinity for the α_1-sites do not alter the α_2-sites. The results are very interesting from several respects. They could open up new ways towards the design of hypotensive drug compounds, further study of the nature α_1-α_2-adrenoreceptors and the structural differences between them.

D. ANALYSIS

HPLC is a paramount method in the detection, purity testing, and assay of prazosin. This is also true for the determination of prazosin in biological fluids (see Table 14). In the above case, the main differences between the methods are in the conditions of extraction, which are often lengthy (involving double extraction) and require 3 to 4 ml of plasma. The method described by Lin and coworkers[278] for the determination of prazosin in human plasma, whole blood, and urine is simple, rapid, and sensitive, without any extraction steps, and requires only 0.2 ml of biological sample. Additionally, this method is suitable for pharmacokinetic studies.

TABLE 11
Blood Pressure-Lowering Activity of 2-Amino-Substituted 6,7-Dimethoxy-4-Quinazolines

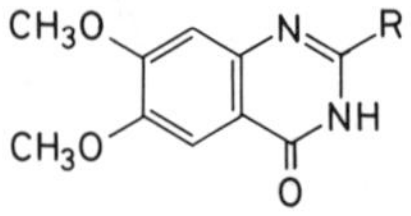

CH_3O, CH_3O, N, R, NH, O

R	Activity at mg/kg 2.5	10.0	40.0
NH_2	0	0	+ +
$N(CH_3)_2$	0	+ +	+ + +
$N(C_2H_5)_2$	+	+ + +	+ + +
$N(C_2H_5)$ $(CH_2{-}CH{=}CH_2)$			
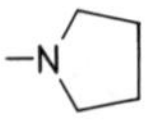 −N	0	+ +	
$N(CH_2CH{=}CH_2)_2$	0	+ +	+ + +
$N(CH_2CH_2CH_3)_2$			0
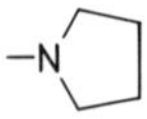 −N		0	+ +
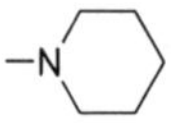 −N		0	+ +
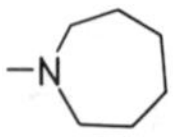 −N		0	+ +
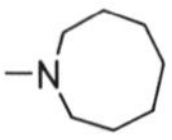 −N	0	+	+ + +
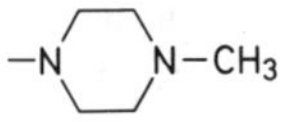 −N N−CH_3	0	+ +	+ + +
$NH{-}C_2H_5$			0
$NH{-}CH(CH_3)_2$	0	+	+ +
$NH{-}CH_2{-}C_6H_5$			0
$N(C_2H_5)_2$ (6,7-diethoxy derivative)	0	+	+ +
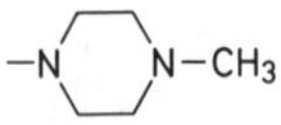 −N N−CH_3 (8-methoxy derivative)			0

TABLE 12
Hypotensive Activity[a] of Piperidinoquinazolines[271]

CH_3O N N R CH_3O N $NH_2 \cdot HCl$

R	Hypotensive[b] activity ED$_{30}$	ED$_{30}$ after 30 min
Prazosin	0.01	0.03
–CO–N	0.2	0.4
–CO–N	0.02	0.08
–CO–N	0.01	0.05
$-CO-(CH_2)_3CH_3$	0.03	0.06
–CO–	0.06	0.09
–CO–	0.07	0.1
$-CH_2-$	0.06	0.07
$-COO-CH_2-C(CH_3)_3$	0.03	2.00
$-CO-NH_2$	0.1	0.8

[a] Urethane-anaesthetized normotensive rats were used.
[b] ED$_{30}$ is the dose (in μmol/kg) that produced a 30% fall in mean blood pressure.

TABLE 13
Adrenergic Affinity and Selectivity of Antagonists with Prazosin-Related Structure

Antagonist	α_1 pA$_2$ (against norepinephrine)	α_2 pA$_2$ (against *clonidine*)	α_1 Selectivity (antilog α_1 pA$_2$ - α_2pA$_2$)
Prazosin	8.74	5.81	851
(14)	8.83	6.29	347
WB 4101			
(15)	8.81	5.22	3890
(16)	7.18	5.37	65
(17)	6.05	6.28	0.59
(18)	7.19	5.85	22

TABLE 14
HPLC Methods for the Investigation of Prazosin

Compound	Column	Eluent	Detection	Detection limit Linearity	Note	Ref.
Prazosin (P)	Porous silica	Methanol-water-glacial acetic acid (7:3:1) + diethylamine 0.2 ml	UV 254 nm	—	—	USP XXII
P + carbamazepine	μBondapack C_{18}	—	UV 254	0.1 μg/ml 1—164 μg/ml	In plasma + urine (after deproteination by adding acetonitrile)	
P	μBondapak pnenyl	1.5 m*M* H_3PO_4-Acetonitrile 77:23	Fluorescence exc: 246 nm emiss: 389 nm	0.2μg/ml	In plasma after treatment with $CHCl_3$	
P + 4-(4-amino-6,7,8-trimethoxy-2-quinazol-1-piperazine carboxylic acid 2-methallyl ester (I.St.)	μBondapak CN	Acetonitrile-water-acetic acid 50:47:3	Fluorescence exc: 246 nm emiss: 389 nm	0.2 μg/ml	In plasma after extraction with alkaline ethyl acetate	
P	LiChrosorb RP_{18}	MeOH-phosphate buffer 61:39	Fluorescence exc: 246 nm emiss: 389 nm	0.2 μg/ml	In plasma after extraction with alkaline ethyl acetate	
P	μBondapak CN	Acetonitrile-water-acetic acid 50:47:3	Fluorescence exc: 246 nm emiss: 389 nm	0.2—0.5 μg/ml 0.5—20 μg/ml	Extraction from alkaline plasma into ethylacetate (metabolites do not interfere)	

Another very simple method was used by Jee and coworkers[279] for sample preparation. Prazosin was then chromatographed as an ion-pair with pentanesulfonic acid. With the thin-layer chromatographic method official in USP, five potential impurities of prazosin can be detected:

I 1-(2-furoyl)-piperazine

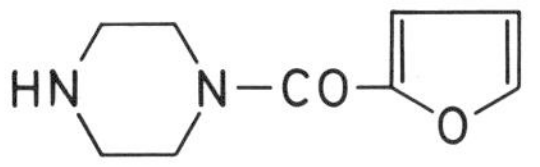

II 1,4-bis-(2-furoyl)-piperazine

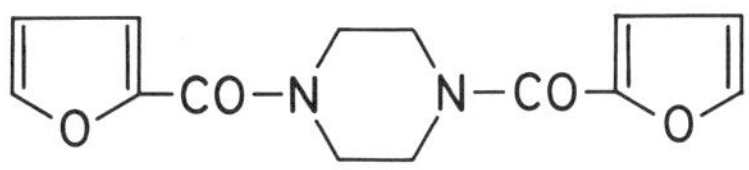

III 2-chloro-4-amino-6,7-dimethoxyquinazoline

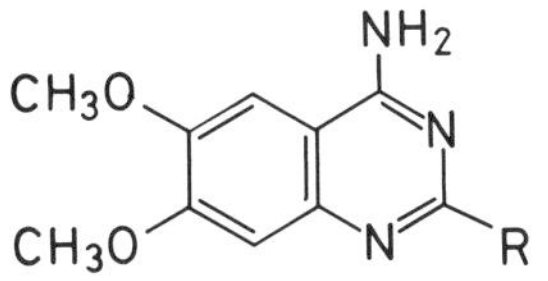

IV 4-amino-6,7-dimethoxy-2-(1-pipearazinyl)-quinazoline

R=—Cl

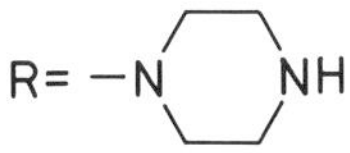

V 1,4-bix(4-amino-6,7-dimethoxy-2-quinazolinyl)-piperazine

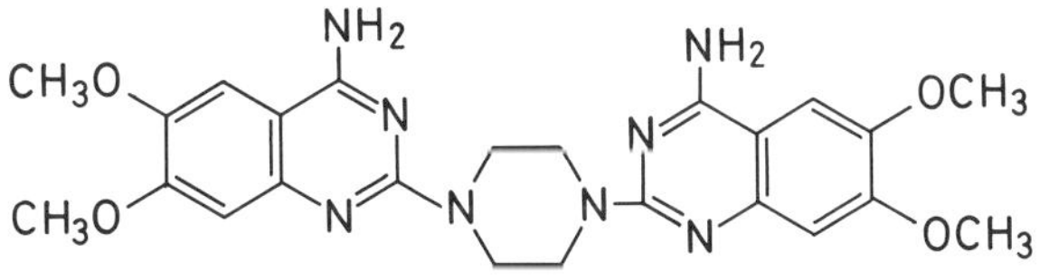

With silica gel as stationary phase, the following mobile phases are prescribed: toluene-ethylacetate-glacial acetic acid 10:10:1 (for III); ethyl acetate-glacial acetic acid-diethylamine, 40:20:3 (for IV); ethyl acetate-diethylamine, 19:1 (for V).

A much simpler way to solve the same analytical task is the HPLC method of Boehman,[285] by which all the five impurities of prazosin can be separated in one development.

Spectrophotometric and fluorometric assays of prazosin hydrochloride in tablet form have been described.[287] The absorbance is measured at 247 nm; Beer's law is obeyed in the interval 1 to 8 μg/ml. The fluorescence intensity was linear in the concentration range 1 to 16 μg/ml. A method for the automatic coulometric titration of prazosin hydrochloride has been published.[288] It is based on the precipitation of Cl^- by Ag^+ ions generated electrochemically at the electrode. The method provides a rapid and simple solution for the microdetermination of prazosin hydrochloride. End-point detection is by potentiometry.

VI. CLONIDINE

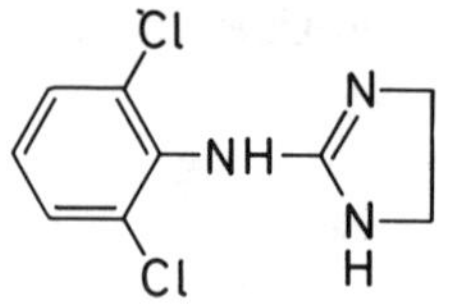

2-(2,6-Dichlorophenylamino)-imidazoline hydrochloride

$C_9H_9Cl_2N_3$ $M_r = 266.56$

$C_9H_9Cl_2N_3 \cdot HCl$ $M_r = 230.1$

A. PROPERTIES

Clonidine is a white crystalline powder that is highly soluble in water. Mp: 305°C, pK_a: 8.05 ± 0.04 (25°C), log P = 0.62 (octanol, pH 7.4).

Since both the free base and the protonated form of clonidine are present under physiological conditions, investigations have been performed to establish the molecular and conformational structures of clonidine. NMR spectroscopy has verified that the clonidine molecule exists predominantly in the imino form in solution,[289-291] and at the receptor site.[292]

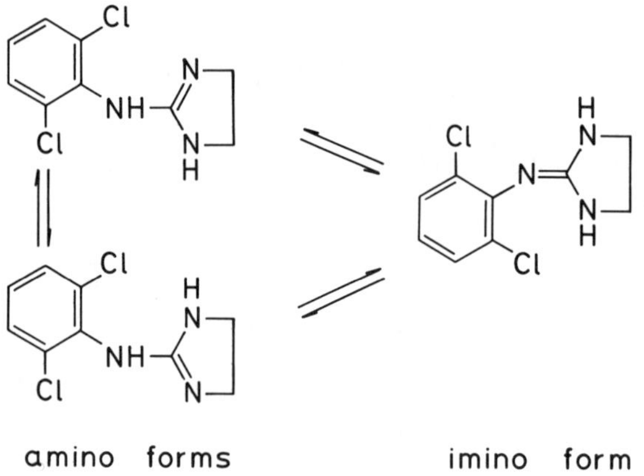

amino forms imino form

Both the UV and the IR spectra prove the presence of the exocyclic double bond, i.e., the imino form. The protonation takes place on the bridging nitrogen atom. In the protonated form of clonidine, the π-electrons of the double bond are delocalized, and the positive charge is dispersed over the three nitrogen atoms:

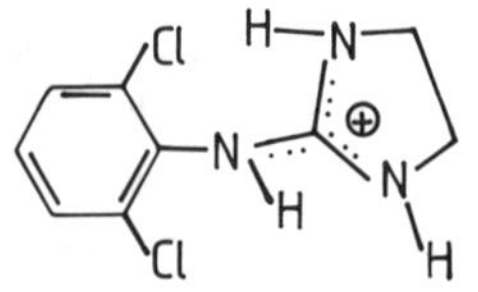

The absorption maxima of clonidine hydrochloride in methanolic solution are at 271 and 278 nm ($A^{1\%}_{1cm}$ = 15.6, 18.1, E = 420 and 480).

B. HISTORY AND STRUCTURE-ACTIVITY RELATIONSHIP

In 1939, Meier[293] reported on the sympatholytic effects of 2-benzylimidazoline (tolazoline), which had been synthetized by Hartmann and Isler.[294] It was shown that tolazoline produces a hypotensive effect by virtue of adrenolysis.

Systematic research on imidazolines led to the development of numerous differently substituted compounds which elicited significant changes in the pharmacological profile of the parent compound.

2-Anilinoimidazoline, produced by Hoefke,[295] had sympathomimetic effects similar to those of tolazoline. Later, the compound St-155 synthetized by Stähle (1962) proved to be a potent therapeutically useful hypotensive product; it was given the generic name clonidine.

In an excellent review, Timmermans et al.[296] gave a very detailed account of the structure-activity relationships in clonidine-like imidazolidines and related compounds.

The most important structural modifications in the molecule that involved change of the action may be summarized as follows.

The hypotensive activity was diminished

— when the 5-membered imidazolidine ring was enlarged into a 6, 7, or 8-membered one
— when the imidazolidine ring was replaced by an oxazolidine or thiazolidine moiety
— when the imidazolidine ring was opened
— when one or two methylene groups were omitted from the open ring
— when the bridge between the two rings was extended, if there was an imino (C=N) function in the ring
— when the 2,6-dichlorophenyl moiety was replaced by quinoxaline
— when the phenyl ring was unsubstituted or was substituted with more than three groups (substances monosubstituted in the *meta* or the *para* position are not effective)
— by alkyl substitution (methyl, ethyl, isopropyl)

The most active derivatives are the 2,6-disubstituted-phenyl-imidazolines. Halogen substitution: chlorine, bromine, CF_3, and fluorine in every combination give highly active derivatives, except for the *ortho-ortho* double fluorine compound.

Of the more than 70 different phenyl-substituted compounds examined, the following molecules had ED_{20} (mg/kg) values under 0.02[297] (Table 15).

In general, it can be said that as the lipophilicity of the molecules increases, the hypotensive activity also increases. As regards the role of the phenyl substituents, an increased lipophilicity and an influence on the elecron density of the aromatic nucleus may be assumed.

Many experiments have been carried out to synthetize new clonidine-like hypotensive compounds.

Numerous useful analogs which became medicines have been found in the imidazoline series. Besides the different halogenated analogs in different positions (see later) other substituted compounds were also active, as follows:

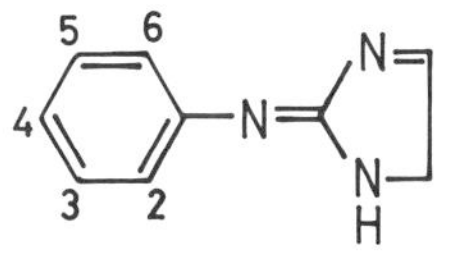

2-Cl, 4-CH_3 St 375, tolonidine Euctan®
2-CH_3,5-F St 600, flutonidine

The doses of these in humans are distinctly higher than the effective dose of clonidine.

TABLE 15
Antihypertensive Activity of Phenylsubstituted Clonidine-Analogs[297]

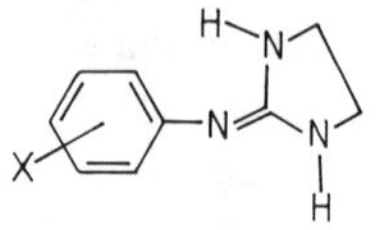

Compound X	ED_{20} mg/kg
2,3-disubstituted	
2,3-di-Br	0.006
2-Cl,3-Br	0.01
2-Me,3-Br	0.01
2-Br,3-Cl	0.015
2,4-disubstituted	
2Cl,4-Me	0.014
2,6-disubstituted	
2,6-di-Cl	0.011
2-Br, -F	0.01
2,4,5-trisubstituted	
2,5-di-Cl,4-Me	0.01
2,5-di-Cl,4-Br	0.015
2,4,6-trisubstituted	
2,6-di-Cl,4-Me	0.01

Replacement of the imidazoline residue by dihydrothiazines, oxazolines, and pyrrolines also yielded highly active analogs:[298]

Bay 1470
Xylazine
Rompune®

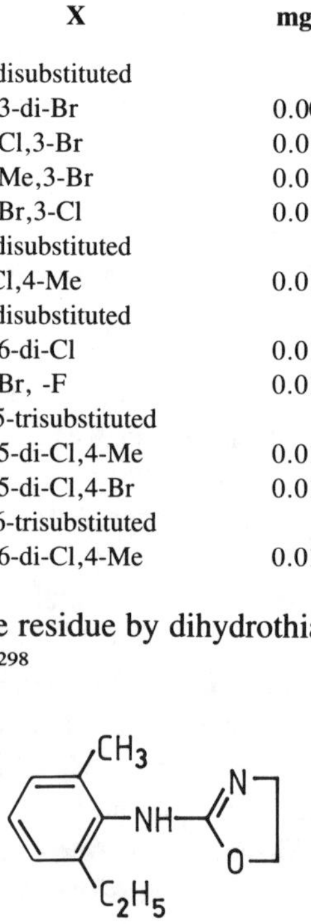

Bay Q 6781

351

Variation of the amino bridge between the phenyl and imidazoline rings brought an especially good result. Lofexidine is an outstanding example from among the compounds mentioned below:

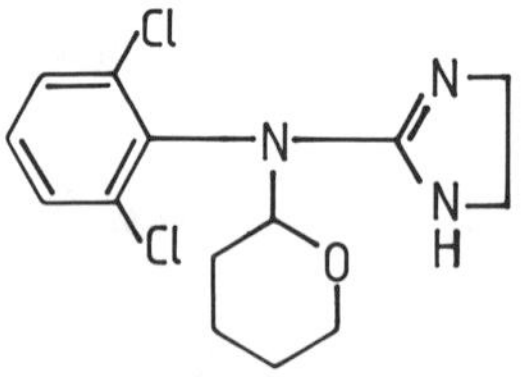

LR -99,853 (Lusofarmaco)

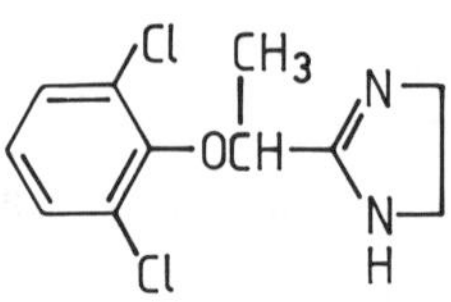

Lofexidine

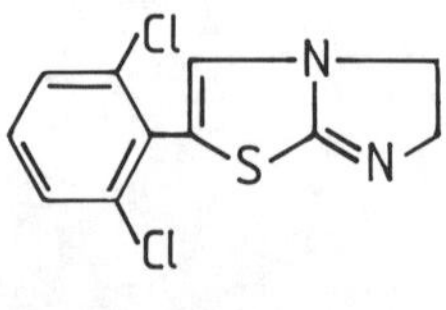

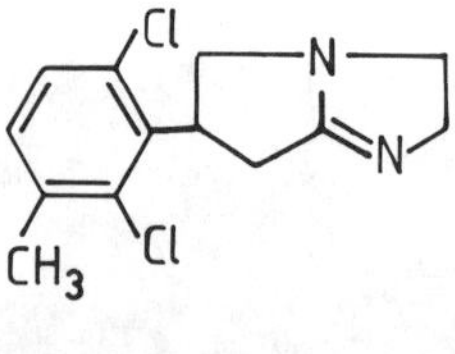

Lofexidine produces a hypotensive effect at oral doses of 0.1 to 0.6 mg.

Clonidine is an antihypertensive agent which causes a transient elevation in blood pressure and many side effects. The mechanism of action of clonidine has been discussed by numerous authors. The mechanism of its hypotensive action proceeds within the CNS due to its lipophilicity. It stimulates the peripheral α_2-adrenergic receptors, inhibiting the release of catecholamines, with resulting stimulation of the postsynaptic α-adrenoreceptors in the vasomotor centre of the brain, and hence the arterial blood pressure and cardiac frequency are diminished. Due to its central action, clonidine also suppresses the plasma renin activity level and acts directly on the kidney. The mechanism of clonidine action has been discussed by numerous authors.[299]

As regards the interaction of the pharmacon with the α-adrenoreceptor, it is very important to know the structural and conformational requirements. The interatomic distances in the calculated, preferred conformation of protonated clonidine have been reported by Timmermans et al.[300]

It was shown that the nitrogen atom is located at a distance of 1.2 Å above the plane of the aromatic moiety, and the distance of this nitrogen atom from the centre of the aromatic ring is 4.9 Å (Figure 3).

Both of these data are very similar to those for the catecholamines, in which the distances are 1.2 to 1.4 Å and 5.1 to 5.2 Å, respectively. In contrast to phenylalkylamines, clonidine does not possess an alcoholic hydroxy group that can participate in hydrogen bond formation. It may be that the bridging NH group occupies this role, or it is also possible that hydrogen bond formation is not necessary for α-adrenoreceptor stimulation.[296]

C. SYNTHESIS

There are two main procedures for the synthesis of clonidine.

In one of them, the parent compound is built up in one step by condensation of the separate ring systems:[301]

$$\text{X-C}_6\text{H}_4\text{-NH}_2 + \text{Y-(2-imidazolinyl)} \longrightarrow \text{X-C}_6\text{H}_4\text{-N=(2-imidazolidinylidene)} + \text{HY}$$

Most frequently, the condensation takes place between stepwise prepared suitable precursors (**1**) and ethylenediamine:[302]

$$\underset{(1)}{\text{X-C}_6\text{H}_4\text{-N(Y)(Z)}} + \text{H}_2\text{N-CH}_2\text{CH}_2\text{-NH}_2 \longrightarrow \text{X-C}_6\text{H}_4\text{-N=(2-imidazolidinylidene)} + \text{HY} + \text{HZ}$$

D. ABSORPTION AND METABOLISM

Orally administered clonidine is almost completely absorbed. About 60% of the drug is excreted unchanged in the urine. The predominant routes of excretion are the urinary and biliary ones. The main metabolites are *p*-hydroxychlonidine and 2,6-dichlorophenylguanidine.

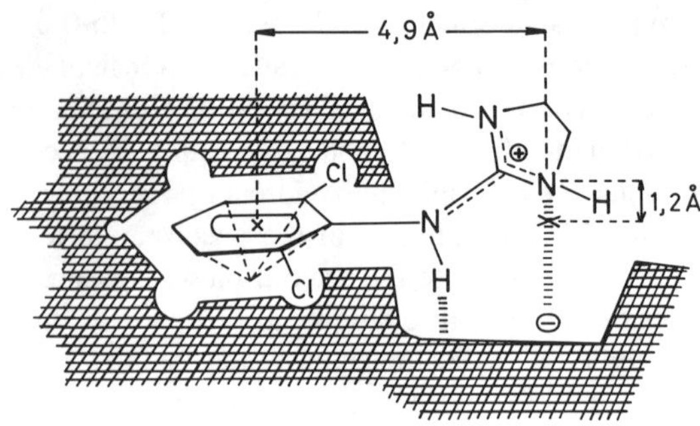

FIGURE 3. Scheme of clonidine-receptor interaction. (From Timmermens, P. B. et al., *Prog. Pharmacol.*, 3, 21, 1980. With permission.)

E. ANALYSIS

Clonidine is official (USP) as its hydrochloride salt. In USP XXII, clonidine hydrochloride is identified through its IR and UV absorption spectra, with concomitantly measured clonidine hydrochloride reference standard.

Similarly, as for tolazoline and related compounds, clonidine is not stable in hot alkaline solution because of hydrolysis:[303]

Lofexidine

USP prescribes thin-layer chromatography on silica gel for the detection of impurities in clonidine. The mobile phase consists of toluene, dioxane, anhydrous alcohol, and ammonium hydroxide (10:8:2:1). The spray reagents are sodium hypochlorite solution and, after drying, starch-potassium iodide test solution.

There are numerous gas-chromatographic methods for the determination of clonidine in plasma and urine. Most often, clonidine is converted into its methyl,[304,305] pentafluorobenzyl,[306-308] heptafluorobutyryl,[309] or 3,5-bis(trifluoromethyl)-benzyl[310] derivatives to alkylate or to acylate the imidazoline N-atoms. The substrate is then subjected to GC and chromatographed, generally on a fused silica column coated with 5% phenylmethylsilicone, or with OV-275 or OV-17. Chromatography on fused-silica capillary columns affords more rapid analysis, better resolution, and greater inertness, as compared with the parameters on the conventional GC methods.[311,312] It is usually operated with N_2 or H_2 or He as carrier gas, with temperature programming between 105 and 330°C.

Suitable detection is provided by electron capture, but when the technique is combined with MS detection, with selected ion monitoring, an even better performance can be attained, particularly for some biomedical applications at the lower detection limits. Clonidine can be determined down to 0.1 pg, with a precision of 3.45% at the level of 1.25 μg ml^{-1}.[306]

For the analysis of different kinds of dosage forms containing clonidine, the HPLC methods on a reversed phase are proposed, with a slightly alkaline eluent, e.g., methanol-water 4:1 containing 0.005% triethylamine,[313] or a buffer solution of pH = 7.9-methanol 7:13[314] has been applied. The calibration plot is linear over the range 2.15 to 44.8 μg/ml.[313]

Spectrophotometric determination was outlined by El Jarli et al.[315] for clonidine base in acetonitrile medium after the addition of 2,3-dichloro-5,6-dicyano-1,4-benzoquinone as reagent; the absorbance was read at 455 nm.

In tablets, clonidine is determined via the reaction with sodium nitroprusside in alkaline (NaOH) medium. The mixture is then treated with aqueous boric acid solution to stabilize the color, and the absorbance is measured at 570 nm. Beer's law is obeyed for solutions containing 200 to 1500 μg/10 ml.[316]

A fluorometric method involving the application of dansyl chloride in acetone-isobutyl methyl ketone solution has been developed. The excitation wavelength is at 345 nm, with measurement at 445 nm.[315]

A precise radioimmunoassay for clonidine in plasma was carried out by raising antiserum in rabbits with the *p*-carboxy derivative of clonidine. The tracer was ^{3}H clonidine. The calibration plot was linear from 0.1 to 10 ng ml^{-1} clonidine.[317] The detection limit of clonidine was 10 pg/ml when ^{125}I-labelled 4-carboxyclonidinetyrosine was used as tracer in radioimmunoassay.[318] A similar RIA method has been devised with the application of an azo derivative of clonidine. The antigen is prepared by reacting 4-hydroxyclonidine with 4-carboxybenzenediazonium chloride. The reaction product was injected into rabbits.[319]

REFERENCES

1. **Geneidi, A. S., Ali, A. A., and Salamo, R. B.,** *Egypt. J. Pharm. Sci.,* 20, 429, 1979.
2. **Roth H. J., Eger, K., Troschütz, R.,** *Arzneistoffanalyse,* Georg Thieme Verlag, Stuttgart, 1981.
3. **Lounasmaa, M., Tolvanen, A., and Kan, S. K.,** *Heterocycles,* 23, 371, 1985.
4. **Schiffl, E. and Pindur, U.,** *Arch. Pharm.,* 319, 443, 1986.
5. **Furlemeier, A.,** *Experientia (Basel),* 9, 331, 1953.
6. **Dorfman, L.,** *Helv. Chim. Acta,* 37, 59, 1954.
7. **Pötter, H. and Voigt, R.,** *Pharmazie,* 22, 436, 1967.
8. **Ahmad, I., Khan, M. A., Usmanghani, K., and Salam, T.,** *Pharmazie,* 34, 402, 1979.
9. **Coffman, H. D., Crabbs, W. C., Kolinski, R. E., and Page, D. P.,** *Am. J. Hosp. Pharm.,* 43, 103, 1986.
10. **Savory, B. and Turnbull, J. H.,** *J. Photochem.,* 24, 355, 1984.
11. **Pfeifer, S., Behnsen, G., and Kuehn, L.,** *Pharmazie,* 27, 639, 1972.
12. **Ljunberg, S.,** *J. Pharm. Belg.,* 14, 115, 1959.
13. **Ljunberg, S.,** *Sven. Farm. Tidskr.,* 62, 693, 1958.
14. **Indemanns, A. W. M.,** *Pharm. Weekbl.,* 94, 1, 1959.
15. **Wright, G. E. and Tang, T. Y.,** *J. Pharm. Sci.,* 61, 299, 1972.
16. **Sanchez, M., Diaz, T., Carmona, M. C., Balon, M., and Hidalgo, J.,** *Oxid. Commun.,* 9, 17, 1986.
17. **Carmona-Guzman, M. C., Balon, M., Gonzalez-Arjona, D., Makaver, J., and Sanchez, M.,** *J. Chem. Soc. Perkin Trans.,* 2, 409, 1986.
18. **Sato, K., Matsumosi, H., and Karama, S.,** *Jpn. Kokai (Tokyo),* 7899, 316, 1978.
19. **Bayer, J.,** *Pharmazie,* 13, 468, 1958.
20. **Bayer, J. and Szporny, L.,** *Magy. Kem. Foly.,* 65, 24, 1959.
21. **Banes, D.,** *J. Am. Pharm. Assoc. Sci.,* Ed., 45, 708, 1956.,
22. **Asker, A. F., Helal, M. A., and Motawi, M. M.,** *Pharmazie,* 28, 526, 1973.
23. **Suzuki, J.,** *Toho Igakkai Zasshi,* 28, 260, 1981.
24. **Stitzel, R. E.,** *Pharmacol. Rev.,* 28, 179, 1976.
25. **Goldenberg, H., Fischman, V.,** in *Metabolism of Psychotropic Agents,* Clark, W. G., Ed., Academic Press, New York, 1970, 179.
26. **Green, A. F.,** *Adv. Pharmacol.,* 1, 161, 1962.
27. **Schittler, E. and Bein, H. J.,** in *Antihypertensive Agents,* Schittler, E., Ed., Academic Press, New York, 1967, 191.
28. **Bender, A. D.,** in *Topics in Medicinal Chemistry,* Rabinowitz, J. L. and Myerson, R. M., Eds., Wiley-Interscience, New York, 1967, 177.
29. **Dhawan, B. N. and Sharma, J. N.,** *Psychopharmacology,* 2, 1101, 1977.
30. **Deligue, P. and Buonodiese, M.,** *Ann. Anesthesiol. Fr.,* 19, 14, 1978.

31. **Kobinger, W.,** *Discoveries Pharmacol.*, 2, 107, 1984.
32. **Finnerty, F. A., Gyftopoulos, A., Berry, O., and MacKenney, A.,** *JAMA*, 241, 579, 1979.
33. **Goodman Gilman, A., Goodman, L. S., Gilman, A., Eds.,** *The Pharmacological Basis of Therapeutics*, 6th ed., Macmillan, New York, 1980, 204.
34. **Kaplan, N. M.,** Systemic hypertension: therapy, in *Heart Disease, A Textbook of Cardiovascular Medicine*, 2nd ed., Braunwald, E., Ed., W. B. Saunders, Philadelphia, 1984, 915.
35. **Muradyan, R. E.,** *Vopr. Onkol.*, 32, 76, 1986.
36. **Lucas, R. A.,** *Prog. Med. Chem.*, 3, 146, 1963.
37. **Klohs, M. W., Draper, M. D., and Keller, F.,** *J. Amer. Chem. Soc.*, 77, 2241, 1955.
38. **Ehrhart, G. and Ruschig, H.,** Arzneimittel: *Entwicklung, Wirkung, Darstellung,* Verlag Chemie, Weinheim, 1972, 194.
39. **Cronheim, G. E., Brown, W., Cawthorne, J., Toekes, M. J., and Ungari, J.,** *Proc. Soc. Exp. Biol. Med.*, 80, 120, 1954.
40. **Ulshafer, P. R., Taylor, W. J., and Nugent, R. H.,** *C. R. Hebd. Seances Acad. Sci.*, 244, 2989, 1957.
41. **Treka, V., Dlabac, A., and Vanecek, M.,** *Life Sci.*, 4, 2257, 1965.
42. **Treka, V., Carlsson, A.,** *Life Sci.*, 4, 2263, 1965.
43. **Siddiqui, S. and Siddiqui, R. H.,** *J. Indian Chem. Soc.*, 8, 667, 1931.
44. **Klohs, M. W., Draper, M., Keller, F., Malesh, W., and Petracek, F.,** *J. Am Chem. Soc.*, 76, 1332, 1954.
45. **Schlittler, E. and Schwarz, H.,** *Helv. Chim. Acta.*, 33, 1463, 1950.
46. **Bader, F. and Schwarz, F.,** *Helv. Chim. Acta* 35, 1594, 1952.
47. **Thomae, F.,** DBP, 966, 024, 1954.
48. **Plummer, A. J., Barrett, W. E., and Rutledge, R. A.,** *Fed. Proc.*, 13, 395, 1954.
49. **Szmuszkovicz, J.,** U. S. Patent 2, 763, 654, 1956.
50. **Szmuszkovicz, J.,** U. S. Patent 2, 763, 655, 1956.
51. **Barbour, B., Irwin, G., Yamahiro, H., Frasher, W., and Maronde, R. F.,** *Am. J. Cardiol.*, 3, 220, 1959.
52. **Ciba,** U.S. Patent 2789, 112, 1955.
53. **Berthaux, P. and Neuman, M.,** *Arzneim. Forsch.*, 14, 1040, 1964.
54. **van Itallie, L. and Steenhauer, A. J.,** *Arch. Pharm. Ber. Dtsch. Pharm. Ges.*, 270, 313, 1932.
55. **Vakil, R. J.,** *Br. Heart J.*, 11, 350, 1949.
56. **Mueller, J. M., Schlittler, E., and Bein, H. J.,** *Experientia*, 8, 338, 1952.
57. **Bein, J.,** *Experientia*, 9, 107, 1953.
58. **Woodward, R. B., Bader, F. E., Bickel, H., Frey, A. J., and Kiersted, R. W.,** *J. Am. Chem. Soc.*, 78, 2023, 1956.
59. **Polniaszek, R. P.,** *Diss. Abstr. Int.* B., 44, 1460, 1983.
60. **White, A. W.,** *Diss. Abstr. Int. B.*, 42, 1901, 1981.
61. **Stahl, E.,** *Dünnschichtchromatographie,* 2. Aufl., Springer-Verlag, Berlin, 1967, 429.
62. **Devred, Y. P., Parmentier, G., and Parmentier, J.,** in Symp. Chromatogr. Electrophor. Lect. Pap., 6th, 1970 (Publ. 1971).
63. **Le Xuan, T., Munier, R. L., and Menuier, S.,** *Chromatographia*, 13, 693, 1980.
64. **Court, W. E. and Habib, M. S.,** *J. Chromatogr.*, 80, 101, 1973.
65. **Court, W. E. and Timmins, P.,** *Planta Med.* 27, 319, 1975.
66. **Court, W. E. and Iwu, M. M.,** *J. Chromatogr.*, 187, 199, 1980.
67. **Court, W. E.,** *Can. J. Med. Pharm. Sci.*, 1, 76, 1966.
68. **Court, W. E.,** *Can. J. Pharm. Sci.*, 3, 70, 1968.
69. **Habib, M. S., and Court, W. E.,** *J. Pharm. Pharmacol.*, 23, Suppl., 230, 1971.
70. **Amato, A., Gagliardi, L., Misto, R., Iafrati, C., Chiavarelli, S., and Porra, R.,** *Riv. Tossicol. Sper. Clin.*, 11, 113, 1981.
71. **Cavazzutti, G., Gagliardi, L., Amato, A., Profili, M., Zagarese, V., Tonelli, D., and Gattavecchia, E.,** *J. Chromatogr.*, 268, 528, 1983.
72. **Stohs, S. J. and Scratchley, G. A.,** *J. Chromatogr.*, 114, 329, 1975.
73. **Mancini, M. A. D., Longo, A., and Hanai, L. W.,** *Rev. Cienc. Farm. Bras.*, 7, 133, 1985.
74. **Huesmann, G.,** *PTA Prakt. Pharm.*, 7, 332, 1978.
75. **Sams, R. A. and Huffman, R.,** *J. Chromatogr.*, 161, 410, 1978.
76. **Tripp, S. L., Williams, E., Wagner, W. E., Jr., and Lucas, G.,** *Life Sci.*, 16, 1167, 1975.
77. **Frijus, J. M.,** *Pharm. Weekbl.*, 106, 605, 1971.
78. **Heusser, D.,** *Planta Med.*, 12, 237, 1964.
79. **Wung, J. and Bonakdar, M.,** *J. Chromatogr. Biomed. Appl.*, 382, 343, 1986.
80. **Owen, J. A., Nakatsu, S. L., Condra, M., Surridge, D. H., Fenemore, J., and Morales, A.,** *J. Chromatogr. Biomed. Appl.*, 342, 333, 1985.
81. **Prasad, T. N. V., Rao, E. V., Sastry, C. S. P., and Rao, G. R.,** *Indian Drugs*, 24, 398, 1987.
82. **Cieri, U. R.,** *J. Assoc. Off. Anal. Chem.*, 70, 540, 1987.

83. **Zhang, S., Du, H., and An, D.,** *Zhonguo Yaoke Daxne Xuabao,* 18, 12, 1987.
84. **Vincent, A. and Awang, D. V. C.,** *J. Liq. Chrom.,* 4, 1651, 1981.
85. **Krugers, D., Klein, E. J. T., and Knuif, A. J.,** *Ziekenhuis Farmacia,* 1, 7, 1985.
86. **Honigberg, I. L., Stewart, J. T., Smith, A. P., Plunkett, R. D., and Hester, D. W.,** *J. Pharm. Sci.,* 63, 1762, 1974.
87. **Honigberg, I. L., Stewart, J. T., Smith, A. P., and Hester, D. W.,** *J. Pharm. Sci.,* 64, 1201, 1975.
88. **Lang, J. R., Stewart, J. T., and Honigberg, I. L.,** *J. Chromatogr.,* 264, 144, 1983.
89. **Lang, J. R., Honigberg, I. L., and Stewart, J. T.,** *J. Chromatogr.,* 252, 288, 1982.
90. **Cieri, U. R.,** *J. Assoc. Off. Anal. Chem.,* 68, 542, 1985.
91. **Cieri, U. R.,** *J. Assoc. Off. Anal. Chem.,* 66, 867, 1983.
92. **Suckow, R. F., Cooper, T. B., and Asnis, G. M.,** *J. Liq. Chrom.,* 6, 1111, 1983.
93. **Gfeller, J. C., Haas, R., Troendlé, J. M., and Erni, F.,** *J. Chromatogr.,* 294, 247, 1984.
94. **Settimj, G., Di Simone, L., and Del Guidice, M. R.,** *J. Chromatogr.,* 116, 263, 1976.
95. **Khayyal, S. E., Ayad, M. M., and Girgis, A. N.,** *J. Chromatogr.,* 285, 495, 1984.
96. **Khayyal, S. E., Ayad, M. M., and Girgis, A. N.,** *J. Drug Res.,* 15, 43, 1984.
97. **Pantarotto, C., Belvedere, G., Frigerio, A., Mennini, T., and Manara, L.,** *Eur. J. Drug Metab. Pharmacokinet.,* 1, 25, 1976.
98. **Horning, E. C., Carroll, D. J., Horning, M. G., Nowlin, J. G., and Stillwell, R. N.,** Proc. Philip Morris Sci. Symp., 1981 (Publ. 1982).
99. **Becker, O., Fuerstenau, N., Knippelberg, W., and Kruger, F. R.,** *Org. Mass Spectrom.,* 12, 461, 1977.
100. **Levin, R. H., Lallemand, J. Y., and Roberts, J. D.,** *J. Org. Chem.,* 38, 1983, 1973.
101. **Almeida, M. and Guzman, C. C.,** *J. Pharm. Biomed. Anal.,* 6, 185, 1988.
102. **Avang, D. V. C., Dawson, B. A., Neville, G. A., Ekiel, I., Deslauriers, R., and Smith, I. C. P.,** *Can. J. Spectrosc.,* 32, 86, 1987.
103. **Perchalski, R. I., Jost, R. A., and Wilder, B. Y.,** *Anal. Chem.,* 55, 2002, 1983.
104. **Haycock. R. P., Sheth, P. B., and Mader, W. J.,** *J. Am. Pharm. Assoc. Sci. Ed.,* 48, 479, 1959.
105. **Undenfriend, S., Duggan, D. E., Vasta, B. M., and Brodie, B. B.,** *J. Pharmacol. Exp. Ther.,* 120, 26, 1957.
106. **Dechene, E. B.,** *J. Am. Pharm. Assoc. Sci. Ed.,* 44, 657, 1955.
107. **Glazka, A. J., Dell, W. A., Wolf, L. H., and Kazenko, A.,** *J. Pharm. Exp. Ther.,* 118, 337, 1955.
108. **Gürkan, T.,** *Mikrochim Acta (Wien),* 1, 173, 1976.
109. **Welman, E., Curry, P. V. L., Krikler, D. M., Rowlands, E., and Callowhill, E. A.,** *Br. J. Clin. Pharmacol.,* 4, 549, 1977.
110. **Balon-Almeida, M., Munoz-Perez, M. A., and Hidalgo-Toledo, J.,** *J. Pharm. Biomed. Anal.,* 4, 505, 1986.
111. **Balon-Almeida, M., Munoz-Perez, M. A., Hidalgo-Toledo, J., Carmona, M. C., and Sanchez, M.,** *J. Photochem.,* 36, 193, 1987.
112. **Hidalgo-Toledo, J., Gonzalez, A. D., Roldan, E., and Sanchez, M.,** *J. Mol. Struct.,* 143, 501, 1986.
113. **Savory, B. and Turnbull, J. H.,** *J. Photochem.,* 23, 171, 1983.
114. **Hidalgo-Toledo, J., Tejeda, P. P., Munoz, M. A., Maestre, A., Balon-Almeida, M., and Sanchez, M.,** *Pharm. Acta Helv.,* 61, 89, 1986.
115. **Indemans, A. W. M.,** *Pharm. Weekbl.,* 108, 641, 1973.
116. **Indemans, A. W. M.,** *Pharm. Weekbl.,* 108, 785, 1973.
117. **Kreienbaum, M. A.,** *J. Assoc. Off. Anal. Chem.,* 59, 289, 1976.
118. **Urbanyi, T. and O'Connell, A. W.,** Adv. Autom. Anal. Technicon Ind. Congr., 1972.
119. **Urbanyi, T. and O'Connell, A. W.,** *Anal. Chem.,* 44, 565, 1972.
120. **Urbanyi, T. and Stober, H.,** *J. Assoc. Off. Anal. Chem.,* 55, 180, 1972.
121. **Kabadi, B. N.,** *J. Pharm. Sci.,* 60, 1862, 1971.
122. **Jung, Z., Jungová, M.,** *Cesk. Farm.,* 22, 195, 1973.
123. **Pindur, U.,** *Arch. Pharm.,* 312, 270, 1979.
124. **Pindur, U., and Schiffl, E.,** *Pharm. Acta Helv.,* 60, 48, 1985.
125. **Grabowska, I., Marcinkowska, K., and Wodkiewicz, J.,** *Farm. Pol.,* 31, 825, 1975.
126. **Stainier, R.,** *J. Pharm. Belg.,* 28, 115, 1973.
127. **Koval'chuk, T. V., Kagan, F. E., Koget, T., and Arzynaeva, E. A.,** *Otkrytiya Izobret.,* 159, 1986; *Farm. Zh. (Kiev),* 67, 5, 1986.
128. **Ruiz, R. E., Alvarez, S. M., and Consuegra, M. S.,** *Rev. Cubana Farm.,* 7, 3, 1973.
129. **Fujita, V., Mori, I., and Kitano, S.,** *Bunseki Kagaku,* 33, E195, 1984.
130. **Karawaya, M. S., Sharaf, A. A., and Diab, A. M.,** *J. Assoc. Off. Anal. Chem.,* 59, 795, 1976.
131. **Dryanowska, L. and Iovchev, I.,** *Pharmazie,* 31, 130, 1976.
132. **Dembinski, B. and Zavadzki, H.,** *Chem. Anal. (Warsaw),* 31, 437, 1986.
133. **Wang, J., Tapia, T., and Bonakdar, M.,** *Analyst (London),* 111, 1245, 1986.

134. **Martens, S. J., Hünther, K., and Schickedanz, M.,** *Arch. Pharm.*, 319, 572, 1986.
135. **Torchiana, M. L., Lotti, V. J., Clark, C. M., and Stone, C. A.,** Arch. Int. Pharmacodyn., 205, 103, 1973.
136. **Cohen, Y., Wepierre, J., Jacquot, C., Rapin, R.,** *Arch. Int. Pharmacodyn.*, 207, 348, 1974.
137. **Waldmeier, P.., Hedwall, P. R., and Mairtre, L.,** *Naunyn-Schmiedebergs Arch. Pharmacol.*, 289, 303, 1975.
138. **Nijkamp, F. P., Ezer, J., and De Jong, W.,** *Eur. J. Pharmacol.*, 31, 243, 1975.
139. **Scriabine, A., Ludden, C. T., Stone, C. A., Wartman, R. J., and Watkins, C. J.,** *Clin. Sci. Mol. Med.*, 51(Suppl.3), 407, 1976.
140. **Finch, L., Hersom, A., and Hicks, P.,** *Br. J. Pharmacol.*, 54, 445, 1975.
141. **Henning, M. and van Zwieten, P. A.,** *Acta Pharmacol. Toxicol.*, 25(Suppl. 4), 25, 1967.
142. **Henning, M.,** *Br. J. Pharmacol.*, 34, 233, 1968.
143. **Henning, M. and van Zwieten, P. A.,** *J. Pharm. Pharmacol.*, 20, 409, 1968.
144. **Henning, M.,** *Acta Physiol. Scand.*, A166 (Suppl.322), 1, 1969.
145. **Scriabine, A.,** Methyldopa, in *Pharmacology of Antihypertensive Drugs,* Raven Press, New York, 1980.
146. **Head, G. and de Yong, W.,** *Clin. Exp. Hypertens.*, Part A, A6, 1984, 2051.
147. **Goldberg, M. R., Tung, Ch. S., Feldman, R. D., Smith, H. E., Dates, J. A., and Robertson, D.,** *J. Pharmacol. Exp. Ther.*, 220, 532, 1982.
148. **Ayitey-Smith, E., and Varma, D. R.,** *Br. J. Pharmacol.*, 40, 186, 1970.
149. **Gaffney, T. E., Sigell, L. T., Mohammed, S., and Atkinson, A. J., Jr.,** *Prog. Cardiovasc. Dis.*, 12, 52, 1969.
150. **Sjoerdsma, A.,** *J. Clin. Pharmacol.*, 13, 45, 1982.
151. **Winn, M., Rasmussen, R., Minard, F., Kynel, J., and Plotnikoff, N.,** *J. Med. Chem.*, 18, 434, 1975.
152. **Zavisca, F. G., Breau, A. P., and Wurtman, R. J.,** *Circulation Res.*, 45, 684, 1979.
153. **Counsell, R. E., Desai, P., Ide, A., and Kulkarni, P. G.,** *J. Med. Chem.*, 14, 789, 1971.
154. **Atkinson, M., Hartley, D., Lunts, L. H. C., and Ritchie, A. C.,** *J. Med. Chem.*, 17, 248, 1974.
155. **Collin, D. T., Hartley, D., Jack, D., Lunts, L. H. C., Press, J. C., Ritchie, A. C., and Toon, P.,** *J. Med. Chem.*, 13, 674, 1970.
156. **Schnettler, R. A., Suh, J. T., and Dage, R. C.,** *J. Med. Chem.*, 19, 191, 1976.
157. **Carter, H. E.,** *Org. React.*, 3, 198, 1946.
158. **Roebel, L. E., Lucas, R. W., Cheng, H. C., and Woodward, J. K.,** *Fed. Proc.*, 39, 1189, (Abstr. 4843), 1980.
159. **Fozard, J. R., Spedding, M., Palfreyman, M. J., Wagner, J., Möhring, J., and Koch-Weser, J.,** *J. Cardiovasc. Pharmacol.*, 2, 229, 1980.
160. **Castagnoli, N., Musson, D., Karashima, D., and Melmon, K.,** Proc. 4th Int. Catecholamine Symp., 1979, 1500.
161. **Taylor, J. B., Lewis, J. W., and Jacklin, M.,** *J. Med. Chem.*, 13, 1226, 1970.
162. **Kew, M. C. and First, R. G.,** *Curr. Ther. Res. Clin. Exp.*, 14, 343, 1972.
163. **Redmond, D. E., Jr., Olander, R., Maas, J. W.,** *Toxicol. Appl. Pharmacol.*, 34, 301, 1975.
164. **Palfreyman, M. G., Danzin, C., Bey, P., Jung, M. J., Ribereau-Gayon, G., Aubry, M., Vevert, J. P., and Sjoerdsma, A.,** *J. Neurochem.*, 31, 927, 1978.
165. **Jung, M. J., Palfreyman, M. G., Wagner, J., Bey, P., Ribereau-Gayon, G., Zraika, M., and Koch-Weser, J.,** *Life Sci.*, 24, 1037, 1979.
166. **Ulm, E. H., Swett, C. S., Duggan, D. E., and Minsker, D. H.,** *Fed. Proc.*, 38, 422 (Abstr. 1021), 1979.
167. **Maycock, A. L., Aster, S. D., and Patchett, A. A.,** *Biochemistry,* 19, 709, 1980.
168. **Bhatkar, R. G. and Almeida, L.,** *Indian Drugs,* 22, 34, 1984.
169. **Rama, R. G. and Reghuveer, S.,** *East Pharm.*, 25, 113, 1982.
170. **Sastry, C. S. P. and Rao, K. E.,** *Indian J. Pharm. Sci.*, 45, 113, 1983.
171. **Shingbal, D. M. and Agni, R. M.,** *Indian Drugs,* 20, 237, 1983.
172. **Emmanuel, J. and Shetty, A. R.,** *Indian Drugs,* 21, 393, 1984.
173. **Shukla, S., Pathak, V. N., and Shukla, I. C.,** *J. Inst. Chem. (India),* 57, 115, 1985.
174. **Wallash, M. I., Abou, O. A., and Salem, F. B.,** *Indian J. Pharm. Sci.*, 45, 223, 1983.
175. **Salem, F. B., and Walash, M. I.,** *Analyst,* 110, 1125, 1985.
176. **Wallash, M. I.,** *J. Assoc. Off. Anal. Chem.*, 68, 91, 1985.
177. **Fujita, Y., Mori, I., Fujita, K., Kitano, S. H., and Tanaka, T.,** *Chem. Pharm. Bull.*, 33, 5385, 1985.
178. **Korany, M. A., Wahbi, A. M., and Abdel-Hay, M. H.,** *J. Pharm. Biomed. Anal.*, 2, 537, 1984.
179. **Wallash, M. I., Ouf, A. A., and Salem, F. B.,** *J. Assoc. Off. Anal. Chem.*, 65, 1445, 1982.
180. **Rao, G. R., Raghuveer, S., and Mohan, K. R.,** *Indian Drugs,* 19, 328, 1982.
181. **Ting, S.,** *J. Assoc. Off. Anal. Chem.*, 66, 1436, 1983.
182. **Ting, S.,** *J. Assoc. Off. Anal. Chem.*, 67, 1118, 1984.
183. **Hoskins, J. A. and Holliday, S. B.,** *J. Chromatogr.*, 230, 162, 1982.

184. **Ong, H., Sved, S., and Beaudoin, N.,** *J. Chromatogr.*, 229, 433, 1982.
185. **Dilger, C., Salama, Z., and Jaeger, H.,** *Arzneim. Forsch.*, 37, 1399, 1987.
186. **Bartlett, W. A.,** *J. Liq. Chromatogr.*, 8, 719, 1985.
187. **Pathak, V. N., Shukla, S. R., and Shukla, I. C.,** Analyst (London), 10, 1086, 1982.
188. **Srinivasan, V. S., Povsic, T. J., and Huntington, J. L.,** *Am. Lab. (Fairfield, CT)*, 15, 57, 1983.
189. **Athanasiou-Malaki, E. M. and Koupparis, M. A.,** *Anal. Chim. Acta*, 161, 349, 1984.
190. **Bishop, E. and Hussein, W.,** *Analyst (London)*, 109, 623, 1984.
191. **Abd el Wahed, M. G. and Ayad, M.,** *Anal. Lett.*, 17 (B53), 205, 1984.
192. **Himeno, A., Kunisado, K., Niwa, M., and Ozaki, M.,** *Jpn. J. Pharmacol.*, 39, 91, 1985.
193. **Beart, P. M., Rowe, P. R., and Louis, W. J.,** *J. Pharm. Pharmacol.*, 35, 519, 1983.
194. **Kamalapurkar, O. S. and Chudasama, J. J.,** *Indian Drugs*, 21, 406, 1984.
195. **Shearer, Ch. M.,** *Anal. Profiles Drug Subst.*, 15, 319, 1986.
196. **Shearer, Ch. M. and De Angelis, N. J.,** *J. Pharm. Sci.*, 68, 1010, 1979.
197. Shell International Research Maatschappij N. V., Fr. 1, 455, 835, 1966.
198. **Houlihan, W. I. and Manning, R. E.,** Sandoz-Wander Inc., U.S. 3, 982, 020, 1976.
199. **Kodama, J. K., Haynes, G. R., and Albert, J. R.,** Shell Oil Company, U.S. 3, 975, 533, 1976.
200. **Scolastico, C. and Tronconi, G.,** *Braz. Pedido* PIBR 7903, 439, 1980.
201. **Bruce, W. F and Baum, T.,** *Ger. Offen.* 1, 802, 394.
202. **Baum, T., Echfield, D. K., Metz, N., Dinish, J. L., Rowles, G., Van Pelt, R., Shropshire, A. T., Fernandez, S. P., Gluckman, M. I., and Bruce, W. F.,** *Experientia*, 25, 1066, 1969.
203. **Baum, T., Shropshire, A. T., Rowles, G., Van Pelt, R., Fernandez, S. P., Echfield, D. K., and Gluckman, M. I.,** *J. Pharmacol. Exp. Ther.*, 171, 276, 1970.
204. **Ozava, H., Uematsu, T., and Chen, Ch.,** *Yakugaku Zasshi*, 95, 966, 1975.
205. **Diamant, S., Agranat, I., Goldbaum, A., Cohen, S., and Atlas, D.,** *Biochem. Pharmacol.*, 34, 491, 1985.
206. **Yokoyama, N., Horisaka, K., Soda, Y., Mori, I., Sokamoto, H., Ohata, K., Shimado, A., Shichino, F., Murai, K., and Tatsumi, H.,** *Iyakuhin Kenkyn*, 13, 1190, 1982.
207. **Meachem, R. H., Kick, C., Ruclins, H., and Sisemvine, S.,** *Fed. Proc. Fed. Am. Soc. Exp. Biol.*, 39, 850, 1980.
208. **Meachem, R. H., Emmett, M., Kyriakopoalos, A. A., Chiang, S. T., Ruclins, H. W., Walker, B. R., Narins, R. G., and Goldberg, M.,** *Clin. Pharmacol. Ther.*, 27, 44, 1980.
209. **Miyamoto, A., Soda, Y., Mori, I., Yokoyama, N., Horisaka, K., Shichino, F., Shimada, A., Koshima, K., Kataoko, T., and Tatsumi, H.,** *Iyakuhin Kenkyn*, 13, 1214, 1982.
210. **Yalcindag, O. N.,** *J. Pharm. Belg.*, 42, 59, 1987.
211. **Knowles, J. A., White, G. R., Kick, C. J., Spangler, T. B., and Ruelius, H. W.,** *J. Pharm. Sci.*, 71, 710, 1982.
212. **von Unruh, G. E. and Hengstmann, Y. H.,** *Quant. Mass Spectrom. Life Sci.*, 2, 295, 1978.
213. **Vio, L., Manolo, M. G., and Furlan, G.,** *Farmaco. Ed. Prat.*, 43, 27, 1988.
214. **Maxwell, R. A., Plummer, A. J., Schneider, F., Povalski, H., and Daniel, A. J.,** *J. Pharmacol. Exp. Ther.*, 124, 127, 1958.
215. **Page, I. H. and Důstan, H. P.,** *JAMA*, 170, 1017, 1959.
216. **Frohlich, E. D. and Freis, E. D.,** *Med. Ann. D. C.*, 28, 419, 1959.
217. **Maxwell, R. A., Mull, R. P., and Plummer, A. J.,** *Experientia*, 15, 267, 1959.
218. **Ciba,** U.S. Patent 2, 928, 829, 1960.
219. **Mull, R. P., Maxwell, R. A., and Plummer, A. J.,** *Nature*, 180, 1200, 1957.
220. **Page, I. H. and Důstan, H. P.,** *JAMA*, 170, 1265, 1959.
221. **Schier, O. and Marxer, A.,** Antihypertensive agents 1962—1968, in *Progress in Drug Research*, Jucker, E., ed., Birkhauser Verlag, Basel, 1969.
222. **Short, J. H., Ours, C. W., and Ranus, W. J., Jr.,** *J. Med. Chem.*, 11, 1129, 1968.
223. **Short, J. H. and Darby, T. D.,** *J. Med. Chem.*, 6, 833, 1967.
224. **Barrett, W. E., Mull, R. P., and Plummer, A. J.,** *Experientia*, 21, 506, 1965.
225. **Baines, N. W., Cobb, D. B., Eden, R. J., Fielden, R., Gardener, J. N., Roe, A. M., Tertiuk, W., and Willey, G. L.,** *J. Med. Chem.*, 8, 81, 1965.
226. **Augstein, J., Green, S. M., Monro, A. M., Potter, G. W. H., Worthing, C. R., and Wrigley, T. I.,** *J. Med. Chem.*, 8, 446, 1965.
227. **Augstein, J., Monro, A. M., Potter, G. W. H., and Scholfield, P.,** *J. Med. Chem.*, 11, 844, 1968.
228. **Kadyrova, R. D., Ikramov, L. T., and Tegisbaev, E. T.,** *Farmatsiya (Moscow)*, 36, 68, 1987.
229. **Nessinger, W. and Auterhoff, H.,** *Dtsch. Apoth. Ztg.*, 119, 1377, 1979.
230. **Casadebaig, F., Dupin, J. T., and Mesnard, P.,** *Ann. Pharm. Fr.*, 37, 313, 1979.
231. **Khamov, V. A. and Spasov, A. A.,** *Khim. Farm. Zh.*, 14, 112, 1980.
232. **Rao, G. R. and Rakhuveer, S.,** *Indian J. Pharm. Sci.*, 42, 141, 1980.
233. **Sakai, A., Inoue, T., and Tanimura, A.,** *Gann.*, 75, 245, 1984.

234. **Kadyrova, R. D., Ikramov, L. T., and Tegisbaev, E. T.,** *Zdravookhr. Kaz.*, 65, 1987.
235. **Kadyrova, R. D., Ikramov, L. T., and Tegisbaev, E. T.,** *Farmatsiya (Moscow)*, 37, 44, 1988.
236. **Zöllner-Iván, É.,** *Acta Pharm. Hung.*, 48, 76, 1978.
237. **Ganescu, I., Papa, I., and Preda, M.,** *Pharmazie*, 40, 495, 1985.
238. **Ganescu, I., Brinzan, G., and Várhelyi, Cs.,** *Chem. Anal. (Warsaw)*, 29, 549, 1984.
239. **Jarsebinski, J. and Gajewska, M.,** *Acta Pol. Pharm.*, 34, 503, 1977.
240. **Shah, M. H. and Stewart, J. T.,** *Int. J. Pharm.*, 20, 65, 1984.
241. **Stewart, J. T. and Clark, S. S.,** *J. Pharm. Sci.*, 75, 413, 1986.
242. **Pellizzari, E. D. and Seltzman, T. T.,** *Anal. Biochem.*, 96, 118, 1979.
243. **Dutt, M. Ch.,** *J. Chromatogr.*, 248, 115, 1982.
244. **Loeffler, L. J. and Pittman, A. W.,** *J. Pharm. Sci.*, 68, 1419, 1974.
245. **McManus, M. E., Davies, D. S., Boobis, A. R., Grantham, P. H., and Wirth, P. J.,** *J. Pharm. Pharmacol.*, 39, 1052, 1987.
246. **Evans, D. B., Bristol, J. A., and Kaplan, H. R.,** Antihypertensive agents, in Verdrame, M., Ed., *Handbook of Cardiovascular and Antiinflammatory Agents*, CRC Press, Inc., Boca Raton, 1986.
247. **Bianco, E. J.,** *Ger. Offen.*, 2, 708192, 1977.
248. **Lehmann, D., Faust, G., Poepel, W., Fiedler, W., and Dietz, G.,** Ger. East 141, 674, 1980.
249. **Schickaneder, H., Grafe, I., and Ahrens, K. H.,** EP 237, 608, 1987.
250. **Taskinen, J., Koivisto, N., and Nore, P.,** *Finn. Chem. Lett.*, 2, 1979.
251. **Fernandes, M., Smith, I. S., Weder, A., Kim, K. E., Gould, A. E., Busby, P., Swartz, C., and Questi, G.,** *Clin. Sci. Mol. Med.*, 48 (Suppl.2), 181, 1975.
252. **Fernandes, M. and Onesti, G.,** *Monogr. Physiol. Soc. Philadelphia*, 2 (New Antihypertensive Drugs), 481, 1976.
253. **Brogden, R. N., Heel, R. G., Speight, T. M. and Avery, G. S.,** *Drugs*, 14, 429, 1977.
254. **Groth, P. E. and Lee, B.,** *Drug Intel. Clin. Pharm.*, 12, 22, 1978.
255. **Löwenstein, W. and Steele, J. M., Jr.,** *Am. Heart J.*, 95, 262, 1978.
256. **Stokes, G. S. and Oates, H. F.,** *Cardiovascular Med.*, 3, 41, 1979.
257. **Kosman, M. E.,** *JAMA*, 237, 157, 1979.
258. **Graham, M. R. and Pettinger, W. A.,** *New Engl. J. Med.*, 300, 232, 1979.
259. **Cavero, I. and Roach, A. G.,** *Life Sci.*, 27, 1525, 1980.
260. **Cavero, I., Fenard, S., Gomeni, R., Lefevre, F., and Roach, A. G.,** *Eur. J. Pharmacol.*, 49, 259, 1978.
261. **Lefevre-Borg, F., Roche, A. G., Gomeni, R., and Cavero, I.,** *Cardiovasc. Pharmacol.*, 1, 31, 1979.
262. **Biagi, D., Giordani, A.,** *Bull. Soc. Ital. Pharm. Osp.*, 25, 295, 1979.
263. **Cavero, I., Fenard, S., Lefevre-Borg, F., and Roche, A. G.,** Alpha-Bloquants Symp. Int., Mason, Paris, 1979 (Published 1981), 7.
264. **Davey, M. J.,** *Clin. Exp. Hypertens.*, 4, 47, 1982.
265. **Scriabine, A., Ed.,** *Pharmacology* of *Antihypertensive Drugs*, Raven Press, New York, 1980, 151.
266. **Hobbs, D. O., Twomey, T. M.,** *Res. Comm. Chem. Pathol. Pharmacol.*, 25, 189, 1979.
267. **Rubin, P. and Blaschke, T.,** *Br. J. Clin. Pharmacol.*, 9, 177, 1980.
268. **Broers, O., Nilsen, O. G., Sager, G., Sandnes, D., and Jacobsen, S.,** *Clin. Pharmacokinet.*, 9 (Suppl. 1), 85, 1984.
269. **Hess, H. J. R.,** *Ger. Offen.*, 2, 457.911, 1975.
270. **Hess, H. J., Cronin, T. H., and Scriabine, A.,** *J. Med. Chem.*, 11, 130, 1968.
271. **Honkanen, E., Pippuri, A., Kairisalo, P., Nore, P., Karppanen, H., and Paakkari, I.,** *J. Med. Chem.*, 26, 1433, 1983.
272. **Alabaster, V. A., Campbell, S. F., Danilewicz, J. C., Greengrass, C. W., and Plews, R. M.,** *J. Med. Chem.*, 30, 999, 1987.
273. **Schickaneder, H., Ahrens, K. H.,** *S. African Z.A.*, 8401, 360, 1984.
274. **Poepel, W., Lehmann, D., Faust, G., and Fiedler, W.,** GER East 203, 911, 1983.
275. **Buchowieczki, W., Nawrot, B., Zjawiony, J., Zajac, H., Magielka, S., and Jarzebski, A.,** POL PL 115, 626, 1982.
276. **Honkanen, E., Pippuri, A., Kairisalo, P., Thaler, H., Koivisto, M., and Tuomi, S.,** *J. Heterocycl. Chem.*, 17, 797, 1980.
277. **Giardina, D., Bertini, R., Brancia, E., Brasili, L., and Melchiorre, C.,** *J. Med. Chem.*, 28, 1354, 1985.
278. **Lin, E. T., Baughman, R. A., and Benet, L. Z.,** *J. Chromatogr.*, 183, 367, 1980.
279. **Jee, Y. G., Rubin, P. C., and Meffin, P.,** *J. Chromatogr.*, 172, 313, 1979.
280. **Reece, Ph. A.,** *J. Chromatogr.*, 221, 188, 1980.
281. **Twomey, T. M. and Hobbs, D. C.,** *J. Pharm. Sci.*, 67, 1468, 1978.
282. **Sohr, R., Wesshuhn, I., Preiss, R., and Scholze, J.,** *Pharmazie*, 38, 496, 1983.
283. **Shen, D. D. and Pitterman, A. B.,** *Methodol. Anal. Toxicol.*, 3, 159, 1985. Sunshine, I., Ed., CRC Press, Boca Raton.

284. **Bhamra, R. K., Flanogan, R. J., and Holt, D. W.,** *J. Chromatogr.*, 380, 216, 1986.
285. **Bachman, W. J.,** *J. Liq. Chromatogr.*, 9, 1030, 1986.
286. **Dokladalova, J., Coco, S. J., Lemke, P. R., Quercia, G. T., and Korst, J. J.,** *J. Chromatogr.*, 224, 33, 1981.
287. **Mohamed, M. E. and Aboul-Enein, H. Y.,** *Pharmazie*, 40, 358, 1985.
288. **Nikolic, K. and Velasevic, K.,** *Arh. Farm*, 38, 3, 1988.
289. **Pook, K. H., Stähle, H., and Daniel, H.,** *Chem. Ber.*, 107, 2644, 1974.
290. **Rouot, D., Leclerc, G., and Wermuth, C. G.,** *Chim. Ther.*, 5, 545, 1973.
291. **Jackman, L. M. and Jen, T.,** *J. Am. Chem. Soc.*, 97, 2811, 1975.
292. **Wermuth, C. G., Schwartz, G., Leclerc, G., Garnier, J. P., and Rout, B.,** *Chim. Ther.*, 1, 115, 1973.
293. **Meier, R. and Müller, R.,** *Schweiz. Med. Wochenschr.*, 69, 1271, 1939.
294. **Hartmann, M., Isler, H.,** *Naunyn-Schmiedebergs Arch. Exp. Pathol. Pharmakol.*, 192, 141, 1939.
295. **Hoefke, W. and Kobinger, W.,** *Arzneim. Forsch.*, 16, 1038, 1966.
296. **Timmermens, P. B., Hoefke, W., Stähle, H., and van Zwieten, A.,** *Prog. Pharmacol.*, 3, 2, 1980.
297. **Timmermens, P. B., Hoefke, W., Stähle, H., and van Zwieten, A.,** *Prog. Pharmacol.*, 3, 31, 1980.
298. **Schier, O. and Marxer, A.,** Antihypertensive Agents 1969—1980, Chemical Research Department, Pharmaceutical Division, Ciba-Geigy Ltd., Basel, Switzerland.
299. **Timmermens, P. B., Hoefke, W., Stähle, H., and van Zwieten, A.,** *Prog. Pharmacol.*, 3, 21, 1980.
300. **Timmermens, P. B., Hoefke, W., Shähle, H., and van Zwieten, A.,** *Prog. Pharmacol.*, 3, 85, 1980.
301. **Najer, H., Giudicelli, R., and Sette, J.,** *Bull. Soc. Chim. Fr.*, 2114, 1961.
302. **Chemie Linz AG.,** Öst. Patent 005474, 1974.
303. **Roth, H. J., Eger, K., and Troschütz, R.,** Pharm. Chemie II., *Arzneistoffanalyse,*, Georg Thieme Verlag, Stuttgart, 1981, 414.
304. **Murray, S., Waddell, K. A., and Davies, D. S.,** *Biomed. Mass Spectrom.*, 8, 500, 1981.
305. **Foerster, H. J. and Staehre, H.,** *Biomed. Mass Spectrom.*, 5, 483, 1978.
306. **Hiltunen, R., Marvola, M., Hirsjarvi, P., and Rajsanen, S.,** *Acta Pharm. Fenn.*, 88, 161, 1979.
307. **Nazarali, A. I., Baker, G. B., and Boisvert, D. P.,** *J. Chromatogr. Biomed. Appl.*, 53, 393, 1986.
308. **Edlund, P. O.,** *J. Chromatogr.*, 187, 161, 1980.
309. **Chu, L. C., Bayne, W. F., Tao, F. T., Schmitt, L. G., and Shaw, J. E.,** *J. Pharm. Sci.*, 68, 72, 1979.
310. **Murray, S. and Davies, D. S.,** *Biomed. Mass Spectrom.*, 11, 435, 1984.
311. **Settlage, J. and Jager, H.,** *J. Chromatogr. Sci.*, 22, 192, 1984.
312. **Haering, N., Salama, Z., Reif, G., and Jager, H.,** *Arzneim. Forsch.*, 38, 404, 1988.
313. **Wilczynska-Wojtulewicz, I. and Sadlej-Sosnowska, N.,** *J. Chromatogr.*, 367, 434, 1986.
314. **Walters, S. M. and Stonys, D. B.,** *J. Chromatogr. Sci.*, 21, 43, 1983.
315. **El Jazli, F. A., Bedair, M., and Korany, M. A.,** *Analyst (London)*, 111, 477, 1986.
316. **Tawakkol, M. S., Jado, A. I., and Aboul-Enein, H. Y.,** *Arzneim. Forsch.*, 31, 1064, 1981.
317. **Arndts, D., Staehle, H., and Struck, C. J.,** *Arzneim. Forsch.*, 29, 532, 1979.
318. **Farina, P. R., Homon, C. A., Chow, C. T., Keirus, J. J., Zavorkas, P. A., and Esber, H. J.,** *Ther. Drug Monit.*, 7, 344, 1985.
319. **Jarrott, B. and Spector, S.,** *J. Pharmacol. Exp. Ther.*, 207, 195, 1979.

Chapter 5

β-ADRENERGIC ANTAGONISTS (β-BLOCKERS)

I. INTRODUCTION

The β-adrenergic antagonists are commercially available as the water-soluble, absorbable salts of the nonsoluble or only slightly soluble bases. They behave as sympatholytics, i.e., they have the ability to interact with β-adrenergic receptors. The β-blocker compounds that are reasonably often used are to be found in the comprehensive Table 1; a few of them will be detailed in this chapter.

Unlike the agonists, the great majority of the β-blocker compounds may be regarded as 1,2-propanediol derivatives. Only sotalol has a skeleton resembling that of adrenergic agonists, including an ethanolamine moiety. On the basis of the elucidated relations, all those physical-chemical properties which are important in the *in vitro* and *in vivo* interactions of the β-blocker compounds can be deduced from the individual features of the main structural units of the molecules (Table 2).

II. PROPERTIES

The commercial salts of β-blocker agents (chloride, tartrate, maleate, etc.) are generally white or off-white odorless crystalline powders.

These salts are very soluble in water and alcohol. Depending upon their hydrophobicity, they may also be soluble in semi-polar solvents. Metoprolol tartrate (USP) is freely soluble in methylene chloride or chloroform, while propranolol hydrochloride (USP) is only slightly soluble, and timolol maleate (USP) sparingly soluble in chloroform. They are insoluble in ether or apolar solvents such as cyclohexane. Due to the poor water-solubility of their corresponding bases, their water-solubility, as expected, is pH-dependent. This may be illustrated by the data for pindolol:

pH	1.5	5.2	7.5
parts of pindolol in 100 parts of water	2	0.03	0.001

The octanol-water partition coefficients of the therapeutic β-blockers show a great variety,[1-3] there being a difference of several orders of magnitude between the values for the most hydrophobic and the most hydrophilic compounds (compare the P values of penbutalol, propranolol, and atenolol in Table 3). On this basis β-blockers may be grouped into three classes of lipophilicity.

In spite of the often significant discrepancies between the values published for the same compounds by different authors, the octanol-water partition coefficient can be regarded as a good indicator of the predictable pharmacokinetic behavior of β-blocker molecules. The differences that may be encountered in the values for a given compound may stem from the different circumstances applied in the determination (pH of the aqueous phase, duration and temperature of equilibration, purity of the substance, etc.). However, the extremely different numerical values of the compounds in Table 3 arise from the different natures of the values. The data in column A are apparent partition coefficients, which relate to the partition of the protonated form of the compounds, while the much higher values in column B are true partition coefficients, which reflect the partition of the non-protonated (base) form.

A good correlation has been demonstrated between octanol-water partition coefficients and the HPLC retention data for the β-blockers. Irrespective of the inconsistency in the

TABLE 1
β-Blockers in Therapeutic Use

General structure: $R_1{-}O{-}CH_2{-}CH(OH){-}CH_2{-}NH{-}R_2$

R_2: i.pr. = $CH(CH_3)_2$ (H$_3$C–CH–CH$_3$); i.b. = $CH(CH_3)_2$ with –CH_3

Generic name	Chemical structure: R_1	Chemical structure: R_2	Proprietary name®	Therapeutic use	Doses
Selective β_1-receptor blocking agents					
Acebutolol	Benzene ring with –CO–CH_3 and –NH–CO–CH_2–CH_2–CH_3	i.pr.	IL-17803, Neptal, Prent, Sectral	Preferred in management of angina pectoris, hypertension, and cardiac arrhythmias. Reduces arterial pressure without reducing cardiac output. Exhibits intrinsic sympathomimetic activity (ISA), correlating with baseline plasma renin activity.	Maximum dose: 300—400 mg/day
Bisoprolol	Benzene ring with –CH_2–O–CH_2–CH_2–O–$CH(CH_3)_2$	i.pr.	Concor	Indicated in hypertension and angina pectoris. Does not exhibit any ISA properties.	Initial dose: 10 mg up to 20 mg once a day
Betaxolol	Benzene ring with –CH_2–CH_2–O–CH_2–$CH(CH_3)_2$	i.pr.	Kerlone	Same as bisoprolol.	

Celiprolol	(p-substituted phenyl) $NH-C(=O)-N(C_2H_5)_2$	i.b.	Selectol.	Has some ISA, β_2- and very slight α_2-receptor blocking effect. Indications similar to those of bisoprolol and betaxolol	Usual dose: 200—300 mg daily.
Atenolol	(p-substituted phenyl) $CH_2-CO-NH_2$	i.pr.	ICI 66082, Tenormin	Three times more potent than practolol. Used in management of hypertension alone or in combination with thiazide-type diuretics.	Initial dose: 50 mg/day (1 tablet) up to 100 mg/day.
Metoprolol Metoprolol tartrate (USP)	(p-substituted phenyl) $CH_2-CH_2-O-CH_3$	i.pr.	Beloc, Betaloc, Lopressor, Selokeen	Has a preferential effect on β_1-adrenoreceptors, but inhibits β_2-receptors too, particularly in higher doses.	Initial dose: 100 mg/day (in single or divided doses) up to a maximum 450 mg/day.
Practolol	(p-substituted phenyl) $NH-C(=O)-CH_3$	.pr.	Cordialine, Dabric, Eralex, Practol, Pralon, Teranol	Acts selectively on β_1-receptors of myocardium only. Therefore possesses fewer side effects than other β-blockers. Has some ISA.	Initial dose: 100 mg up to 400 mg/day.
Non-selective β-receptor (β_1 and β_2) blocking agents					
Oxprenolol	(o-substituted phenyl) $O-CH_2-CH=CH_2$	.pr.	Coretal, Trasicor, Trazitensine	Intensity of β-blocking effect similar to that of propranolol. Indicated for all grades of hypertension. Does not cause orthostatic collapse, hypersensitivity, weakness, or fatigue. Cardioprotective. Combination with diuretics gives better hypotensive result and diminished side effects.	Usual dose: 2 × 80 mg/day
Sotalol	$CH(OH)-CH_2-NH-CH(CH_3)_2$ on phenyl, p-$NH-SO_2-CH_3$		Beta-Cardone, Sotacor, Sotalex, Sotaper	Has screening effect against sympathomimetic stimulation, and weak negative inotropic and chronotropic effect. Indicated in therapy of hypertension, angina pectoris, and tachycardiac arrhythmia.	Usual dose: 300 mg/day

TABLE 1 (continued)
β-Blockers in Therapeutic Use

Generic name	Chemical structure R$_1$	Chemical structure R$_2$	Proprietary name®	Therapeutic use	Doses
Cloranolol	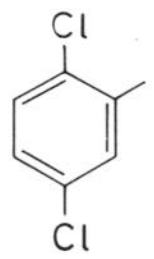 Cl, Cl	i.b.	Tobanum	Three times more active than propranolol. Does not possess ISA. Useful in mild and moderate hypertension.	Initial dose: 2—3 times 2.5 mg daily, up to a maximum 45 mg/day
Bunitrolol	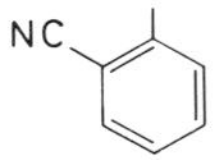 NC	i.b.	Stresson, Bunitrolol hydrochloride, Kö 1366	Devoid of marked negative inotropic action. Possesses ISA.	
Alprenolol	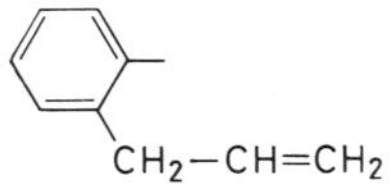 $CH_2-CH=CH_2$	i.pr.	Aptin	Exerts only very light agonistic effect. Used mainly in management of arrhythmia, heart neurosis, and angina pectoris.	
Propranolol Propranolol hydrochloride (USP)	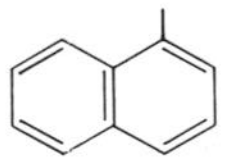	i.pr.	Alindol, Anaprilin, Betadren, Camcor, Dociton, Inderal, Indobloc, Ipranol, Propranolol hydrochloride, Stobetin, Sumial, Tonum	Blocks both β_1 and β_2-receptors. Does not exhibit any ISA. Can be used as alternative to thiazide in mild hypertension. Its antihypertensive activity is related to plasma renin activity.	Initial dose: 40 mg (once daily) up to 320 mg/day
Pindolol	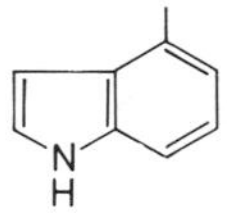 N H	i.pr.	Carvisken, LB-46, Visken	A nonselective β-blocker, which has definite ISA.	Initial dose: 5 mg up to a maximum 60 mg/day
Bunolol	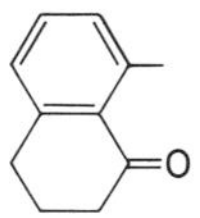 O	i.b.	W 6412		

Nadolol		i.b.	Corgard, Solgol	Approximately as potent as propranolol. Essentially devoid of agonist or local anaesthetic activity.	Initial dose: 40 mg (once a day) up to 320 mg/day
Penbutolol		i.b.	Betapressin	Similar in action to timolol. Has moderate ISA.	Usual dose: 40 mg once a day
Timolol Timolol maleate (USP)		i.b.	Blocadren, MSD, Blocanol, Temserin, Timacar, Timolol maleate	As effective as pindolol. Indicated for patients with mild or moderate hypertension, but not in hypertension emergencies. Usually used in combination with thiazide diuretics.	Initial dose: 10 mg twice a day, up to 480 mg/day
α_1- and β-blocker					
Labetalol			Ibidomite, Trandate, Normodin	Has selective α_1 and non-selective β-adrenergic blocking activity. Possesses ISA. Potent in treatment of essential hypertension and other cardiovascular effects associated with pheochromocytoma.	Initial dose: 100 mg twice daily, orally

Note: The β-blockers are contraindicated in manifested heart insufficiency, AV-block, sinus bradycardia, asthma bronchiale, and diabetes mellitus. Their general side effects: bradycardia, hypotension, fatigue, weakness, gastrointestinal disturbances, and various CNS effects (hallucinations, nightmares, insomnia, lassitude, and depression).

TABLE 2
Structures and Corresponding Features of β-Blockers

$$R-OCH_2-\underset{\displaystyle OH}{\underset{|}{CH}}-CH_2-\overset{\oplus}{NH}-R' \qquad X^{\ominus}$$

Structural unit	Aryl	Sec[a]-alcohol	Sec-ammonium	Alkyl (isopropyl, *t*-butyl)
Corresponding features	Hydrophobic; nucleophilic interacetion	Hydrophilic; H-bond formation; oxidizability	Proton acception (electrophilic interaction); H-bond formation; nitrosamine formation	Lipophilic

[a] Secondary.

TABLE 3
Partition Coefficients of β-Blockers

Compound	Partition coefficient A[1]	Partition coefficient B[2]	HPLC retention[4] ('min)
Penbutolol	185.9	14,200.0	—
Propranolol	13.41	1,640.0	16.2
Oxprenolol	2.36	235.0	8.77
Timolol	1.274	82.0	6.20
Metoprolol	1.133	76.0	6.49
Acebutolol	1.008	59.0	6.41
Pindolol	0.884	—	4.20
Nadolol	0.089	8.5	4.32
Sotalol	—	0.24	3.20

Correlation between partition and HPLC data:

	1	2
b	= −4.788	b = −603.2
m	= 1.045	m = 126.1
r	= 0.96	r = 0.94

Note: A = apparent partition coefficient (P) determined by octanol-phosphate buffer (pH 7.4); t = 37° C; equilibrization: 6 h. B = true partition coefficient (distribution coefficient, D) determined by octanol-phosphate buffer (pH 7.38); t = 35° C; equilibrization: 5 h; calculated by the equation $P = D\left(1 + \frac{1}{\text{antilog}(pH\text{-}pK_a)}\right)$.

methods used and the nature of the partition coefficients, the quality of the correlation is very satisfactory in both cases (Table 3), indicating the general applicability of HPLC for the determination of *in vitro* lipophilicity. The reliability of HPLC retention as a measure of lipophilicity is convincingly proved by the data of Vila et al.: at three different values (7.5, 6.2, and 3.5), they found a significant linear correlation between log K′ and log D. The log D values were obtained from four bibliographic sources, while the log K′ data were estimated in a reversed-phase system, with a methanol-aqueous buffer mixture as mobile phase, adjusted to the pH values mentioned above.[5] It is also worth noting that the partition coefficients for β-blocker compounds of aryloxypropanol type may be calculated with very

TABLE 4
Comparison of Observed and Calculated log P Values for Propranolol

log P observed	log P calculated R[6]	K[7]	L[8]
3.56	3.66	3.53	2.80

Note: R: by Rekker method;[6] K: by Klopman method;[7] L: by Leo method.[8]

TABLE 5
Basicities of Therapeutic β-Blocker Compounds

Compound	pK_a 16	17	In presence of urea[17] 8 *M*	16 *M*
Isoproterenol	9.70	9.42	9.88	10.85
Propranolol	9.41			
Alprenolol	9.19			
Oxprenolol	9.21	9.40		
Pindolol	9.26	9.46		
Timolol		9.40		
Nadolol		9.50	9.91	10.95
Practolol	9.50	9.44		
Metoprolol		9.53		
Pronethalol		9.53	9.80	10.70
Sotalol	9.86	9.55		
Di-*n*-propylamine	10.97			
Di-*n*-butylamine		10.80	10.80	10.80
Tripropylamine	10.64			

good accuracy by the hydrophobic fragmental method of Rekker[6] and the atomic charge density method of Klopman,[7] but a rather great difference arises when the log P of propranolol is calculated by the fragmental method of Leo[8] (Table 4).

Although the octanol-water partition coefficient may generally serve as a basis for QSAR study and drug design, the octanol lipophilicity under *in vivo* circumstances can be modified and enhanced by the formation of ion pairs in the human organism. This idea was proposed by Higuchi,[9] and later by other authors.[10-12] The membrane permeability of medicinal compounds,[13,14] including propranolol, was found to be enhanced in the presence of certain counter-ions. Gasco et al.[15] established that the *in vitro* diffusion rate constants of β-blockers were higher in the presence of taurodesoxycholate ions. On the basis of this finding, ion pairing was propounded as a tool for influencing the rate and extent of absorption.

The melting points of the bases and salts lie in the range 100 to 200° C, and may be used for identity and purity testing.

The pK values (Table 5) indicate that, under physiological conditions, the β-blockers are organic bases of medium strength, which are completely protonated and dissolved in the stomach. Although the ratio base/ammonium form increases at higher pH values even in mildly alkaline medium, e.g., at pH = 8, approximately 96% of propranolol is in the protonated form. Consequently, the interacting form of β-blockers must be considered to be the ammonium form. The side-chain secondary amine group of therapeutic β-blockers makes these compounds medium-strong bases. Although the published pK values of β-blockers differ somewhat, pK_a 9.4 to 9.6 may be taken to be an average range. Table 5 lists pK_a values for therapeutic β-blocker compounds from different papers. These values indicate that at physiological pH (7.4) the β-blockers exist as singly charged cations. This has a strong influence on the conformation and reactivity of β-blockers under both *in vitro* and *in vivo* circumstances.

In this respect, attention should be paid to the investigations by Laxer et al.,[17] who assumed intramolecular hydrogen bonding in β-blocker compounds. Their conclusion was based in part on the fact that the pK_a values of the β-blockers are more than one unit smaller than those of the dialkyl secondary amines (pK_a of diethylamine, dipropylamine, and dibutylamine = 10.98, 10.97, and 10.80, respectively). On the other hand, on the addition of urea, the pK_a values of the β-blockers increased and became almost identical with those of the alkylamines (see Table 5). It is important to note that the pK_a of dibutylamine did not change on the addition of urea. It was assumed that urea disrupted the intramolecular hydrogen bonds, increasing the proton-acceptor ability of the secondary amine group. This indirect evidence is strongly supported by IR spectroscopic data, but some contradiction arises from NMR sources.

The PMR spectroscopic investigation by Jen and Kaiser[18] indicates hydrogen bonding only in acidic solution (i.e., for the protonated form) of propranolol. The large PMR shift of the β-proton in a $CDCl_3$ solution of propranolol hydrochloride relative to the shift of the same proton in the propranolol base has been interpreted in terms of hydrogen-bond and chelate bicycle formation. In this confermer **(1)** the Hβ atom is bonded equatorially and is located relatively distant from other protons:

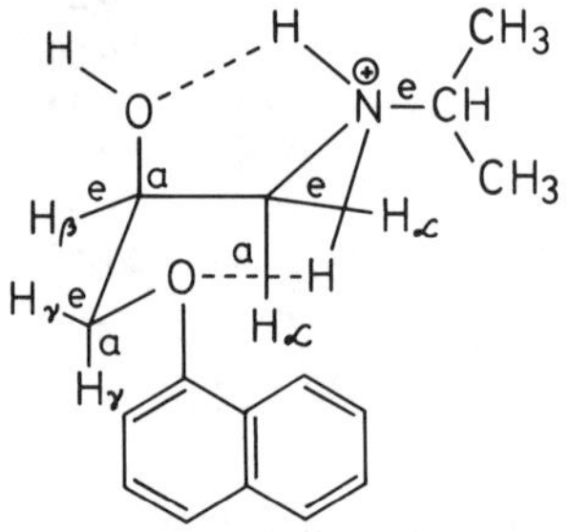

(1)

This form requires a conformation **(2)** where the ammonium group is in the gauche position relative to the aryloxymethylene groups:

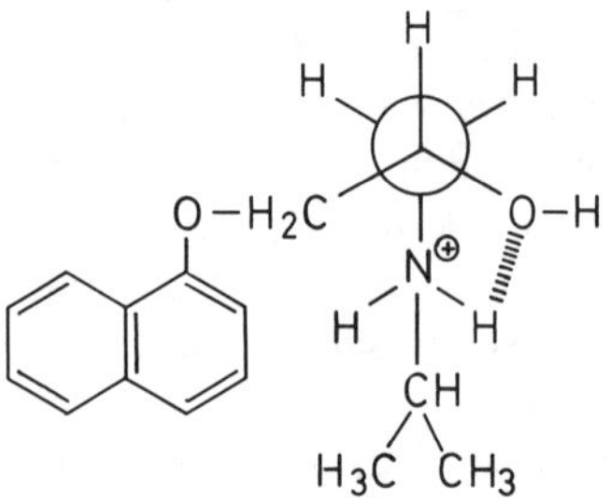

(2)

In accordance with the conclusions of PMR and other investigations, the structure of isoproterenol may be interpreted similarly.[18] Basically the same conclusions were drawn from the more recent PMR investigations by Balsamo,[19] who utilized the PMR spectra of INPEA 1-(4-nitrophenyl)-1-hydroxy-2-isopropylaminoethane and isoproterenol to determine the conformation of the β-adrenergic agent of the model compounds:

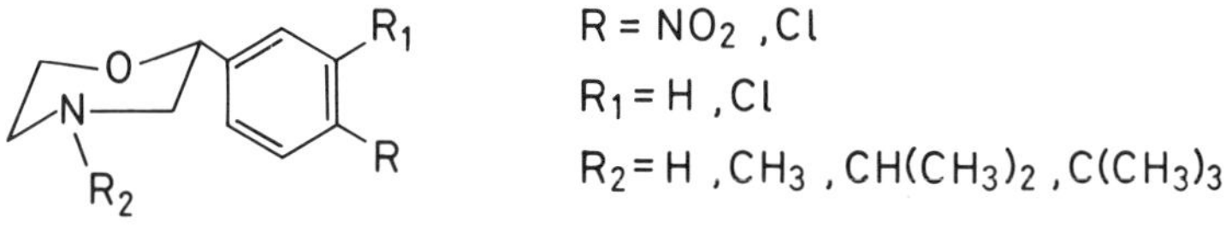

$R = NO_2$, Cl
$R_1 = H$, Cl
$R_2 = H$, CH_3, $CH(CH_3)_2$, $C(CH_3)_3$

(3)

The conformer **(3)**, in which the aryl and N-R_2 groups are anti and which allows the formation of an intramolecular hydrogen bond between OH and NHR_2, proved to be predominant;[20] this is illustrated by the Newman projection for isoproterenol **(4)**.

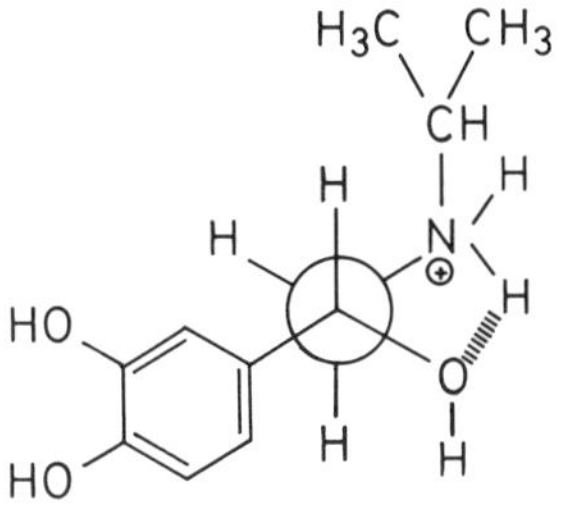

(4)

The^{13} NMR resonances of pindolol, timolol, penbutolol, and nadolol were assigned by Aboul-Enein et al.[21] From the chemical shifts, it can be concluded that the aromatic ring systems in these compounds have negligible effects on the 2-propanol side chain. This finding gives an explanation for the almost identical pK_a values of the different β-blockers, irrespective of their different aromatic moieties (see Table 5).

Crystallographic investigations of (aryloxy) propanolamines have shown that the side chain is in an extended conformation, which may be considered to be the preferred conformation of β-adrenoreceptor ligands in the solid state.[22,23]

A basic principle in the pharmacopeial control of β-blockers is the application of their UV and IR spectra, which are compared with the spectra of standard reference substances. As a consequence of the benzene nucleus in the phenylethylamine and phenylpropylamine β-blocker derivatives, their spectra are characterized by a UV absorption maximum at about 250 (±20) nm, while the coexistence of certain auxochromic groups or/and the extension of the aromatic moiety (see propranolol, timolol, pindolol, etc.) give rise to a bathochromic effect accompanied by a shift in the maximum absorbance to near or over 300 nm (see Table 6).

As expected, acidification (ammonium salt formation) exerts very little influence on the UV spectra of β-blockers; it appears as a hypsochromic shift of the maximum by 1 to 4 nm. However, a very significant bathochromic effect can be observed when amphiphilic

TABLE 6
UV Absorption of β-Blocker Agents

Compound	Absorption: λ_{max}, nm	Absorption: Molar absorbance
Sotalol	232	14,560
Toliprolol	274	555
	279	7,340
	272	1,430
	218	1,360
Practolol	248	16,240
Atenolol	227	10,000
	275	1,430
Labetalol	305	3,490
Bunitrolol	231	9,180
	291	3,950
Acebutolol	235	2,640
	328	2,580
Nadolol	271	1,100
	279	1,130
Alprenolol	271	1,840
	277	1,650
Oxprenolol	273	2,330
Propranolol	289	6,120
	319	1,950
Pindolol	265	7,950
	288	4,200
Timolol	299	9,070

compounds are made alkaline. This is illustrated by the data for labetalol, which contains a phenolic group:

	λ_{max}, nm: methanolic solution	λ_{max}, nm: 0.1 *N* NaOH solution
Labetalol	305	333

As for other series of compounds which are related to each other in structure, examination of the IR spectra is one of the most reliable tools for the identification of β-blockers. The IR spectrum of the compound and that of the official standard must agree in all respects, i.e., as regards both the positions and relative intensities of the bands. As an example, the IR spectra of oxprenolol and pindolol will be briefly interpreted (see also Figures 1 and 2). The broad and intense bands to be found at 3200 to 2700 cm^{-1} and 3500 to 2850 cm^{-1}, respectively, in the spectra indicate the presence of a hydroxy group (νO–H stretches) involved in hydrogen bond formation. The positions and shapes of these bands suggest intramolecular hydrogen-bonding, though intermolecular interaction has also been assumed.[24] The bands of the corresponding ν C–O stretches appear at about 1100 cm^{-1}. The sharp bands at 3200 cm^{-1} and 3340 cm^{-1} originate from ν N–H stretches. The different methyl and methylene groups exhibit peaks at 2950 to 2550 cm^{-1}, and also at 1460 cm^{-1}. The isopropyl group yields a characteristic absorbance at 1390 to 1350 cm^{-1}. The aromatic structure of the molecules gives rise to intense absorbance at 1580 and 1515 cm^{-1} (ν C–C vibration). Bands due to the ether anti-symmetric stretching vibration (1255 cm^{-1}) and the terminal vinyl group (ν C=C 1655, ν C–H 1000 and 930 cm^{-1}) may be found in the spectra of both oxprenolol and its hydrochloride.

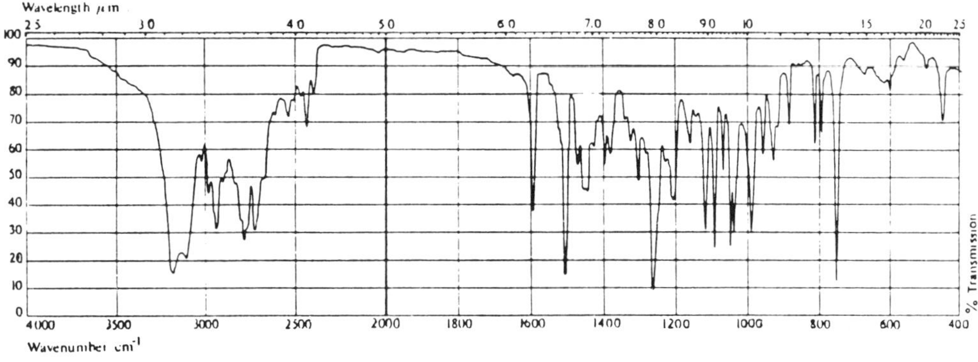

A.

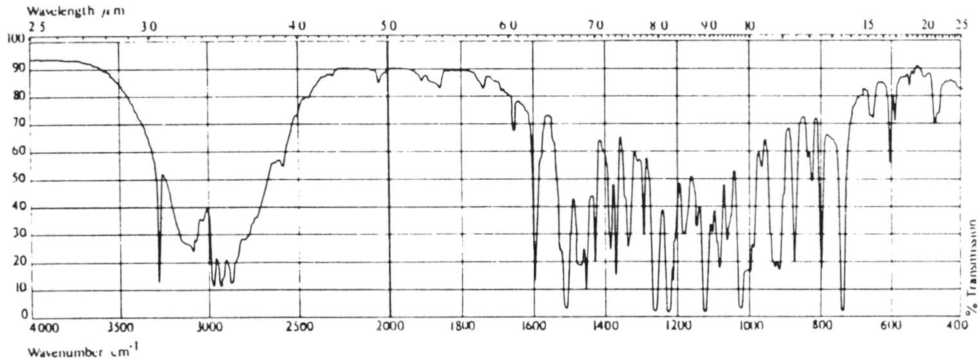

B.

FIGURE 1. A. IR spectrum of oxprenolol. B. IR spectrum of oxprenolol hydrochloride.

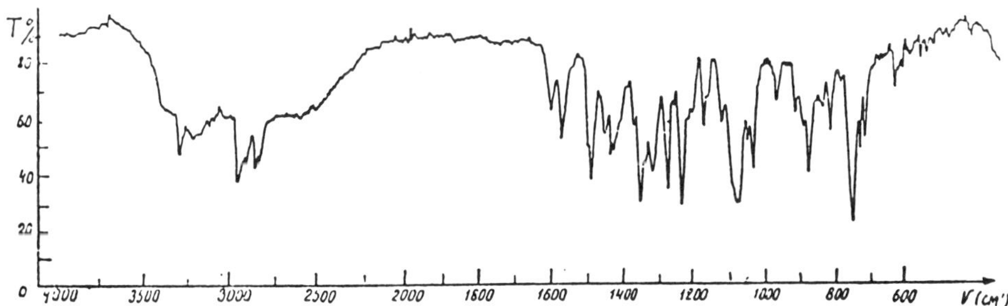

FIGURE 2. IR spectrum of pindolol.

Just as for other pharmacological groups, the optical isomers of β-blockers are not equipotent. In general, the levorotatory enantiomers of the β-adrenergic blockers are 50 to 100 times more active than the corresponding dextrorotatory isomers.[25] In the (aryloxy) propanolamine series they possess the (*S*)-configuration, and in the ethanolamine series the (*R*)-configuration. Regardless of the difference in biological activity, the great majority of the substances available commercially are racemates. To check the optical purity or to resolve the pure enantiomers from the racemate mixtures, mainly HPLC is used. The methods are based on direct separation or on the use of derivatization agents. Petterson and Schill[26] have demonstrated the usefulness of (+)−10-camphorsulfonic acid as a chiral ion pair reagent for the direct resolution of some β-blockers. The methods including derivatization are more frequent for the separation of β-blocker enantiomers (see p. 221).

III. MECHANISM OF ACTION

The pharmacology of β-blockers has been investigated extensively. Several papers have dealt with the antihypertensive property of β-blockers,[36-40] and reviews have also been published on the pharmacological and pharmacokinetic properties in general,[41] and on individual β-blockers (propranolol,[42] labetolol,[43] atenolol,[44-46] and nadolol[47]).

The mechanism of action of β-blockers can be explained only in terms of the cooperative effects of several factors. The majority of opinions focus on the inhibitory effects within the central, cardiac, renal, vascular, and peripheral mechanisms. Staroukine et al.[48] have emphasized the hormonal features of the effect. It was established that β-blockers (propranolol, sotalol, atenolol, metoprolol, and labetalol) caused a greater decrease of the blood angiotensin in hypotensive than in normotensive animals, but the depressive effect on the cardiac activity was also taken into consideration.

IV. STRUCTURE-ACTIVITY RELATIONSHIP

A. HISTORICAL

Åhlquist[27] postulated the existence of different kinds of adrenergic receptors in 1948. His concept, i.e., the differentiation or classification of the adrenoreceptors into α and β subspecies, was experimentally supported only 10 years later, when Powell and Slater[28] observed the selective β-blocking effect of dichloroisoproterenol (DCI, **1**), which had been synthetized by Mills.[29] The development continued when Lands et al.[30,31] classified β-receptors into β_1 and β_2 subgroups. The following 30 years has been marked by the almost unbroken propagation of β-blocker agents.

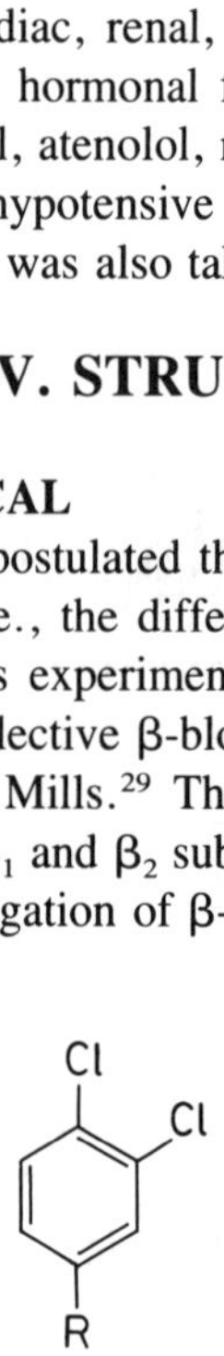

Cl Cl R (1)   R (2) $-OCH_2-R$ (3)

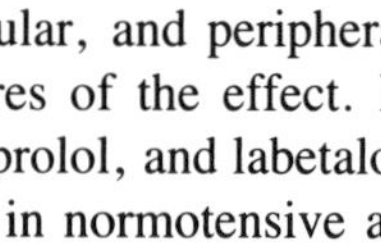

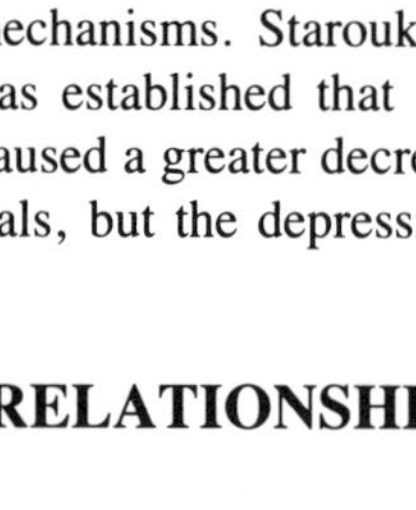

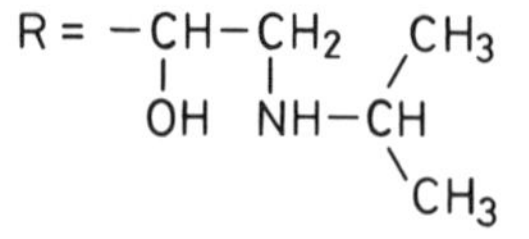

$R = -CH(OH)-CH_2-NH-CH(CH_3)_2$

Although none of the hypotheses put forward to explain their effects is entirely satisfactory, β-blockers are indispensable drugs in the armament of cardiovascular (antihypertensive) therapy. The structural similarity between DCI and isoproterenol (IPT) led to DCI and IPT being made the prototypes in the search for new β-blockers with antihypertensive features. Though DCI exhibits rather strong β-adrenergic receptor-inhibiting properties, its partial agonist character *ab ovo* prevented even its clinical testing. In 1962, Black et al.[32] synthetized promethalol **(2)**, a congener of DCI, which proved to be equipotent with DCI in β-blocking, but to have a much lower agonist potency in clinical trials. Unfortunately, this highly promising compound displayed carcinogenicity in animal experiments. Whereas substitution on the phenylethyl skeleton on DCI or IPT did not result in compounds with more valuable pharmacological features than those of DCI, the elongation of the side chain provided a breakthrough when propranolol **(3)** was prepared and introduced into cardiovascular therapy.[33-35] This compound was ten times more potent as a β-blocker than promethalol, but devoid of the carcinogenic property of the latter. It may now be regarded as a matter of drug history that propranolol was initially marketed as an antiarrhythmic agent under the name Inderal, and that after 25 years it still maintains its place among β-blocker drugs.

The number of β-blockers available at present which have a significant role in cardiovascular therapy may be estimated as around a dozen. Of these, sotalol is the only one with a phenylethanolamine skeleton; all the others are aryloxypropanolamine derivatives. The main fields of their therapeutic application are in the treatment of cardiac arrhythmias, angina pectoris, and hypertension.

B. RECENT DEVELOPMENTS

The results obtained on the pharmacological features and mechanism of action of propranolol led to the recognition of general structural requirements for the design and synthesis of new β-sympatholytic agents with extended strengths and selectivities of action. Bercher et al.[49] surveyed the literature data and performed a comprehensive quantitative structure-activity analysis of 415 β-adrenergic blocking drugs. More recently, the structure-activity relationships and thermodynamics were studied by Testa et al.[50]

From the beginning it was a very logical strategic research aspect that competitors of β-adrenergic sympathomimetics must bear a structural resemblance to propranolol. Consequently, it was a quite understandable trend during the past 25 years that the aromatic and the amino moieties, the two presumable receptor binding parts of isoproterenol, were subjected to structural modification in the search for new antihypertensive β-blockers:

$$R-OCH_2-\underset{\displaystyle OH}{\underset{|}{CH}}-CH_2-\overset{\oplus}{NH}-R' \qquad X^{\ominus}$$

The secondary amino group, the alcoholic hydroxy group, and even the ether oxygen in the aminopropanediol skeleton could not be replaced by other groups or atoms without partial or total loss of efficacy.

When the substitution on the aromatic group ring involves the fusion of another aromatic ring, the result is an increased β-blocker effect and a lower sympathomimetic effect. Although replacement of the naphthalene group by polycyclic ring structures led to a decreased activity, the compounds containing indolyl and benzofuryl moieties retained the β-blocking effect and also some sympathomimetic activity. The β-blocker activity of pindolol is 10 to 20 times stronger than that of propranolol, and its initial intrinsic β-adrenergic agonist activity also surpasses that of other known blocking agents.[55]

In all of these cases the extension of the aromatic moiety is equivalent to an extension of the π-electron system, i.e., to strengthening of the Lewis base nature of the aromatic nucleus. The same effect seems to be achieved by substitution of the aromatic ring with groups able to interact with the π-electron system, as in the case of oxprenolol, alprenolol, bunitrolol, toliprolol, and cloranolol. While alprenolol was equipotent with propranolol in antagonizing positive chronotropic and inotropic effects, oxprenolol, though slightly less potent than propranolol, has several advantageous properties. Even at present, more than 20 years after its discovery, oxprenolol has a distinguished place among hypertensive β-blockers.

TABLE 7
Structural Types of β-Blockers

Aryl	R	Compound	Ref.
Compounds with fused aromatic rings			
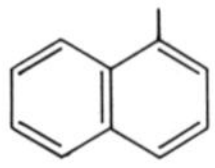	i.pr.	Propranolol	33,35
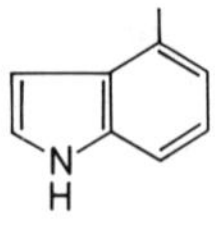 N H	i.pr.	Pindolol (2-methyl:mep-indolol)	51,52
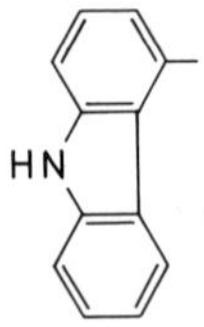 HN	i.pr.	Carazolol	53
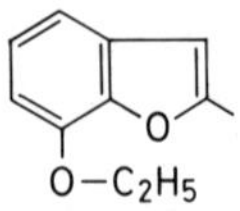 O; $O-C_2H_5$	i.b.	Bufuralol	54
Compounds with ring activator substituents			
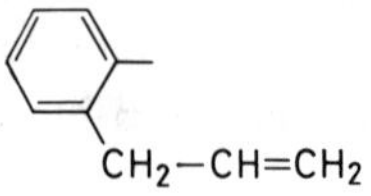 $CH_2-CH=CH_2$	i.pr.	Alprenolol	56,57
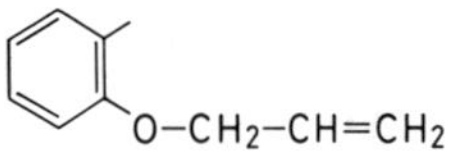 $O-CH_2-CH=CH_2$	i.pr.	Oxprenolol	58—60
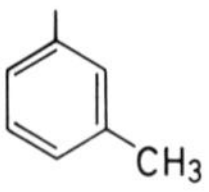 CH_3	i.b.	Toliprolol	61,62
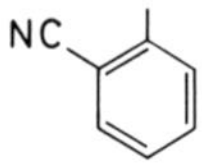 NC	i.b.	Bunitrolol	63,64
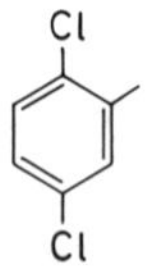 Cl; Cl	i.b.	Chloranolol	65,66

TABLE 7 (continued)
Structural Types of β-Blockers

Aryl	R	Compound	Ref.
Compounds with cycloaliphatic substituents on aromatic ring			
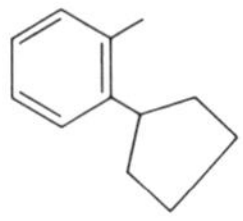	i.b.	Penbutolol	67,69
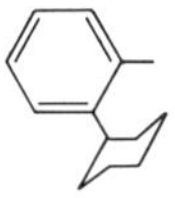	i.pr.	Exaprolol	70
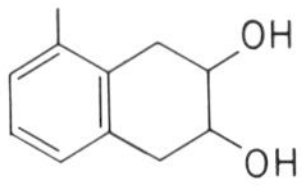 OH, OH	i.b.	Nadolol	71,72
Compounds with polar substituents on C-3 or C-4			
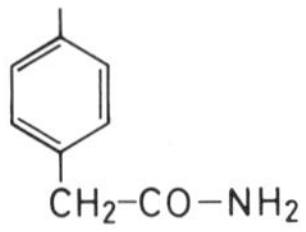 $CH_2-CO-NH_2$	i.pr.	Atenolol	73,75
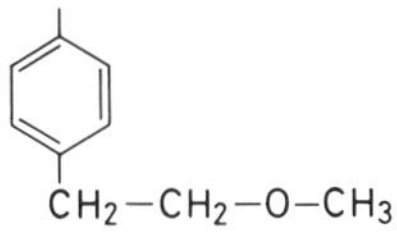 $CH_2-CH_2-O-CH_3$	i.pr.	Metoprolol	76,77
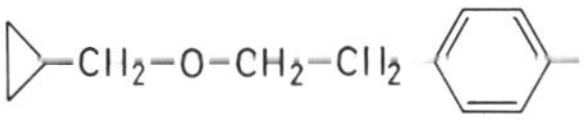 $-CH_2-O-CH_2-CH_2$	i.pr.	Betaxolol	78
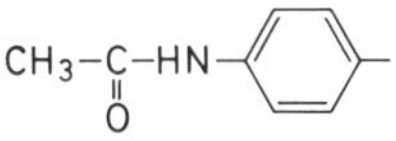 $CH_3-\underset{\underset{O}{\parallel}}{C}-HN-$	i.pr.	Practolol	79,80
Compounds with special substituents on side-chain amino group			
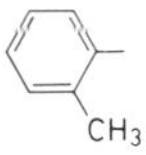 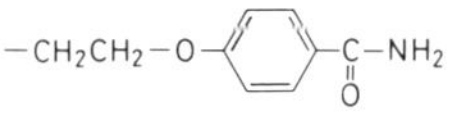 CH_3 ; $-CH_2CH_2-O-$ … $-\underset{\underset{O}{\parallel}}{C}-NH_2$		Tolamolol	81,82
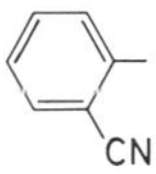 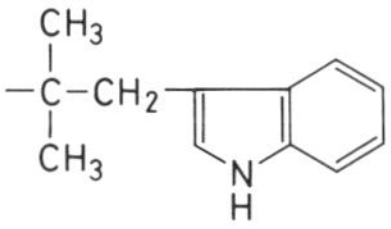 CN ; $-\underset{CH_3}{\overset{CH_3}{C}}-CH_2-$ (N, H)		Bucindolol	83
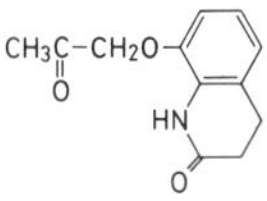 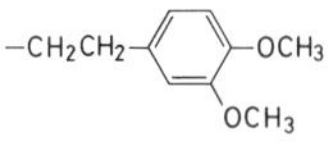 $CH_3\underset{\underset{O}{\parallel}}{C}-CH_2O-$ (HN, O) ; $-CH_2CH_2-$ … $-OCH_3$, OCH_3		Bometolol	84

It does not fit into the above picture that effective β-blockers are known with a cycloalkyl substituent in the *ortho* position to the propanolamino side chain. The best known compound of this type, penbutolol, proved to be 26 times more effective as a β-blocker than propranolol.[69] Penbutolol and exaprolol are strongly lipophilic; they are therefore metabolized and excreted rapidly, and have relatively short plasma half-lives. The strongly lipophilic drugs (propranolol, penbutolol, etc.) penetrate more easily and quickly into the cerebrospinal fluid, and in general into the central nervous system, causing CNS side effects more often than the hydrophilic β-blocker compounds (atenolol, nadolol, etc.). Nadolol is one of the most hydrophilic β-blockers, due to its cycloalkene (tetrahydronaphthalene) diol structure.

Distinct categories of β-blockers are formed by (1) bearing polar substituents at C-3 and/or C-4 of the aromatic ring, and (2) compounds with substituents other than isopropyl or *t*-butyl on the side-chain amino group. Atenolol, metoprolol, betaxolol, and practolol are examples of the first type; tolamolol and bucindolol are examples of the second type; and bometolol is of both structural types. These latter derivatives were at the focus of interest during the past decade, and drug research in this field led to promising results relating to cardioselectivity and/or antihypertensive effects.

Although atenolol, betaxolol, and metoprolol are potent cardioselective β-blocker antihypertensive agents, the rather high doses used to treat hypertension mean that part of this selectivity is lost, due to the β_2-mediated side effects. Efforts have been continued since the seventies to prepare β-blockers with higher cardioselectivity and significant antihypertensive activity. In many of these studies, practolol served as a structural model, and it appeared that the *p*-acylamino moiety is responsible for its cardioselectivity.[85] The benzamido analog **(2)** of practolol **(1)** displayed β-blocker effect and a complete or (in the case of the corresponding lactam analog **(3)**) a partial loss of cardioselectivity.[85] This was explained by the opposite electronic (electron-releasing) effect of the benzamido group from that of the acetamido group in practolol (electron withdrawing). The numbers in brackets under the formulas are the vascular/myocardial ED ratios, as a measure of cardioselectivity:

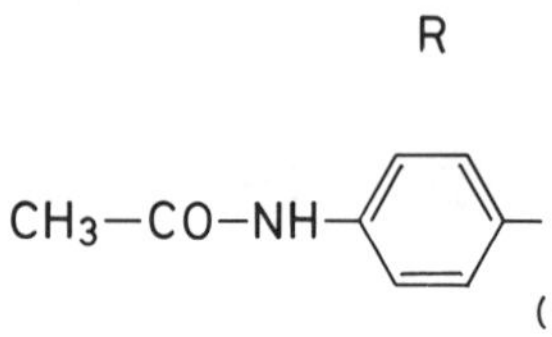

R

$CH_3-CO-NH-C_6H_4-$ (1) (practolol)

(31)

$R-OCH_2-CH(OH)-CH_2-NH-CH(CH_3)_2$

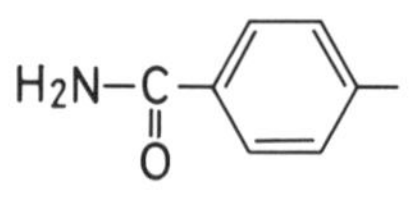

$H_2N-C(=O)-C_6H_4-$ (2)

(0,8)

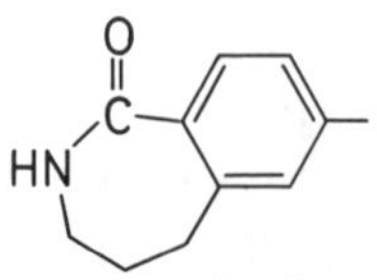

(3)

(2,6)

TABLE 8
Comparative Data of β-Blocker Activity for 1-Substituted 3-Amino-2-Propanol Derivatives[70]

O–CH₂–CH(OH)–CH₂–NH–R on the aromatic ring; ring substituents R₁ and NH–C(=O)–NH–R₂

Compound, substituents			Dose μg/kg, giving 50% inhibition of tachycardia	Inhibition %, of depressor response
Practolol			167	8
Propranolol			62	85
R	**R_1**	**R_2**		
H	*t*-Bu	H	101	0
CH_3	*t*-Bu	H	74	0
CH_3	*t*-Bu	*n*-Pr	93	7
C_2H_5	*t*-Bu	I	23	6
C_2H_5	*t*-Bu	C_2H_5	80	2
C_2H_5	t-Bu	$CH_2CH{=}CH_2$	38	30
n-Pr	*t*-Bu	H	33	0
n-Pr	*t*-Bu	CH_3	44	0
n-Bu	*t*-Bu	H	48	0
n-Bu	*t*-Bu	Cl	8	0
n-Bu	*i*-Pr	CH_3	46	0
n-Bu	*t*-Bu	CH_3	12	11
n-C_8H_{17}	*t*-Bu	CH_3	80	11
i-Pr	*t*-Bu	CH_3	49	5
$CH_2{-}CH{=}CH_2$	*t*-Bu	H	89	6
$CH_2{-}CH{=}CH_2$	*i*-Pr	CH_3	107	12

The results indicated the influence of the charge distribution exhibited by the aromatic part of the cardioselectivity of practolol-type β-blockers. After the cardioselective β-blocking property of practolol had been established,[80] a series of practolol analogs was prepared in which the acylamino moiety in the aryl residue was replaced by a ureido moiety.[86] Compounds with significant cardioselectivity and strength of effect were obtained only following further substitution at position 2 (**4**).

O–CH₂–CH(OH)–CH₂–NH–R on the aromatic ring; ring substituents R₁ and NH–C(=O)–NH–R₂

(4)

When their data were compared with those of practolol and propranolol, many of the 60 investigated compounds proved to be potent cardioselective β-blockers (Table 8).

It is concluded that the effect of R_2 in increasing the cardioselective potency increases with the lipophilicity of the substituent, but the selectivity of action is not apparently related to any single physical parameter, and no general conclusion can be drawn as concerns this property.

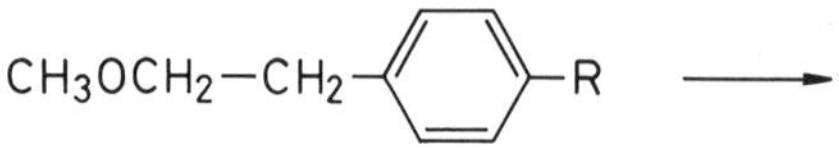

$CH_3OCH_2-CH_2-C_6H_4-R \longrightarrow HOOC-CH_2-C_6H_4-R \longrightarrow$

(5) 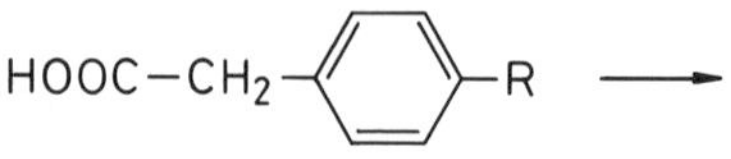 (6)

$\longrightarrow R_1OOC-CH_2-C_6H_4-R$

$R_1 =$ CH_2-CH_2- (adamantyl)

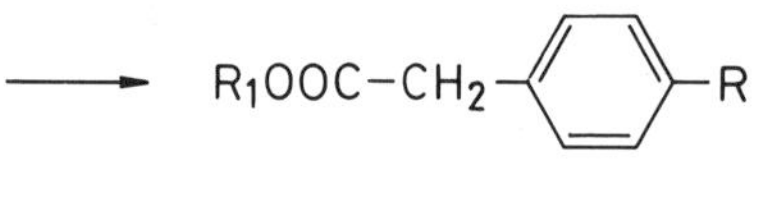

(7) 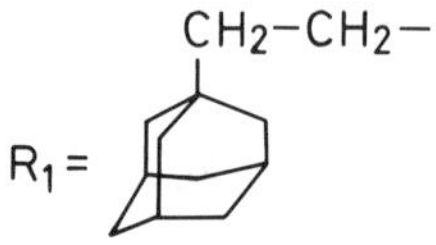 (8)

Bodor et al.[87,88] applied the "inactive metabolite approach" principle to the metabolites of metoprolol **(5)**, and designed and synthetized ester derivatives with β-blocker activity. In a recent paper,[89] they report that esterification of the inactive metoprolol-metabolite carboxylic acid **(6)** led to the preparation of compounds with an oculoselective β-blocking activity **(7)**. The "approach" is based on the strategy that the inactive metabolite can be transformed to derivatives with the same pharmacodynamic properties, but different pharmacokinetic features. The esters of the metoprolol-metabolite carboxylic acid will be degraded quickly to the inactive free acid by esterases in the plasma, but they will remain in the intraocular fluid for a relatively long period of time. Consequently, these "soft" "β-adrenoreceptor antagonists[88] exhibit an (ultra) short duration of nonselective systemic action, but due to a selective distributional feature they exert an ocular hypotensive (antiglaucoma) activity of normal or occasionally prolonged duration. The most promising of a number of compounds tested, the adamantylethyl ester **(8)**, emerged as the best potential candidate for ophthalmic (antiglaucomatous) use. Compared with the effect of timolol maleate, it was reported to induce a prolonged and significant reduction of intraocular pressure, and it was degraded very rapidly in human blood.

Substitution of the aromatic nucleus of the phenoxypropanolamine skeleton with further aromatic moieties resulted in potent cardioselective agents with marked antihypertensive activity.[90] From a series of imidazolyl-phenoxypropanolamine derivatives, many compounds proved to be more cardioselective than the known therapeutic agents. This is reflected in the values of the cardioselectivity ratio (Table 9).

As demonstrated in other reports, the *(S)*-isomers were found to be the more efficient. It is also worth mentioning that the 2-thienyl group as imidazole substituent, and 2-(3,4-dimethoxyphenyl)ethyl as *N*-terminal moiety, greatly enhanced the cardioselectivity. This latter finding resulted from systematic work which revealed that it is possible to increase the cardioselectivity by appropriate choice of the amino substituent;[91] moreover, the cardioselectivity was optimum when the "traditional" *N*-isopropyl or *t*-butyl moiety was replaced by the *N*-(3,4-dimethoxyphenetyl) group.[91] The unique nature of this group was substantiated by the fact that replacement of the 4-methoxy with a hydroxy, ethoxy, benzyloxy, or chloromethyl substituent decreased the selectivity; similar results were obtained when the positions of the methoxy substituents on the pnenyl ring were changed. The "magic" feature of (3,4-dimethoxyphenethyl)amino is apparent from a consideration of the data in Table 10.

TABLE 9
Comparative Cardioselectivity Ratio Values[a]

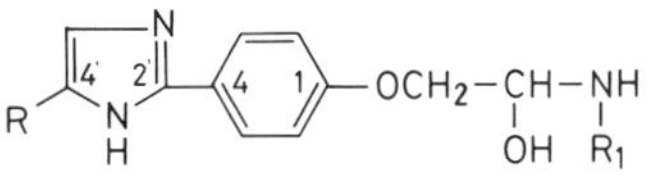

(Structure: 2-aryl imidazole with R at 4′ position; phenyl 4-substituted with $OCH_2-CH(OH)-NH-R_1$)

Compound R	R_1	Ratio[b]
H	$-CH_2CH_2-C_6H_3-3,4-(OCH_3)_2$	3236
CH_3	$-CH_2CH_2-C_6H_3-3,4-(OCH_3)_2$	7762
2-thienyl	$-C(CH_3)_2-CH_2-C_6H_3-3,4-(OCH_3)_2$	3981
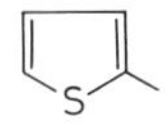 2-thienyl	$-(CH_2)_2-C_6H_3-3,4-(OCH_3)_2$	8709
	(S)-isomer	
	Practolol	43
	Atenolol	49
	Betaxolol	60
	Metoprolol	11
	Propranolol	12

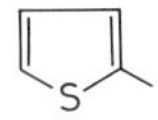

[a] *In vitro* values obtained from guinea pig tissue.
[b] Obtained by taking antilog of (pA_2 (β_1)-pA_2 (β_2)) relation.

TABLE 10
Comparative Selectivity Data for (3,4-Dimethoxyphenethyl) Amino Derivatives[91]

Compound	Apparent K_B Atrium	Apparent K_B Trachea	Ratio K_B
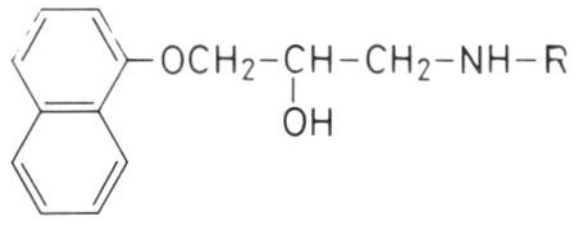 1-Naphthyl-$OCH_2-CH(OH)-CH_2-NH-R$			
R = $CH(CH_3)_2$ = propranolol	$1.3 \cdot 10^{-6}$	$0.8 \cdot 10^{-8}$	0.6
R = (DPE) 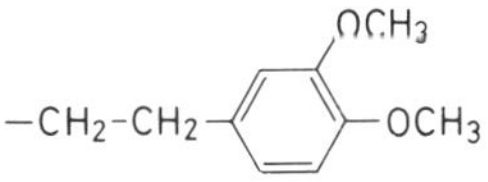 $-CH_2-CH_2-C_6H_3-3,4-(OCH_3)_2$	$1.1 \cdot 10^{-8}$	$4.8 \cdot 10^{-8}$	4.5
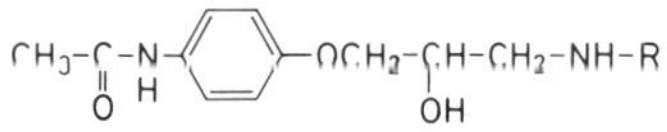 $CH_3-C(=O)-NH-C_6H_4-OCH_2-CH(OH)-CH_2-NH-R$			
R = $CH(CH_3)_2$ = practolol	$1.4 \cdot 10^{-6}$	$2.5 \cdot 10^{-5}$	18
R = DPE	$2.8 \cdot 10^{-7}$	$1.1 \cdot 10^{-4}$	393
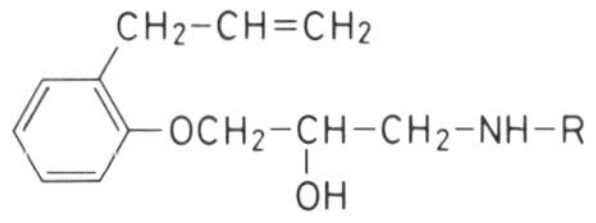 2-($CH_2-CH=CH_2$)$C_6H_4-OCH_2-CH(OH)-CH_2-NH-R$			
R = $CH(CH_3)_2$ = alprenolol	$8.6 \cdot 10^{-9}$	$1.1 \cdot 10^{-8}$	1.3
R = DPE	$7.6 \cdot 10^{-9}$	$9.2 \cdot 10^{-8}$	12

In connection with the continued efforts to prepare cardioselective β-blocking agents by substitution of the phenyl nucleus on the phenoxypropanolamine skeleton, the synthesis and biological testing of pyrazolyl **(9)** and triazolyl **(10—12)** derivatives[92] may be mentioned.

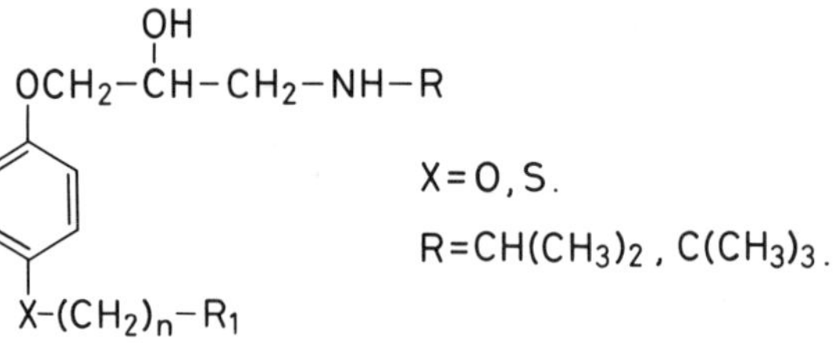

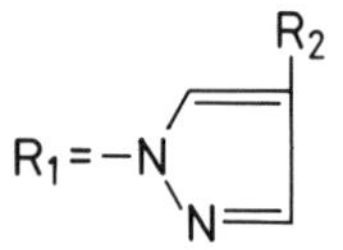

$X = O, S.$

$R = CH(CH_3)_2, C(CH_3)_3.$

(9)

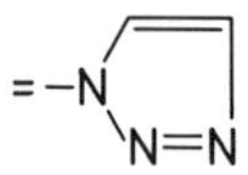

(10) 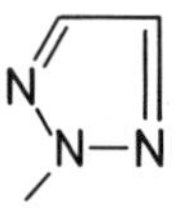 (11) 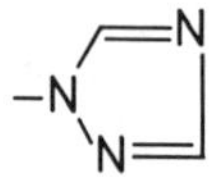 (12)

Some of the above derivatives exhibited a high level of β_1-antagonism in rats, with a selectivity ratio of more than 100:1. Taylor et al.[93,94] enlarged the number and types of azolyl-linked phenoxypropanolamine derivatives by the synthesis of 6-arylpyridazinones. An interesting compound from this series was prizidolol **(13)**, in which the structural elements of a direct-acting vasodilator (e.g., hydralazine) and a β-adrenoreceptor antagonist (e.g., propranolol) were united; this was shown to be a promising tool in the therapy of hypertension.[95] Because of the adverse toxicity exhibited by prizidolol, which was attributed to the hydrazino moiety of the compound, the research was continued with new azo-linked phenoxypropanolamine antihypertensives. This research led to the recent paper by Slater et al.,[96] who reported on the synthesis and biological evaluation of a series of 4,5-dihydropyridazinone derivatives of the traditional β-blocker skeleton. Another novel structural feature of these compounds was the acylaminoalkyl moiety inserted between the 4-pyridazinonyl-substituted phenyl and the secondary amino group **(14)**. Further, the structure-activity relationships were established between the electronic and steric parameters (van der Waals volume, etc.) of substituent R_1 and the β-blocker vasodilator effect. As a result, one of the compounds **(15)** was selected for continued trials.[96,97]

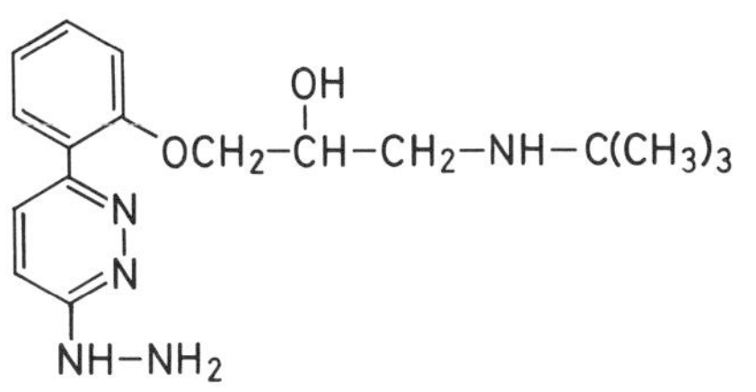

(13)

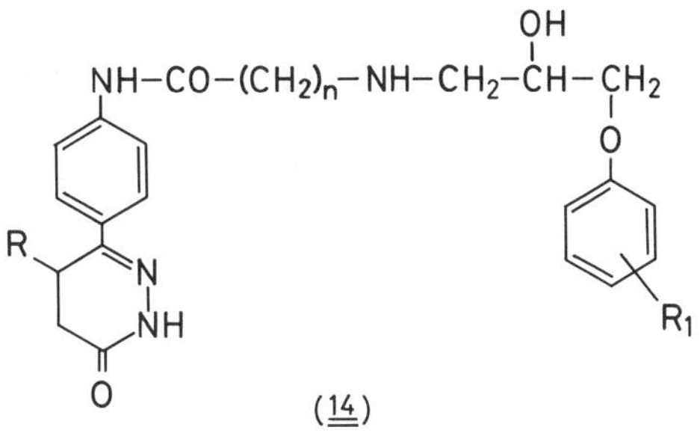

(14)

n=2 ., R_1=2CH_3 ., R=H

(15)

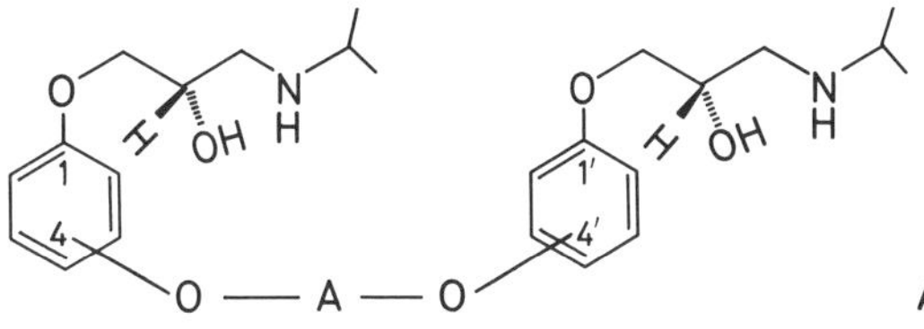

A=(alkyl)n or (ethylenoxy)n

(16)

β_1-Selective antagonists found among the binary (aryloxy) propanolamines consisting of two **(S)**-(phenyloxy)propanolamine pharmacophores **(16)**. In general, the compounds with 4,4′-linkages proved to be β_1-selective.

The insertion of an ester group into a suitable structural moiety basically having β-blocker activity may result in β-blocker compounds with an ultrashort duration of action. The experience stemming from this line of research is collected in several papers.[98-102] These agents are mainly needed to avoid the often dangerous side effects (cardiac failure during surgery or after myocardial infarction, etc.). The designing principle, bearing some common features with the "inactive metabolite approach",[87] led to the synthesis of compounds with a very unusual side chain: an extra planar function could be inserted between the ring and the ether oxygen of an (aryloxy) propanolamine molecule without the abolishment of β-blocker activity. This is illustrated by compound ACC-9089 **(17)**, which has been selected for toxicological and clinical study.

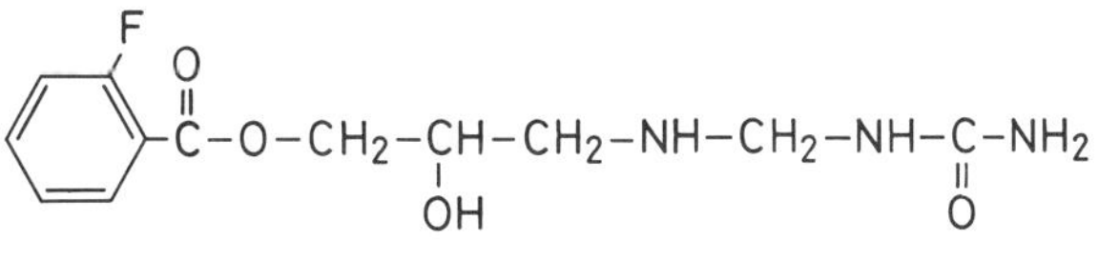

(17)

Essentially new structural types of β-blockers were discovered by the synthesis and biological evaluation of aliphatic and alicyclic oxime ethers.[103-105] Although the compounds that have been prepared so far are more active on the tracheal than on the cardiac β-receptor, analysis of the structure-activity relationship data provided bases for new concepts and initiated further research on the drug-adrenergic receptor interaction. One of the most important conclusions drawn from this work is that the aromatic moiety need no longer be regarded as an essential moiety for β-blocker activity. Despite the lack of an aromatic system, a number of oxime ether compounds **(20)** and also 3-(acyloxy)-propanolamine derivatives **(21)** exhibited marked competitive β-blocking activity.[105] Together with phenylethanolamines **(18)** and phenoxypropanolamines **(19)** there are nowadays four classes of β-blocker agents. These have a common aminoethanol portion, –CH(OH)–CH_2NMR, which is present in all the four structural classes **(18-21)** and also in the skeleton of phenylethylamine agonist molecules.

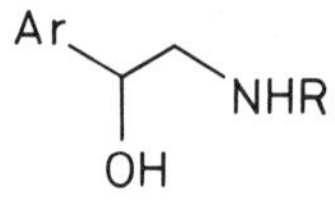

Ar, NHR, OH (18)

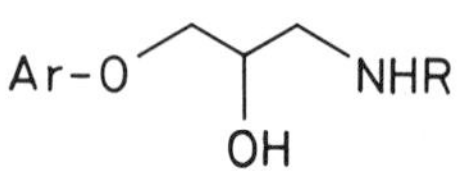

Ar–O, NHR, OH (19)

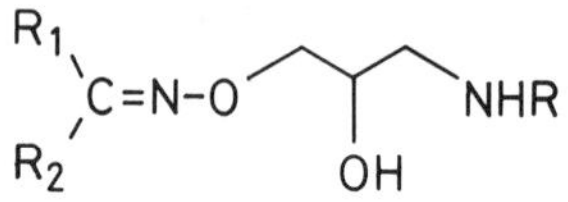

R_1, R_2, C=N–O, NHR, OH (20)

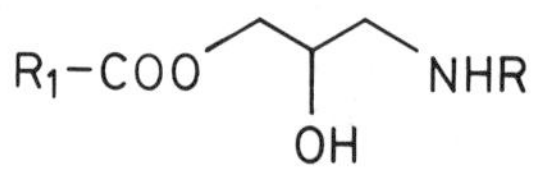

R_1–COO, NHR, OH (21)

This aminoethanol portion may therefore be associated with the ability of these molecules to bind to the β-adrenergic receptors. The remaining part of the molecules will determine the adrenergic (1) stimulant or (2) blocking properties, and the cardioselective or/and vasculoselective features of the drugs. In light of the new findings, it can be concluded that a specific stereoelectronic distribution is required for the interaction with β-receptors, and this must not be generated by the presence of an aromatic ring(s). The molecular electrostatic potential (MEP) has recently been suggested as a good tool for characterizing the electronic features of drug molecules,[106-110] and a reactivity index for phenylethanolamines was derived from the MEP.[111,112] Testa et al.[113] have carried out MEP calculations for β-blocker compounds belonging in the different structural classes. For the calculations, based on crystallographic results and theoretical conformational studies, the side chains were placed in an extended position and in the plane of the aromatic ring system. When the MEPs of β_1-selective, β_2-selective, and nonselective (aryloxy) propanolamine antagonist compounds were compared, a local negative minimum could be observed in the vicinity of the ether oxygen atom. This minimum is localized above and below the plane of the aromatic ring. The β_1-selective (aryloxy) propanolamines consistently generate a negative minimum centered 2 to 3 Å or more away from the *meta* position, and a positive maximum at the *para* position. The latter is replaced by a negative zone or minimum in the β_1-selective partial antagonists. On the basis of this observation, it was suggested[113] that the agonist-like activity of the β_1-selective partial antagonists is induced by an electrostatic interaction between this zone of negative potential and a complementary area in the β_1-adrenoreceptor. The experimental facts show that change in the *para* substituent from a negative potential to a positive one eliminates the intrinsic sympathomimetic (agonist) activity (ISA), the β_1-selectivity being retained, as indicated by the simultaneous presence of a negative minimum (see Figure 3, M3) located away from the *meta* position and a minimum (M_4) or maximum (P_4) in the *para* position. The positive or negative sign of the latter electrostatic feature is postulated to reflect the absence or presence, respectively, of an ISA. The β_2-activity (nonselective character) is enhanced by an increase of the lipophilicity (aromatic ring-fusion, etc.) and the possibilities of hydrophobic interactions.

Whether the knowledge accumulated from the structure-activity relationship studies of

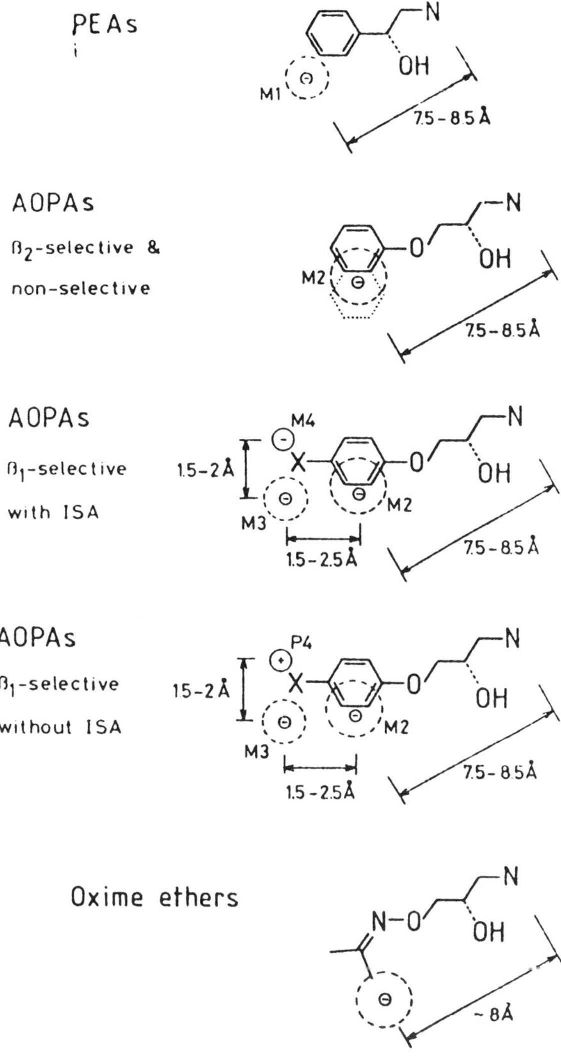

FIGURE 3. MEP-based pharmacophoric elements of various chemical and pharmacological classes of β-adrenoreceptor ligands. (From El-Tayar, N. et al., *J. Med. Chem.*, 31, 2072, 1988. With permission.)

the past decade will give rise to a new era in β-blocker design and put an end to the dominancy of (aryloxy) propanolamines is a question to be answered in the future.

V. SYNTHESIS

The synthesis of propranolol may be regarded as the initiating step for the systematic structure-activity relationship investigations in this field, for it led to the development of several general synthesis routes. One of the most frequently applied methods is the reaction of the corresponding phenolic compound with epichlorhydrin **(23)**. For the synthesis of propranolol, oxprenolol, and pindolol, 1-naphthol, benzcatechol **(22)**, and 4-hydroxyindole, respectively, are used. The synthesis of oxprenolol[104] led to the formation of the glyceryl ether **(24)**. The latter is condensed with isopropylamine **(25)** to give oxprenolol **(26)**.

OH, OH (22) — $BrCH_2{-}CH{=}CH_2$, Acetone K_2CO_3 → OH, $OCH_2{-}CH{=}CH_2$ — $ClCH_2{-}CH{-}CH_2$ (epoxide, O) → (23)

→ $OCH_2{-}CH{-}CH_2$ (epoxide, O), $OCH_2{-}CH{=}CH_2$ (24) — $H_2N{-}CH(CH_3)_2$ (25) → OH $NH{-}CH(CH_3)_2$, $OCH_2{-}CH{-}CH_2$, $OCH_2{-}CH{=}CH_2$ (26)

A. SOTALOL

The synthesis of β-blockers of phenylethanolamine type will be illustrated by the method for the preparation of sotalol. In the reaction between methanesulfonalilide **(27)** and chloroacetyl chloride **(28)**, 4-chloroacetylmethanesulfonanilide **(29)** is formed; on reaction with isopropylamine **(30)**, this product is transformed to 4-(isopropylaminoacetyl)-methansulfonanilide **(31)**. In the final step of the synthesis, the carbonyl group is catalytically reduced to a secondary alcohol group, yielding sotalol **(31)**.

$NH{-}SO_2{-}CH_3$ (27) — $ClCOCH_2Cl$ (28) → $NH{-}SO_2{-}CH_3$, $C(=O){-}CH_2Cl$ (29) — $H_2N{-}CH(CH_3)_2$ (30) →

$NH{-}SO_2{-}CH_3$, $C(=O){-}CH_2{-}NH{-}CH(CH_3)_2$ (31) — H_2 catal. → $NH{-}SO_2{-}CH_3$, $CH(OH){-}CH_2{-}NH{-}CH(CH_3)_2$ (32)

The synthesis route of practolol-like agents involves as preliminary step the introduction of a substituted ureido group, with continuation by the reaction with epichlorhydrin.

VI. METABOLISM

The two main routes in the metabolism of β-blockers are phenol oxidation in the aromatic nucleus and side-chain oxidation. A general review on this subject (the metabolism of 15 β-blockers) may be found in the paper by Pfeiffer.[115]

A. PROPRANOLOL

Propranolol **(33)**, one of the most often investigated substances, was found to metabolize fairly quickly and completely. It has been shown that more than 95% of the administrated propranolol is metabolized by the liver in humans.[116] The initial reports[117] suggested naphthoxylactic acid and 4-hydroxypropranolol as main end products of propranolol metabolism. From the use of a systematic HPLC assay procedure, Pritchard et al.[116] proposed a more detailed scheme for the metabolism of propranolol. The structures of six metabolites and their conjugates are described, and their quantitative determination is also reported. It is worth considering that the formation of an intermediate aldehyde **(35)** through the formation of the *N*-desisopropyl derivative **(34)** is assumed. Besides naphthoxylactic acid **(36)**, the formation of naphthoxyacetic acid **(37)**, propranolol glycol **(38)**, and α-naphthol **(39)** was established.

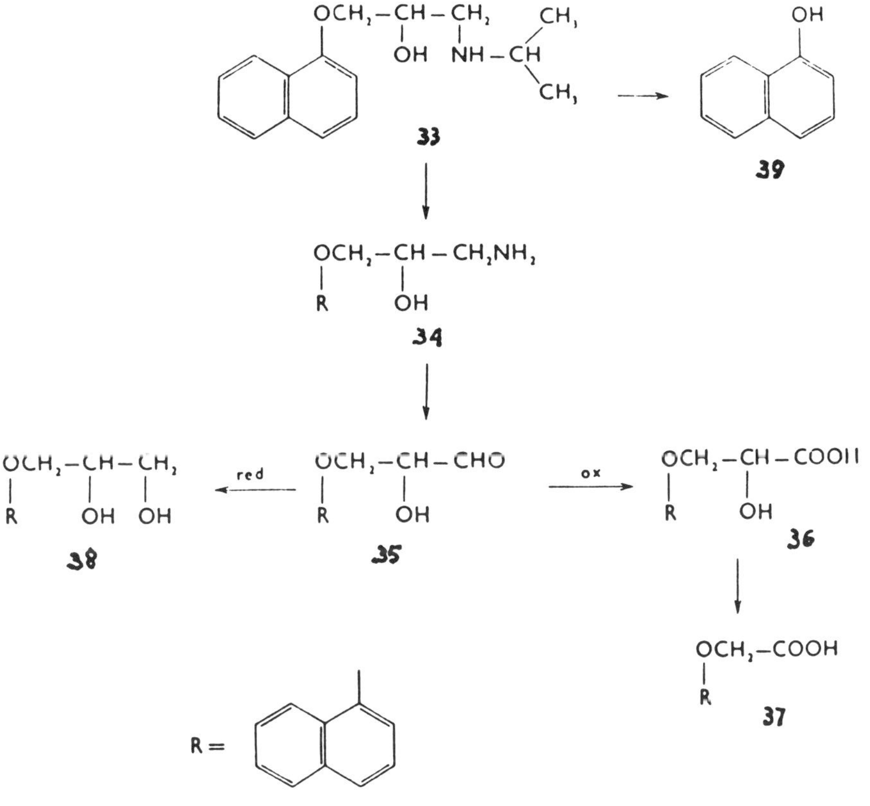

$OCH_2-CH(OH)-CH_2-NH-CH(CH_3)_2$ (R) **33** → OH (R) **39**

33 ↓

$R-OCH_2-CH(OH)-CH_2NH_2$ **34**

↓

$R-OCH_2-CH(OH)-CH_2(OH)$ **38** ←red— $R-OCH_2-CH(OH)-CHO$ **35** —ox→ $R-OCH_2-CH(OH)-COOH$ **36**

36 ↓

$R-OCH_2-COOH$ **37**

R = 1-naphthyl

Formation of the 4-hydroxy and 5-hydroxy derivatives as main metabolites has also been found for oxprenolol.[118]

The preparation and identification of two propranolol metabolic products was accomplished by Chen and Nelson.[119] 1,1-Diethoxy-3-(1-naphthoxy)-2-propanol and two corresponding ring-closure products were characterized. A fraction of previously unrecognized metabolite exhibiting a highly polar nature was found by Bargar et al.[120]

The metabolic oxydation of β-adrenoreceptor antagonists (metoprolol, timolol, bufuralol, and bopindolol) has been reviewed,[121] a survey being made of the evidence for polymorphic oxidation of β-blockers influencing their pharmacokinetic and pharmacodynamic parameters. It is an interesting observation that in "poor metabolizers" of metoprolol the half-life of *(R)*-metoprolol is longer than that of *(S)*-metoprolol. It was found that α-hydroxylation is greatly impaired in "poor metabolizer" individuals, and that there are also significant but smaller differences in *O*-demethylation: *(R)*-metoprolol was more rapidly *O*-demethylated than *(S)*-metoprolol. α-Hydroxylation in poor metabolizers was selective for the *(S)*-enantiomer. The (−)/(+)-enantiomer ratios for unchanged propranolol (1.50) and for propranolol glucuronide (1.76) indicate a selective oxidative metabolism towards the (+)-enantiomer.[122] The analysis of metabolites also showed that the overall ring oxidation strongly favored (+)-propranolol, leading to a greater bioavailability of the (−)-enantiomer of propranolol. The four dihydroxylated propranolol metabolites were identified by Nelson et al.[123] by GC-MS. Each of them arose stereoselectively from the (2*R*)-enantiomer of propranolol.

The formation of nitrosamine derivatives in the reaction of β-blockers and sodium nitrite was proved under *in vitro* conditions when oxprenolol was incubated for 1 to 4 h at 37° C and pH 1.3 to 3.8 in aqueous solution; the amount of oxprenolol converted to nitrosamine derivative was found to be considerable.[124] This and other similar findings[125] should serve at least as a warning for consideration of the carcinogenic risks accompanying the use of β-blockers (or almost any secondary amine drug substance).

VII. ANALYSIS

A review of the publications relating to the examination of β-receptor blockers reveals that the armaments of analytics include simple colour reactions, titrimetry, spectroscopy (UV and IR spectroscopy, and fluorimetry), electrochemical methods (polarography), the different kinds of chromatography, NMR, MS, and RIA. These methods can be classified according to the purpose of their use: detection; purity testing; determination of the compound; assay in biological samples, in pharmaceutical dosage forms or in the presence of other β-blockers or metabolites, or chemically different substances; and separation of enantiomers. For each of the methods, the above sequence of discussion will be followed below.

A. COLOR REACTIONS (IN COMBINATION WITH TLC SEPARATION)

For the identification of β-adrenergic blockers extracted from either dosage forms or biological specimens, Abdel-Hamid et al.[126] describe a TLC separation with subsequent color reactions. For the latter aim they propose indication of the spots through the use of (1) certain well-known reagents (see Table 11), (2) acetaldehyde *p*-chloranil or *p*-fluoranil in isopropanol; tetracyanoquinodimethane (TCNQ) or 2,3-dichloro-5,6-dicyano-*P*-benzoquinone (DDQ) in acetonitrile. It is presumed that the green to bluish-green colors elicited by the aldehyde-anil reagents are based on enamine formation via interaction of the secondary amino group of the β-blockers and acetaldehyde, which reacts further with the electron-acceptor reagent to produce colored substituted vinylamine *P*-benzoquinone derivatives. TCNQ gives rise to a green, and DDQ to a red color, due to the presence of radical anions.

B. CHROMATOGRAPHY

1. High-Performance Liquid Chromatography

When β-blockers are to be determined in biological samples, the technique involves either a direct injection, omitting deproteinization and purification,[127] or previous extraction and sometimes special treatment of the drugs. The extraction or cleaning can be performed by applying:

1. Solid-phase extraction (S) by:
   - Cartridge (Sep-Pak G_{18}, C_{18} Bond-Elut GDX 502 resin, etc.);[135,142-144,188,194,196,215,235,236]

TABLE 11
Colors[a] by β-Adrenergic Blockers with Various Color Reagents[126]

β-Adrenergic blocker	Zwikker I & II reagent	Marquis reagent	Vanillin conc H_2SO_4	Mandelin reagent	Methanol conc H_2SO_4	Froehde reagent
Acebutolol	Bl	Y	R	Gn	Y	Y
Alprenolol	Bl	R	Y	V	R	Gy
Atenolol	Bl	no	no	V	no	no
Carazolol	Bl	Gn	V	Gy	Y	Gn
Nifenalol	Bl	no	Y	no	no	no
Oxprenolol	Bl	V	Y	Br	V	V
Pindolol	Bl	Gn	V	Gy	V	Gy
Practolol	Bl	no	r	Gn	R	no
Propanolol	Bl	Bl	V	Gy	V	Gy
Toliprolol	Bl	R	V	R	no	R

[a] Bl: blue; Br: brown; R: red; V: violet; Gy: grey; Gn: green; Y: yellow; no: no color.

- Pre-column (Hypersil MOS, the column is filled with the same adsorbent as the separating one);[152,176,184,186,222,228]
- A system consisting of two or three columns;

2. Traditional extraction (T):
   - After deproteinization with trichloracetic acid in an organic solvent at alkaline pH;
   - From alkaline medium (pH = 8 to 13) with an organic solvent or a solvent mixture: the organic phase can be reextracted into an aqueous acidic phase (H_2SO_4, HCl, $HClO_4$), the organic phase is usually centrifuged, the supernatant is evaporated to dryness, and the residue is dissolved in an organic solvent or (generally) in the mobile phase.

The urinary metabolites of β-blockers are present as conjugates (glucuronide or sulfate), which are cleaved by hydrolysis with β-glucuronidase or arylsulfatase before the determination.

The generally used stationary phases are mainly reversed phases (C_8, C_{18}, cyano-amino, e.g., cyano[129] and phenyl), and rarely silica gel or ion-exchange adsorbents.[130] Accordingly, the mobile phases mostly contain aqueous buffer solutions (pH = 2 to 6) and organic modifier(s). For the detection, fluorometric, UV spectrophotometric, or electrometric detections[131,132] are used.

For the separation of β-blocker enantiomers, pre-column derivatization is most frequently proposed, the compounds being treated with:

- Phosgene (in toluene);[170,177]
- 2,3,4,6-Tetra-*O*-acetyl-β-β-glucopyranosyl isothiocyanate[164,166,167,240]
- *t*-BOC-L-Ala-anhydride[242,243]
- *t*-BOC-L-Leu-anhydride[242,243]
- 2,3,4-Tri-*O*-acetyl-α-D-arabinopyranosyl iso-2,3,4-thiocyanate[167]
- *N*-Trifluoroacetyl-*(S)*-(−)prolyl chloride[242,243]
- *(S)*-(−)-phenylethyl isocyanate[179]
- *(S)*-(−)-α-methylbenzyl isocyanate[202,230,231]
- *(R)*-(+)-1-isocyanatephenylethane[165]
- *(R)*-(−)-1-(1-naphthyl)-cthyl isocyanatc[168]
- *O,O*-disubstituted tartaric acid anhydride[176]
- Dansyl chloride[156,210]
- *O*-(isopropylcarbamoyl)isopropyl urea

TABLE 12
HPLC Methods for the Analysis of β-Blocking Agents

Compounds	Stationary phase	Mobile phase	Detection, λ	Sensitivity/linearity	Note	Ref
Propranolol (Pr)	YWG-C_{18}	Me-OH-H_2O-phosphate buffer (pH = 9.3) 92:8:1	UV, 232 nm	50 ng 0.05—2 μg/ml	B (T)	134
Pr	μBondapak C_{18}	MeOH-(50 m*M* KH_2PO_4-2.5 m*M*. Na pentane-1-sulfonate) (pH = 5) 1:1	Fluorimetry (F) exc: 296 nm emiss: 338 nm	— 5—400 μg/l	B (S)	135
Pr Pronethalol (I.St)	RP-C_{18}	CH_3CN-H_2O-CH_3COOH 30:69:1	F exc: 235 nm emiss: 335 nm	0.5 ng/ml (I)	B (T) For pharmacokinetic studies	136
Pr *N*-ethyl propranolol (I.St)	μBondapak-Phenyl (10 μm)	CH_3CN-H_2O-AcOH-Triethylamine 9000—12000:200:3 (pH = 3.5)	exc: 228 nm	0.5—1 ng/ml —	B (T)	137
Pr Metoprolol tartrate (I.St)	CN cartridge	50 m*M* NaH_2PO_4 (pH = 3, adjd with 85% H_3PO_4)-CH_3CN-CH_3-OH (190:39:21)	exc: 250 nm emiss: 336 nm		B (T)	138
Pr	Spherisorb S5W silica	CH_3OH-2,2,4-trimethylpentane, containing 1 m*M* camphorsulfonic acid	F	2.5 μg/l —	B	139
Pr	Hitachi gel 3031 (5 μm)	65% EtOH in 0.02 *M*-$HClO_4$-$NaClO_4$ buffer (pH = 2)	exc: 285 nm emiss: 340 nm	— 1—100 ng	B (T)	140
Pr Procainamide (I.St) 4-Hydroxypropranolol *N*-desisopropylpropranolol Propranolol glycol α-Naphtol	μBondapak C_{18} (1 μm)	CH_3CN-MeOH-CH_3COOH 35:5:1	F emiss: 295 nm			141

1-Naphtoxylactic acid 2-Naphtoxyacetic acid 1-Naphtoxyacetic acid						
Pr 4-Hydroxypropranolol (II) *N*-desisopropylpropranolol (III) Propranolol glycol (IV)	µBondapak C_{18}	CH_3CN-H_2O 30:70	—	1 ng (Pr)/ml plasma 6 ng (II) 1 ng (III) 2.5 ng (IV)	B (S)	142
Pr 1-Naphtyloxylactic acid	Nova-Pak C_{18}(5 µm)	CH_3CN-0.1% H_3PO_4 3:7	F exc: 220 nm	50—1000 ng/ml	B (S)	143
Pr 4-Hydroxypropranolol (II) *N*-desisopropyl-Pr (III) Propranolol glycol (IV)	µBondapak C_{18} (10 µm)	OH_3CN-0.1% H_3PO_4 (23:77)	F	1.0, 6.0, 1.0 and 2.5 ng Pr II-III-IV, resp	B (S)	144
Pr 4-Hydroxy propranolol (II) 1-Amino-3-(-1-naphtyloxy)propan-2-ol (I.St)	µBondapak C_{18} (10 µm)	H_2O-CH_3CN-H_3PO_4 1400:1000:10 (containing 1.6 g of Na dodecyl sulfate)	F exc: 308 nm emiss: 398 nm	2 ng (Pr)/3 ml plasma 1 ng (II)/3 ml p.	B (T)	145
Pr (1) 4-Hydroxy propranolol (II)	ODS coated with 6% bovine serum-albumin after equilibration of the column with 0.1 *M* phosphate solution (pH = 3.0)	CH_3CN-phosphate buffer (pH = 7.4) 22:78	F exc: 297 nm 327 nm emiss: 347 nm 427 nm I and II resp	100—400 p*M*	B	146
Pr 4-Methylpropranolol(I.St) 4-Hydroxy-Pr (II)	ODS	CH_3CN-H_2O-CH_3COOH-MeOH triethylamine 7000:1400:200:8000:1 (pH = 3.4)	F exc: 238 nm emiss: 360 nm	1 and 2 ng/ml Pr and (II) resp	B (T) As antioxidant ascorbic acid was added.	147

TABLE 12 (continued)
HPLC Methods for the Analysis of β-Blocking Agents

Compounds	Stationary phase	Mobile phase	Detection, λ	Sensitivity/linearity	Note	Ref
Pr 4-Hydroxypropranolol sulfate	C_{18}	10 m*M*-ammonium acetate soln.-CH_3CN 85:15	UV, 280 nm	—	B (T) Tetrabutyl ammonium ion is added (for ion pairing). Corresponding glucuronide does not interfere.	148
Pr Basic, acidic, and neutral metabolites	C_{18}	0.004 *M* phosphate buffer (pH = 2.7)-THF-CH_3OH-CH_3CN with the addition of *n*-butylamine as a competing base	F	In the low ng to sub ng range was detectable	Basic and neutral metabolites at pH = 10.5 with 2% *n*-butanol in dichlormethane, acidic metabolites into ether *n*-butanol-dichlormethane or CCl_4 were extracted.	149
Pr Glucuronide naphthyl-acetic acid	μBondapak C_{18}		F. exc: 242 nm emiss: 370 nm	—	In plasma and breast milk.	150
Pr Indalpine (I.St.) 4-Hydroxypropranolol Propranolol glycol *N*-desisopropylpropranolol	RP-C_8 (5 μm)	5 m*M* Na-acetate + 5 m*M* tetramethylammonium hydroxyde (pH = 4.3 adjd with 25% H_3PO_4)CH_3CN 10—45% (nonlinear gradient elution)	F exc: 295 nm emiss: 360 nm	— 1.25—20 μg/ml	B	151
Pr 4-Hydroxy-propranolol (II)	Hypersil MOS (5 μm)	0.1 *M* H_3PO_4-MeOH (1:1)	UV 219 nm (Pr) and 208 nm (II)	0.1 and 0.08 (Pr) and (II) resp 20—1000 μg/ml	B (S)	152
Pr 4-Hydroxypropranolol (II) pronethalol hydrochloride (I. St.)	μBondapak Phenyl (pre-column 37—50 μm, Bondapak Phenyl Corasyl)	CH_3CN-H_2O-anhyd-CH_3COOH 200:293:3 (containing 0.5 g/l each of Na dodecyl sulfate and Na heptane sulfonate)	F exc: 212 nm	1 ng/ml (Pr) 2 ng/ml (II) 10—200 ng/ml	B (T)	153

Pr 4-Hydroxypropranolol (II) *N*-desisopropranolol (III)	RP-phenyl, or octyl or ODS	CN_3CN-MeOH-H_3PO_4 (combination depending upon the column)	F	0.2 ng/ml (Pr) 1.0 ng/ml (II) 0.2 ng/ml (III)	B (T) Extraction: in the presence of *ascorbic acid*	154
Pr 4-Hydroxypropranolol (II)	Lichrosorb CN	MeOH-0.03 *M* acetate buffer 65:35	F exc: 285 nm emiss: 405 nm	10 ng/ml (both of Pr, II) 10—250 ng/ml	B (T)	155
Pr Acetoacetic acid Pyruvic acid	Partisil PXS 5	$CHCl_3$-HCOOH 97.85:2.15	UV, 360 nm	0.5 ng/ml	B The two acids (endogenous) were converted into 2,4-dinitrophenylhydrazones, propranolol (exogenous) amino group was modified with dansyl chloride	156
Pr 1-Naphtyloxylactic acid (II) Quinidine sulfate (I.St)	TSK gel LS 410 (5 μm)	MeOH-5% ammonium acetate sol (7:3)	F exc: 295 nm emiss: 337 nm	0.5 ng/ml (Pr) 2 ng/ml (II) 20—100 ng/ml (Pr) 0.2—2 μg/ml (II)	B (T)	157
Pr Metoprolol (II) Atenolol (III)	LiChrosorb C_{18} (10 μm)	MeOH-H_2O mixture, containing alkyl sodium sulfate and 1.3% of phosphoric acid	F exc: 280 nm emiss: 330 nm	2 ng/g (Pr, II) 5 ng/g (III)	B	158
Pr Pindolol (II) Oxprenolol (III)	GYQG-C_{18}	MeOH-H_2O 9:1 aq 2% CH_3COOH	UV, 260 nm	2 ng (Pr) 2 ng (II) 12 ng (III)	B	159
Pr Metoprolol (II) Atenolol (III)	Si 60 RPC_{18}	CH_3CN-aq-phosphate buffer (pH = 4) CH_3CN-H_2O in different ratio	F	1—2 ng/ml in plasma 1 ng/ml(Pr) 10 ng/ml(II) 5 ng/ml(III) in urine	B	160
Pr Metoprolol Atenolol	LiChrosorb	CH_3OH-H_2O-phosphoric acid-alkyl sulfate Na salt	—	2—5 ng/ml from plasma 5—10 ng/ml from tissue	B (T)	161

TABLE 12 (continued)
HPLC Methods for the Analysis of β-Blocking Agents

Compounds	Stationary phase	Mobile phase	Detection, λ	Sensitivity/linearity	Note	Ref
Pr Nadolol (II) Prazosin (III) 4,5,7,8-tetramethyl-aporphine hydro-chloride (I.St)	Partisil 10-SCX (10 μm)	CH_3CN-H_2O-diethylamine-85% H_3PO_4 (400:1600:4:3)	F exc: 225(Pr), 205(II), 246(III) nm emiss: 370 nm	5—200 ng Pr (serum) 100—1000 ng Pr (urine) 20—500 ng II (serum) 2—50 ng III (serum)	B	162
Pr Metoprolol (II) Atenolol (III) Bisoprolol (IV)	(a) Silicagel Si 60 (b) RP-8	(a) aq ammonium phosphate buffer (pH = 4)-CH_3CN 3—7% (b)CH_3CN-H_2O mixtures		2 ng/ml(II), (IV), in plasma 10 ng/ml(II), (IV), in urine 1 ng/ml (Pr) in plasma 5 ng/ml(III) in plasma 50 ng/ml(III) in urine		163
Pr	Spherisorb ODS-5	0.02 *M* $NH_4H_2PO_4$-CH_3CN 21:29	UV, 230 nm		B (T) Enantiomer separation Derivatization: 2,3,4,6-tetra-*o*-acetyl-β-D-glucopyranosyl isothiocyanate	164
Pr 4-Hydroxypropranolol	C_{18}, or silica (10 μm)	MeOH-H_2O or MeOH-$CHCl_3$	UV, 313 nm		(T) Enantiomer separation Derivatization: P-(+)-1-isocyanatophenylethane was added. In urine, the glucuronic acid conjugates were enzymatically hydrolyzed; the sulfate conjugates of metabolites	165

					were extracted as their ion pair with tetrabutylammonium salt	
Pr 4-Hydroxypropranolol sulfate	Spherisorb ODS (5μm)	CH_3CN-MeOH-H_2O-CH_3COOH (35:5:59:1) in 0.05 *M* ammonium acetate (pH = 4)	UV, 313 nm	20 ng (each enantiomer) 100—5000 ng	B (T) Derivatization: 2,3,4,6-tetra-*o*-acetyl-β-D-glucopyranosyl isothiocyanate	166
Pr 10 Related compounds	Ultrasphere ODS	CH_3CN 37—75% in 0.02 *M*-ammonium phosphate			Enantiomer sepn. Derivatization: 2,3,4-tri-*o*-acetyl-α-D-arabinopyranosil isothiocyanate or (preferable) 2,3,4,6-tetra-*o*-acetyl-β-D-glucopyranosyl isothiocyanate	167
Pr Bunitrolol Metoprolol Alprenolol	RP-18	CH_3OH-water 70:30 or 60:40	UV, 290 nm F exc: 285 nm emiss: 335 nm		B Enantiomer separation Derivatization: R-(−)-1-(1-naphthyl) ethylisocyanate	168
Pr	Silica, coated with cellulose-tris-(3,5-dimethyl-phenylcarbamate polymer)	Hexane-propanol-*N*,*N*-dimethyl-octylamine 92:8:1	F exc: 290 nm emiss: 340 nm		B (T) Enantiomer separation	169
Pr Pronethalol (I.St)	Chiral phase with (*R*)-*N*-(3,5-dinitrobenzoyl)-phenylglycine as chiral discriminator	Hexane-propan-2-ol-CH_3CN 97:3:1	F exc: 290 nm emiss: 335 nm	0.5—100 ng/ml	B (T) Enantiomer separation Derivatization: phosgene	170
Pr Timolol Acetobutolol Metoprolol Alprenolol Betaxolol	Partisil ODS (10μm)	20—30 or 40% CH_3CN in 20 m*M* H_3PO_4, 0.2 or 4 m*M* in Na heptane-1-sulfonate	UV, 210 nm	—	—	171

TABLE 12 (continued)
HPLC Methods for the Analysis of β-Blocking Agents

Compounds	Stationary phase	Mobile phase	Detection, λ	Sensitivity/linearity	Note	Ref
Pr Quinidine Lidocaine Procainamide	RP-C_8	CH_3CN-aq phosphate buffer	UV, F			172
Pr Verapamil (I.St)	μBondapak C_{18}	MeOH-CH_3COOH 86:1 containing 0.2 *M* ammonium formate	UV, 270 nm	0.8—2.5 μg	Ph	173
Pr Hydralazine hydrochloride (II) Pyridoxine hydrochloride (I.St)	μBondapak C_{18}	MeOH-CH_3CN-H_2O-KH_2PO_4-NaH_2PO_4 150:30:14:3:3	UV, 254 nm	0.6 (Pr) and 1.5 (II) μg 1.5—22.5 μg (Pr) 1.5—11.5 μg (II)	Ph	174
Pr Atenolol Practolol Metoprolol Oxprenolol Alprenolol Acebutolol Pronethalol	LiChrosorb Si 60	Phosphate buffer (pH = 2.1) with 0.008 *M* octylsulfate and 0.0014 *M* *N,N,N*-trimethyloctylamine + KBr cc of 0.008 *M* of bromide (I = 0.1)	UV, 270 nm or refractive index		Study on dynamically modified silica with ion-pairing modifiers	175
Pr 13 β-blockers	Polyogosil RP-18 (7 μm) Pre-column: LiChrosorb RP-18 (25 μm)	(a) 0.2 *M* phosphoric acid adjd the actual pH with ammonia-methanol 1:1 (b) 0.2 *M* ammonium phosphate buffer (pH = 6)-MeOH 1:1 (c) 2% acetic acid in H_2O (pH = 3)-CH_3OH 1:1 or 35:65	UV, 254 nm		Enantiomer separation Derivatization: tartaric acid anhydrides	176

Pr Alprenolol Oxprenolol Metoprolol Pindolol	α_1 acid glycoprotein bonded silica (10 μm)	Phosphate buffer (pH = 7) containing 4—15% of propan-2-ol	—	—	Enantiomer separation Derivatization: phosgene soln in toluene	177
Pr	Cyanopropyl-silica	Dichlormethane-hexane-acetonitrile 79:20:1 + 5 m *M* *d*-10-camphorsulfonic acid + 2.5 m *M* tert-butylamine			Enantioner separation	178
Metoprolol (M)	Zorbax ODS	MeOH-H_2O 75:5	F exc: 265 nm emiss: 313 nm	2 ng/ml	B (T) Enantiomer separation Derivatization: (*S*)-(−)-(Phenylethylisocyanate)	179
M	RP- C_{18}		F exc: 222 nm emiss: 320 nm	5—10 ng/ml 20—160 ng/ml	B	180
M 4-Methylpropranolol (I.St)	RP-C_{18}			5 ng/ml 10—300 ng/ml	B (T)	181
M Alprenolol (I.St)	μBondapak C_{18}	CH_3CN-MeOH-H_2O 3:6:11 (pH = 3) (with H_3PO_4 adjd)	F exc: 275 nm emiss: 300 nm	2 ng/ml 5—500 ng/ml	B (T)	182
M α-Hydroxy metoprolol	RP-C_8	CH_3CN-0.088 *M* acetate buffer 75:25	F exc: 225 nm emiss: >320 nm		B (S)	183
M α-Hydroxymetoprolol(II) 1-(4-butyramido-2-butirylphenoxy-(-3-isopro-pylamino)-propan-ol (I.St.)	(a) Hypersil 5-ODS (b) Spherisorb phenyl	(a) 20% CH_3CN containing 1% of triethylamine (for plasma) pH = 3 (b) Aqu 45% MeOH containing 5 m*M* Na-heptane sulfonate + 0.1% of CH_3COOH (for urine)	F exc: 222 nm emiss: 320 nm	5 ng/ml (M) 0.2 μg/ml (II) 10—400 ng/ml in plasma 0.5—40 μg/ml in urine	B (T)	184

TABLE 12 (continued)
HPLC Methods for the Analysis of β-Blocking Agents

Compounds	Stationary phase	Mobile phase	Detection, λ	Sensitivity/linearity	Note	Ref
M Nadolol (I.St) α-Hydroxy metoprolol (II) *o*-Dealkylated metab (III) Carboxylic acid metab (IV)	(1) Radial-Pak cartridge containing Nova-Pak C_{18} (5 μm) C_{18}-precolumn (2) Radial-Pak cartridge containing Partisil SCX CN-precolumn (for carboxilic acid metab)	(1) Aq 12% CH_3CN containing 1% of triethylamine (pH = 3) (2) 0.05 *M* NaH_2PO_4-MeOH 49:1 (pH = 5)	F exc: 193 nm (no emission filter)	0.05 μg/ml (M) 0.01 μg/ml (II) 0.02 μg/ml (III) 0.5—1 μg/ml (IV)	B (T)	185
M α-Hydroxymetoprolol Pronethalol (I.St)	μBondapak C_{18}	CH_3CN-H_2O-aqu 1% CH_3COOH-5m *M* Na heptane-1-sulfonate 39:38:1:2	F exc: 230 nm emiss: 300 nm	5—600 ng/ml for serum, 15—150 ng/ml for cerebrospinal fluid	B (T)	186, 187
M Atenolol (II)	(1) μBondapak C_{18} (2) Spherisorb nitrile silica	CH_3CN-0.1% H_3PO_4 23:77 (1) CH_3CN-0.05 *M* phosphate buffer 1:9 of pH = 7 (2)	F exc: 193 nm (no emission filter)	20 ng/ml (M) 10 ng/ml (II), 1—1000 ng/ml (M) 1—500 ng/ml (II)	B (S)	188
M α-Hydroxy metoprolol (II) Pafenolol (I.St) Metoprolol-carboxylic acid (III)	Microspher C_{18} (3 μm)	1)CH_3CN-10 m*M*-*N*,*N*-dimethyloctylamine (in phosphate buffer pH = 3) 6:5:39 2)With gradient elution of 11—24% of CH_3CN in MeOH-propan-2-ol-0.1 *M* Na-dodecyl sulfate in phosphate buffer of pH = 2:132:33:100	F exc: 228 nm emiss: 306 nm	—	B (T) (S)	189
M α-Hydroxy metoprolol (II)	LiChrosorb RP-8	CH_3CN-acetate buffer soln 3:1	F exc: 225 nm miss: 320 nm	— 1—1462 n*M* (M) 1—1188 n*M*(II)	B (T)	190

Alprenolol (I.St) M α-Hydroxy metoprolol (II) Inactive amino-acid metab (III)	LiChrosorb RP-8 (5 μm)	70% CH_3CN in 2.9 m*M* Na acetate-40 m*M* CH_3COOH (pH = 3.5)	UV, 222 nm	>14.6 μM (5 μg/ml) *M* 19.9 μM (II) 16.5 μM (III) 11.9—237.5 μM (II) 16.5—329.5 μM (III)	B Without any purification	191
M Hydrochlorothiazide (II) Furosemide (I.St)	μBondapak NH_2	1 m*M* Na-acetate in MeOH (pH adjd to 4.6 with CH_3COOH)	UV, 254 nm	1 μg (M) 0.25 μg (II), 1—30 μg (M) 0.25—2.5 μg (II)	Ph	192
M Hydroxymetoprolol (II) (±)ethyl-2-(4-(3-isopropyl-amino-2-hydroxy-propoxy-phenyl)ethyl carbamate (I.St)	Silicagel B/5 (5 μm)	Hexane-isopropanol-MeOH-NH_3 850:100:50:1	F exc: 224 nm (no emission filter)	3 ng/ml (M) 12.5 ng/ml (II)	B (T)	193
Oxprenolol (O)	ODS	MeOH-0.05 M/l aq dodecyl sulfonate 90:10 90:10	UV, 275	0.03 mg/l	B (S) In urine by HPRC	194, 195
O Propranolol (I.St)	Spherisorb 3 ODS (3 μm)	0.03*M* KH_2PO_4 buffer (pH = 2.2)-CH_3CN 1:1 containing 0.02% of triethylamine (pH = 3)	UV, 275	17,6—282,2 ng	B (S)	196
O Timolol Pindolol	Hypersil C_{18}	(a)Phosphate buffer (pH = 3)-25% CH_3CN (b)Phosphate buffer (pH = 3)-0.001 *M* dimethyloctylamine and 21% CH_3CN	UV, 270 and 254 nm	—	—	197

TABLE 12 (continued)
HPLC Methods for the Analysis of β-Blocking Agents

Compounds	Stationary phase	Mobile phase	Detection, λ	Sensitivity/linearity	Note	Ref
Atenolol Practolol *o*-Desmethylmetoprolol Prenalterol Pafenolol Acebutolol	LiChrosorb Si 60	0.008 *M* trimethyloctyl-ammonium-0.008 *M* octylsulphate in phosphate buffer (pH = 2.2) (I = 0.1)				
O Alprenolol (I.St)	LiChrosorb RP-8 (10 μm)	CH_3CN-buffer 3:7-1 m*M* of Na octanesulfonate-1 m*M* anhyd Na acetate, 470 ml of H_2 O 550 ml of EtOH-0.6 ml anhyd acetic acid	UV, 222 nm	66 n*M* rectilinear up to 3310 n*M*	B (T)	198
O *N*- and *O*-dealkylated metabolites, their oxidation and acylation derivs. ring-hydroxylated metabolite	LiChrosorb RP-8	A mixture of sodium octanesulfonic acid-sodium acetate-H_2O-MeOH-CH_3COOH in acetonitrile		20—744 ng/ml O		199
O	Radial-Pak μBondapak CN	MeOH-CH_3CN-53.9 m*M* sodium phosphate buffer soln (pH = 3.0) 21:39:190	UV, 272 nm	0.04—0.8 mg/ml	Ph	200
O 4-Hydroxy-*O* (II) Deisopropyl-*O* (III) Methylamino-*O* (IV) Carboxylic acid of O (V) Glycol (VI) Acetamide (VII)	LiChrosorb Si 60	For O,II,III,IV dichloroethane-propanolbutanol-aq NH_3 25:25:5:1 For V,VI,VII dichloroethane-tetrahydrofuran-dioxane-formic acid-H_2O 80:10:10:4:1	UV, 275 nm	4 ng/ml	B (T) Isotope dilution with ^{14}C-labelled O	201

O 9-Metabolites	LiChrosorb Si 60	1.2-dichlorethane-*n*-heptane-methanol-2-propanol 95:60:2:1	UV, 275 nm	5 or 1 ng/ml 600—1300 μg	B Enantiomer separation Derivatization: with racemic *S*(−)-α-methylbenzyl isocyanate	202
Labetalol (L)	Biophase ODS	CH_3CN-0.05 *M* phosphate buffer 35:65 (pH = 6)	Amperometry		B	203
L	RP-8	CH_3CN-phosphate buffer (pH = 3)	UV, 207 nm	10—400 ng/ml	B	204
L *O*-benzyllabetalol	LiChrosorb 5 RP-8	0.05 *M* Na-citrate (pH = 6.5)-propanol-CH_3CN 25:14:11	F exc: 311 nm emiss: 400 nm	1 ng/ml, 10—350 ng/ml	B (T)	205
L Propranolol (I.St)	Ultrasphere octyl (5 μm)	0.01 *M* Ammonium phosphate-MeOH at pH = 3	UV, 216 nm	10 ng/ml, rectilinear up to 1000 ng/ml	B (T)	206
L *O*-benzyllabetalol (I.St)	LiChrosorb 5 RP-8	0.05 *M* Na citrate soln (pH = 6.5)-2-propranolol-CH_3CN 50:28:22	F emiss: 400 nm	1 ng/ml	B	207
L 5-1-Hydroxy-2-(4-*p*-tolylbut-2-ylamino)ethyl-salicylamide HCl (I.St)	PRP-1 macroporous (Poly(styrene-divinylbenzene) (10 μm)	0.05 *M* Ammonium carbonate buffer of pH = 9.5-CH_3CN-tetrahydrofuran 99:25:8	F exc: 370 nm emiss: 415 nm	4-500 ng/ml	B (T) The method is also suitable for determining dilevalol, the (RR) isomer of labetalol.	208
L 5-(2-(4-Chlorophenyl)-ethyl)salicylamide hemihydrate (I.St)	PRP-1 column, packed with poly(styrene-divinylbenzene) copolymer	50 m*M*-phosphate buffer (pH = 4)-CH_3CN 3 1	F exc: 340 nm emiss: 412 nm	1 μg/l, 1-250 μg/l	B (T)	209
L	LiChrosorb Si60	Hexane-ethylacetate-MeOH-H_2O 100:100:4:1	UV, 270 and 360 nm		Enantiomer separation Derivatization: dansyl chloride	210
L	μBondapak C_{18}	—	UV, 254 nm	0.25 μg/ml	Ph	211

TABLE 12 (continued)
HPLC Methods for the Analysis of β-Blocking Agents

Compounds	Stationary phase	Mobile phase	Detection, λ	Sensitivity/linearity	Note	Ref
Nadolol (N)	Hypersil ODS (5 μm)	H_2O-acetonitrile-triethylamine 800:200:1 (pH = 3) with H_3PO_4 adjd	F exc: 265 nm emiss: 295 nm	0.1—0.5 ng/ml, 40—300 ng/ml in serum 0.5—10 μg/ml in urine	B (T)	212
N Pindolol (I.St)	Sperisorb ODS (5 μm)	MeOH 0.05 *M*-$NH_4H_2PO_4$ (pH = 4.6) 3.7	F exc: 227 nm emiss: 299 nm	5 ng/ml, 5—250 ng/ml	B (T)	213
N Acebutolol (I.St)	Hypersil ODS	Aq 57% MeOH containing 0.1% of octanesulphonic acid	UV, 220 nm	Rectilinear up to 400 ng/ml	B (T)	214
N *N*-butyl-analog of nadolol (I.St)	PRP-1 packed with poly(styrenedivinylbenzene) resin (10 μm)	Aq 17% CH_3CN added 0.5 ml/l $HClO_4$ + 0.5ml/l of methanolic tetramethylammonium hydroxide 160 g/l	F. exc: 265 nm emiss: 305nm	10—400 μg/l	B (S)	215
Timolol (T) Phenacetin (I.St)	Hypersil 5-ODS (5 μm)	Aq 13% acetonitrile containing triethylamine (pH = 3)	UV, 295 nm	2 ng/ml in plasma 5—200 ng/ml in plasma 50—2000 ng/ml in urine	B (T)	216
T Propranolol or pindolol (I.St)	Whatman PXS/25 Partisil ODS 3 (5 μm)	MeOH-0.2 *M* Na-H_2PO_4-88% H_3PO_4-H_2O 500:200:3:297	TL5A thin layer cell assembly with an LC2A controller	2 ng/ml rectilinear for ≤100 ng/ml	B (T) In plasma and breast milk For pharmacokinetic studies	217
T Maleic acid (II)	Zorbax-C_8	0-25% CH_3CN in 0.04 *N* H_3PO_4 containing 0.75 g Na hexanesulfonate (gradient elution)	UV, 235 nm (for 5 min) and then 294 nm	~500 pg	Ph	218

T Iso-timolol 4-hydroxy-3-morphol-ino-1,2,5-thiadi-azol(I) I-oxide	Zorbax Sil (5—6 μm) μBondapak C_{18}	Methanolic 5.4 m*M*-hexadecyl trimethyl-ammonium bromide-H_2O-aq 150 m*M* NaH_2PO_4 (adjd to pH = 7 with NaOH) 19:16:15 H_2O-acen-tonitrile-acetic acid 37:12.5:5 + Na-hexane sulfonate	UV, 280 nm	5—100 μg/ml (50 ng—5 μg/ml the potential degradates)	Ph	219
Sotalol (S) Atenolol	Hypersil ODS (5 μm)	5 m*M*-heptane-sulfonic acid-0.5 m*M*-Na dodecylsulfate in H_2O-CH_3CN-anhyd acetic acid 79:20:1 (pH = 2.5)	F exc: 235 nm emiss: 290 nm	0.1—50 μg/ml in plasma 0.2—100 μg/ml in urine	B (T)	220
S	Shandon ODS	MeOH-H_2O-CH_3CN 11:9:4 containing 1% of CH_3COOH and 5 m*M* Na-dodecylsul-fate	UV, 227 nm	10 ng/ml in plasma 40 ng/ml in tissue 44-440 ng/ml	B (T)	221
S Chloro-derivative of sotalol (I.St)	μBondapak C_{18} (10 μm) (precolumn Li-Chro-sorb RP 8)	H_2O-MeOH-CH_3COOH-methane sulfonic acid 3640:340:20:1 (pH = 3)	UV, 235 nm	25 ng/ml, rectilinear up to 4 μg/ml	B (T)	222
S	Altex ODS	Phosphate buffer (pH = 2.4) containing 2 m*M* nonylamine	UV, 235 nm	0.15—7.5 μg/ml	B	223
S	ODS	0.01 *M* phosphate buffer (pH = 3.2)-CH_3CN 20:80 con-taining 3 m*M* octyl-sodium sulphate	UV, 226 nm	0.03 μ*M*/l	B (T)	224
S	μBondapak C_{18}		F exc: 240 nm emiss: 310 nm	10—6000 ng/ml in plasma 0.5—100 μg/ml in urine	B (S)	225

TABLE 12 (continued)
HPLC Methods for the Analysis of β-Blocking Agents

Compounds	Stationary phase	Mobile phase	Detection, λ	Sensitivity/linearity	Note	Ref
Pindolol (Pn)	LiChrosphere ODS (5 μm)	MeOH-0.01 *M* $HClO_4$ 1:4	Amperometric (vitreous-carbon electrode operated at 1.0 V vs. Ag-AgCl)	0.5 ng/ml rectilinear: ≤200 ng/ml	B (T) No I.St is required. No interference was observed from six drugs tested	226
Pn	Supelcosil LC-8 (5 μm)	Propan-2-ol-CH_3CN-0.3% H_3PO_4 1:10:9	F exc: 255 nm emiss: 315 nm	2 ng/ml	B (T)	227
Pn	Supelcosil C_8 (5 μm)	CH_3CN-0.3% H_3PO_4-propan-2-ol 10:9:1	UV, 220 nm	2—100 ng/ml	B (T)	228
Pn	Micropac CN-10	0.01 *M* monobasic potassium phosphate in water (1.36 g/l) adjusted to pH = 2.6 with H_3PO_4			B (T) Quinidine, *n*-acetylprocainamide, lidocaine interfered with pindolol peak	229
Pn	Ultrasphere ODS	Aq 62% methanol	F exc: 209 nm emiss: 320 nm	2 ng/ml 100—2500 ng/ml	B (T) Derivatization: *S*(−)-methylbenzyl isocyanate, Diastereoisomers of atenolol and acebutolol were also separated	230
Pn Bopindolol	Resolvosil BSA 7	0.1 *M* Na_2HPO_4-0.1 *M* NaH_2PO_4-0.5% propan-2-ol	UV, 220 nm (GCMS)		Enantiomeric separation. Derivatization: *o*-(isopropylcarbamoyl) isopropyl urea	231
Atenolol (A)	Ultrasphere ODS (5 μm)	Heptane-1-sulfonic acid in CH_3COOH-0.1 *M* triethanolamine-MeOH 1:50:950	F	1.0 μg/ml in urine 20—50 ng/ml in plasma	B (T)	232, 233
A	μBondapak C_{18} (10 μm)	H_2O-CH_3CN anhyd acetic acid 50:50:1	F exc: 280 nm emiss: 300 nm	5 ng/ml 5—300 ng/ml	B + Ph (T) It is used to a single dose pharmacokinetics	234

A	Zorbax CN (6 μm)	CH_3CN-0.0125 *M* $NH_4H_2PO_4$ triethylamine 4:96:0:25 (pH = 5.5)	UV, 224 nm	<10 ng/ml 25—1000ng/ml	B (S)	235
A	Spherisorb CN (5 μm)	CH_3CN-0.05 *M* phosphate buffer (pH = 7) 23:77	F exc: 235 nm	rectilinear <100 μg/ml	B (S)	236
A benzimidazole	Spherisorb 5 μm silica	Methanolie 1 m*M* (+)-camphor-10-sulfonic acid monohydrate soln	F exc: 195 nm (no emission filter)	20 ng/ml rectilinear up to 1 μg/ml	B (T) Plasma, serum, breast milk, urine	237
A 4-(Butyrylamino)phenol (I.St)	Supelcosil LC-8-DB (5 μm)	Aq 25% CH_3CN containing [illegible] m*M*-ammonium acetate + 2 m*M* octanesulfonate (pH = 3.5)	UV, 225 nm	0.02—1.2 μg/ml	Ph	238
A Chlorthalidone	RP-18	Aq 10% CH_3COOH-CH_3CN (gradient elution)	UV, 275 nm		Ph	239

Note: B: assay from biological medium; (T): extraction by traditional method; (S): solid phase extraction; Ph: assay from pharmaceutical preparations p: plasma, aq: aqueous, anhydr: anhydrous, derivs: derivatives.

Another means of enantiomer separation by HPLC is the use of a chiral stationary phase, such as:

- Silica coated with cellulose-tris-(3,5-dimethylphenyl-carbamate) polymer[169,241]
- *(R)-N*-(3,5-dinitrobenzoyl)-phenylglycine (as chiral discriminator)
- α-Acid glycoprotein bonded silica[177]

A rarer mode of enantiomer separation is when the chiral reagent is included in the mobile phase

- *N,N,N*-trimethyloctylamine[175] (and octylsulphate)[197]
- 10-D-camphorsulfonic acid[178]
- *N,N*-dimethyloctylamine[169,197]

The separation and determination of β-blockers from related compounds, their enantiomers, and also their metabolites may be performed by ion-pairing HPLC, using:

- Na-alkylsulfonate[161,177]
- Alkyl sodium sulfates[158,161,175,189]
- Tetraalkylammonium cations[148,165,175]

Ion pairing has the advantage of providing more chance for the separation of structurally similar phenoxypropanolamines with different polarities, and concentration on peak broadening too.

Besides the very extended analytical use, HPLC separation is also applied on a semi-preparative scale. Cyclic carbamates of β-blocking agents have been separated by using swollen microcrystalline triacetylcellulose stationary phase.[123]

Gas chromatography (see Table 13) is employed in another large group of methods of β-blocker analysis on bulk material or after extraction from biological samples. Yamay et al.[244] have determined the retention index data of β-blockers, and have compared the selectivities and sensitivities of different detectors.

The methods of extraction are similar to or even simpler than those used in relation to HPLC, but the separation has to be followed by derivatization in every case. The following derivatization reagents have been used most often:

- Trifluoroacetic anhydride[250,251,254,257,262,270,274,275]
- Pentafluoropropionic anhydride[253]
- Heptafluorobutyric anhydride[259,268,271-273]
- Hexafluorobutyric anhydride[266]
- *N-O*-bis-(trimethylsily)-trifluoroacetamide[256,264,267]
- Trifluoro-*N*-methyl-*N*-(t-butyldimethylsilyl)acetamide[252]
- Methyl-bis-(trifluoroacetamide)[269]
- *N*-trimethylsilylimidazole[276]
- *n*-Butylboronic or phenylboronic acid[255,279,280]

In the separation of enantiomers, phosgene[260,261,264,267] has been used (formation of oxazolidinone derivatives), with trimethylsilylation of the remaining OH and COOH groups. Capillary gas chromatographic separation of β-blockers on the enantioselective stationary phase XE-60-L-valine-*(R)*-phenylethylamide was described by Koenig and Ernst,[245] and enantiomer separation of perfluoroacylated β-blockers has also been achieved.[246,247]

The stationary phases employed cover a broad spectrum (see Table 13). For the detection,

TABLE 13
Gas Chromatographic Methods for the Analysis of β-Blocking Agents

Compounds		Stationary phase	Carrier, gas	Detection	Sensitivity/linearity	Note	Ref
Propranolol (Pr)		3% OV-17 or 5% Apiezon L-10% KOH		FID		B	249
Pr				Selective N-P-thermoionic detection		B Derivatization: trifluoroacetic anhydride	250
Pr		OV-101 on chromosorb W-HF	N_2	^{3}H ECD	2 ng/ml	B (T) Derivatization: trifluoroacetic anhydride	251
Pr 7 Monohydroxy-derivatives of (Pr)		Glass column wall-coated with SP-2100	He	Selective-ion monitoring 70 eV	5—250 ng/ml	In bulk and in rat urine Derivatization: with trifluor-*N*-methyl-*N*-(*t*-butyldimethylsilyl) acetamide	252
Pr Acebutolol Alprenolol Metoprolol Oxprenolol Penbutolol Sotalol and their metabolites		Fused silica column SE-30	H_2 (temp programming 140°—260°C)	ECD	5—250 ng/ml	B (T) Derivatization: pentafluoropropionic anhydride	253
Pr Atenolol Oxprenolol Metoprolol Alprenolol Pindolol	I II III IV V VI	CP Sil 5-CB	He (temp programming 140°—280°C)	(a) *N* specific detection (b) 70 eW MS	(a) 500 ng/ml (I—V) 1000 ng/ml (VI) (b) 10 and 20 ng/ml(MS)	B (S) Derivatization: trifluoroacetic anhydride	254
Pr Alprenolol Bufetalol Bupuranolol	I II III IV	(a) Silanized glass column packed with (b) 2% OV-17 on Gas-chrom Q	He (column, inj, det,temp. 265, 320, 300°C, resp)	N-P selective detection	0.5 ng/ml (I—VII), 1—500 ng/ml (I—VII) 2—500 ng/ml (VIII)	B Derivatization: *n*-butylboronic or phenylboronic acid	255

TABLE 13 (continued)
Gas Chromatographic Methods for the Analysis of β-Blocking Agents

Compounds	Stationary phase	Carrier, gas	Detection	Sensitivity/linearity	Note	Ref
Carteolol V Nadolol VI Oxprenolol VII Pindolol VIII	(I—VII) 2% Dexsil 410 GC at 260°			Minimum detectable 1.5—4 pg (Pr)		
Pr β-blockers			GC-MS online computer was used for ionchromatography with the ions of m/e 72, 86, 98, 140, 151, 157, 200, 355		B	256
Pr 15 β-blockers	OV-101 or OV-17 on Chromosorb W		(a) N-P-D (b) FID (c) ECD	(a) and (c) were the most sensitive and selective; (a) limits were 0.1—1 ng; (c) 0.01—0.1 ng	Derivatization: trifluoroacetic anhydride	257
P Alprenolol Oxprenolol Timolol Pindolol	Precolumn: 3% SE-30 on Supelcoport Analytical column: glass capillary wall-coated open tubular CP Sil 5	He	(a) FID (b) GC-MS	Limit: 5 ng (packed column) 500 pg (precolumn capillary system), 5—100 µg/ml	Derivatization: *N-O*-bis (trimethylsilyl)-trifluoroacetamide	258
Metoprolol (M) Pindolol Propranolol (I.St)	Fused silica capillary column coated with OV-1	N_2 (temp programming 180°—230°C)	^{63}Ni ECD	5—800 ng/ml	B (T) Derivatization: heptofluorobutyric anhydride	259
M Ethoxyethyl analog of (M) (I.St)	(a) OV-17 on Gas chrom. Q (b) capillary column: coated with Carbowax 20 *M*	He (temp programming 150-240°C)	Up to 10 n*M* in plasma		B (T) Derivatization: phosgene	260

M α-Hydroxylated metoprolol Metoprolol acid	Chiral capillary column		FID		Enantionmer sepration Derivatization: phosgene	261
M	Column was coated with methylphenyl polysiloxane	He	^{63}Ni ECD	Rectilinear up to 0.8 M	B (T) Derivatization: trifluoroacetic anhydride	262
M	3% of SP-2100 on Supelcoport	He	FID		B (T)	263
M α-Hydroxy methoprolol	Chrompack CP-Sil 8	N_2 (temp. programming)	(a) MS electronimpact (key ion for all comps at m/e 336) (b) *N*-selective detn for the metabolite in urine		B Derivatization: phosgene, then with bis-(trimethylsilyl)acetamide	264
M *N*-alkyl derivatives of M; acid analog of M; 2-(-2-hydroxy-3-(isopropylamino)propoxyphenyl) acetic acid	Capillary column coated with chiral stationary phase			Derivatization: phosgene, and metabolites having hydroxy group(s) derived into their TMS-derivatives. Acid analogs converted to its Me-ester. Enantiomeric separation was achieved of all of the M-metabolites and analogs	265	
M	3% OV-1 on Gaschrom Q	He	Negative ion MS m/e 488, 494, 491	20—1500 n*M*	B (T) Derivatization: hexafluorobutyric anhydride	266
M Hydroxy-M Acid-M	Se-54-coated capillaries (temp programming 180°-240°)		FID	4 μ*M* for M. 20 μ*M* fpr (I) 5.5—37 μ*M* 18—71 for (I) 56—300 for (II)	B (T) Derivatization: Phosgene and bis(trimethylallyl)acetamide	267

TABLE 13 (continued)
Gas Chromatographic Methods for the Analysis of β-Blocking Agents

Compounds	Stationary phase	Carrier, gas	Detection	Sensitivity/linearity	Note	Ref
M Oxprenolol (I.St)	3% OV-1 Gas chrom Q (100 mesh)	Ar-CH_4 9:1	^{63}Ni ECD	10—500 ng/ml	B (T) Derivatization: heptafluorobutyric anhydride	268
M *O*-demethyl-M α-Hydroxy-M	OV-17		MS-selected ion monitoring m/e 266	1 n*M* 1 = 0.3 ng/ml	B (T) Derivatization: methylbis(trifluorocetamide)	269
M Propranolol (I.St)		Ar-CH_4 9:1	ECD	0-700 ng/ml	B (T) Derivatization: trifluoroacetic anhydride	270
M Pindolol (II) Propranolol (I.St)	Fused silica capillary column OV-1	N_2	^{63}Ni ECD	5 ng/ml 5—800 ng/ml	B (T) Derivatization: heptafluorobutyric anhydride. Retention factors of alprenolol, atenolol, metipranolol, oxprenolol are given	271
Oxprenolol (O) Propranolol (I.St.)	3% of SE-30 on Gas chrom Q		ECD	33.2 n*M* 10—1000 ng	B (T) Derivatization: heptafluorobutyric anhydride	272
O (I) ($^{13}C_3$) Oxprenolol (I.St) (II) ($^2H_6$) oxprenolol (III)	3% OV-17 on Supelcoport	He	Negative ionized MS, N_2O as reagent gas m/e 488, 491, 494 for I, II, III, resp.	20 n*M*	B Derivatization: heptafluorobutyric anhydride	273
Nadolol (N)	CP Sil-5 CB column		GCMS m/e 86		B Derivatization: trifluoroacetic anhydride	274-275
N *N*-methylnadolol (I.St) deuterated analogue (II)	3% of SP-2100 DB on Supelcoport	He	Selected ionmonitoring m/e 86.1 95.2 and 100.2	1 and 0.5 ng/ml for N and II resp.	B (S) Derivatization: *N*-trimethylsilylimidazole	276

N Cinnarizine (I.St)	Glass column with 2% Dexsil 300 on gas chrom Q	Ar	FID	0.8—3 mg/ml	Ph (T)	277
N	methylphenyl-poly-siloxane coated silica		GCMS m/e ion 86	0.2 ng/ml 0.6—60 ng/ml	B (S)	278
Labetalol (L)	(1) Fused silica Chrompak WCOT CP Sil 5 capillary column (2) OV-17 on GCQ (3) SE-30 on GCQ	He N_2 N_2	FID		Enantiomer separation as cyclic boronate derivatives	279
Mixture of R, S-L S, R-L R, R-L S, S-L Prochlorperazine maleate (I.St) was added into urine	1.5% of OV-1 Chromosorb W	N_2			Enantiomer separation in urine Derivatization: butane-boronic acid in pyridine	280
Pindolol (P)	OV-101	H_2 and air flow	FID	1—3 mg/cm^3	Ph	281

Note: B: assays from biological medium; (T): extraction by traditional method; (S): solid phase extraction; Ph: assays from pharmaceutical preparations; adjd: adjusted; resp: respectively; I.St.: internal standard; sepn: separation; metab: metabolite; anhyd: anhydrous; derivs: derivatives; temp: temperature; inj: injector; det: detector; comps: compounds; detn: detection; r: reagent(s); soln: solution.

electron capture, flame ionization, N-P selective thermoionic and mass-spectrometric detectors are frequent, with pg or ng sensitivity.

The temperature of gas chromatographic examinations ranges from 100 to 300° C.

In certain cases, especially before the gas chromatographic determination of labile β-adrenoreceptor blocking drugs, it is reasonable to add an esterase inhibitor to the investigated sample. Holm et al.[248] applied dodecyl sulfate for this purpose.

2. Thin-Layer Chromatography

The TLC technique is used in the main for the identification and separation of β-blockers from related compounds or ingredients in pharmaceutical products, either in the bulk or in dosage forms. TLC is also, but rarely, used for analysis in biological samples. The preliminary methods of extraction from biological tissues in this case are generally similar to those used in HPLC and GC (see Table 14).

The adsorbent is usually silica gel. Because of the polar features of the compounds, reversed-phase TLC is very rarely employed.[286]

The detection can be achieved by spraying with color reagents; a very good tabulative summary is given by Abdel-Hamid et al.[126] (Table 11). Irradiation with UV light is also a frequent method of spot indication.

Bernhard et al.[290] recently reported the TLC separation of nine β-blocker compounds, and also gave data on the color of the fluorescence arising after their derivatization with dansyl chloride.

C. OPTICAL METHODS

Spectrophotometric and fluorimetric methods for the determination of β-blockers were reviewed by Marko.[291] Due to their aromatic nucleus, β-blockers can be determined by direct UV spectrophotometry. Timolol in hydrochloric acid solution was measured by spectrophotometry (λ_{max} = 295 nm),[292,293] and in ophthalmic forms[294] by dual wavelength spectrophotometry (λ = 252 and 281 nm). The same compound was determined[295] in tablets in the presence of hydrochlorothiazide by using orthogonal function spectrophotometry (for timolol λ = 306 nm, for hydrochlorothiazide λ = 272 nm). In the method of Vetuschi et al.,[296] derivative UV spectroscopy allowed the simultaneous determination of oxprenolol and chlorthalidone.

An interesting approach for the quantitation of propranolol was made by Castleden et al.,[297] who applied photoacoustic spectroscopy in both the UV-visible and near-IR regions to powdered samples. As a phenolic compound, labetolol could be determined on the basis of the difference in absorbance of its acidic and alkaline solutions (λ = 332 nm).[298]

Due to their amine (base) property, spectrophotometry based on ion pair formation is a general method for the determination of β-blockers. Ion pairing with an acidic indicator dye,[279] and especially with bromothymol blue, has been utilized fairly often for such purposes. The yellow ion pair complex is extracted from the slightly acidic[300] or slightly alkaline[301,302] aqueous solution, mostly with chloroform. The absorbance is measured at about 400 nm; Beer's law is obeyed in the concentration interval up to 20 to 25 μg/ml, and the limit of detection is approximately 1 μg/ml. Timolol in ophthalmic solutions, propranolol in tablets and different pharmaceutical preparations, and oxprenolol in tablets have been determined. A series of β-blockers have been extracted with octanol or decanol, and spectrophotometrically determined as ion pairs with chloride or acetate anions.[303]

With propranolol[304] and metoprolol,[305] 2,4-dinitrofluorobenzene forms yellow condensation products, which are measured at 370 and 380 nm, respectively.

The blue color appearing in the reaction of propranolol,[306,307] atenolol, or metoprolol[308] with acetaldehyde + chloranil reagent can also be subjected to spectrophotometric determination (λ_{max} = 680 nm) in the ranges 5 to 60 μg/ml, and 25 to 75 μg/ml, respectively.

Similarly, a blue reaction product is formed in the reaction of labetolol with 2,6-dichloroquinone chlorimide.[309]

The phenol reagents such as the Folin-Ciocalteu reagent, aminoantipyrine, and ammonium ferric sulfate could be used for the selective determination of labetolol.[310]

The simple and fast reaction with ferric chloride in hydrochloric acid solution was found to be suitable for the determination of metoprolol in tablets (λ = 380 nm; linearity between 40 and 200 μg).[311]

A sensitive kinetic procedure has been described for the spectrophotometric determination of propranolol in tablets, based on oxidation with $K_2Cr_2O_7$ in 5 *M* sulfuric acid solution at 90° C for 20 min. The applicable concentration range is 0.13 to 1 m*M* propranolol.[312]

The determination of pindolol with special reagents (2,3-dichloro-5,6-dicyano-p-benzoquinone[313] and 3-methyl-2-benzothiazolone hydrazine·HCl[314]) has been reported (λ = 460 nm and 550 nm, respectively).

A sensitive and more selective quantitation of β-blockers is available by means of fluorimetry. Some of these compounds exhibit intrinsic fluorescence, e.g., pindolol in 96% ethanol,[315] or propranolol in methanol[316] solution. Certain β-blockers (propranolol and labetolol) have been measured in sulfuric acid solution:

	Excitation wavelength (nm)	Emission wavelength (nm)	Linearity range (μg/ml)	Note
Propranolol	317	340	10—200	In plasma, 4-hydroxy-P does not interfere
Labetolol	301	425	0.5—5	
Pindolol	263	305		

Propranolol was extracted from tablets with ethanolic potassium iodide solution, and the solution was spotted on to filter paper discs. The measurement of fluorescence (using comparative discs) allows the determination of 20 to 300 μg/ml propranolol (λ_{exc} = 306 nm; λ_{emiss} = 492 nm) in pharmaceutical formulations.[317]

After extraction (benzene) from serum, metoprolol may be measured directly at 305 nm λ_{exc} = 275 nm), or after derivatization with fluorescamine in hydrochloric acid solution[318] (λ_{exc} = 390 nm; λ_{emiss} = 480 nm).

D. NUCLEAR MAGNETIC RESONANCE SPECTROSCOPY

As in other fields of drug analysis, efforts have been made to introduce NMR spectroscopy into the standard methods of β-blocker drug analysis. Earlier, Chiarelli et al.[319] recommended NMR spectroscopy in DMSO-d_6 for the detection of propranolol with isosorbide dinitrate in tablets and pharmaceutical mixtures. Iorio et al.[320] recently determined several β-blockers in the bulk and in tablets by means of NMR spectroscopy in non-deuterated solvents. The method was based on the signals of the methyl groups attached to either the side chain or aromatic carbon atoms.

Anodic rotating-disc electrode voltammetry has been found to be suitable for the rapid determination of oxprenolol (up to 5 m*M*), pindolol, practolol, metoprolol, and sotalol.[321] The electrode kinetic parameters for mass and charge transfer were determined, and the reaction mechanism was also discussed. Propranolol and β-blocking agents were determined via a differential pulse polarographic method[322] based on the conversion of the drugs to *N*-nitroso derivatives. The method can be applied to both pure forms and tablets. The conductivity, solubility, and several other parameters of the propranolol- and oxprenolol-tetraphenylborate ion pairs were studied.[323]

Oxidimetry with the use of *N*-bromosuccinimide is preferred within the titrimetric quan-

TABLE 14
Thin-Layer Chromatographic Methods for the Analysis of β-Blocking Agents

Compounds	Stationary phase	Mobile phase	Detection	Sensitivity/linearity	Note	Ref
Propranolol (Pr)	Silica gel G	Benzene-acetone-propan-2-ol 5:2:3	UV radiation or Dragendorff r. or by IR spectrophotometry (3290 cm^{-1})		Purity and quality control (impurities from photo-oxidation)	282
Pr Verapamil Talinolol Nifedipine Oxprenolol Ethacizine Moricizine Metoprolol	Silufol UV_{254}	Toluene-$CHCl_3$-96% ethanol-diethylamine 46:10:5:1, or toluene-acetone-$CHCl_3$-propan-2-ol 30:15:5:1, or hexane-acetone-aqu 25% NH_3 20:20:1, or hexane ethylacetate-96%-ethanol-aq 25% NH_3 30:10:5:1	(a) Marquis, Froehde, Maadelin and Erdman reagents and with cc. H_2SO_4, HNO_3, H_2SO_4-$NaNO_2$, and $HClO_4$ (b) Dragendorff r. and thiourea soln	1-10 μg (A) 1-5μg (B)	B	283
Pr Oxprenolol Metoprolol	Silica gel 60	Toluene-acetone 10:1 in atmosphere of NH_3	F exc: 313 nm emiss: 365 nm	~100 ng/ml Linear up to 30 μg/ml (PR) and (II) 40 μg/ml (III) in urine	B Derivatization: S-(+)-benoxoprofen chloride	284
Pr Acebutolol Atenolol Metoprolol Nadolol Oxprenolol Pindolol Sotalol Timolol	Toxi-grams-A (silica gel)				B (S)	285
Metoprolol and three of its impurities (I, II, III)	KC_{18} RP plates	10% NH_3 in H_2O-MeOH 3:7	UV, 280 nm and 300 nm	15.0; 5—10; 10—20 ng for I-II-III.	Ph HPTLC in tablets	286

Nadolol	Silica gel 60	Chloroform-MeOH-CH_3COOH 75:20:5	Fluorodensitometry exc: 265 nm emiss: 313nm	5 ng/spot	B (T) The fluorescence is increased 2× by spraying with a mixture of 10% citric acid in water-ethylene glycol (1:1) and about fivefold by dipping into a soln of 4% nujol in cyclohexane, which is based upon the nonpolarity and viscosity of the solvent	
Pindolol (Pn) Nadolol (I.St)	Silica gel 60	$CHCl_3$-MeCH-CH_3COOH 15:4:1	F (After dipping the chromatogram into 4% of kerosene in cyclohexane) exc: 265 nm emiss: 313 nm	Limit: 0.5, 1 ng Linear: <350 ng/ml in plasma <4000 ng/ml in urine	B	288
Pn Clopamide (II)	Silica gel G_{60} F_{254}	CH_2Cl_2-MeOH-HCOOH 150:47:3	UV, 254 nm	Ph		289
β-Blockers	Silica gel 60 F_{254}	Ethylacetate-saturated aq methanol-formic acid 800:200:1	Scanned at 254 nm, then treated with three successive sprays to identify the β-blockers. For low dose β-blocker the plate was treated with dansyl chloride after sample application, and then heated. Mobile phase in this case was: aceton-$CHCl_3$ 9:1. Detection fluorimetric exc: 366 nm emiss: 410 nm	100—200 ng	B (T)	290

titation of β-blockers, and mainly propranolol. Semimicro amounts of propranolol were determined by direct titration. The end point detection in the method of Abbasi et al.[324] is based on utilization of the heat generated in the reaction of propranolol and *N*-bromosuccinimide. The stoichiometry of the latter reaction was established by TLC. The method involved the addition of the reagent in excess and determination of the unconsumed reagent iodometrically;[325] a similar procedure was described by Shukla et al.[326] Direct titration with *N*-bromosuccinimide, with methyl red as indicator, was used on a semimicro scale for the determination of propranolol in the presence of methyldopa.[327]

The applicability of RIA, EIA, and radioreceptor assay methods for the determination of β-blocking agents has been reviewed by Marko.[328] Of the individual β-blockers, propranolol has been repeatedly examined by RIA. The antiserum is produced by the use of propranolol metabolites,[329,330] or propranolol itself,[331] conjugated to bovine serum albumin and labelled with ^{123}I. The various methods allow the quantitation of propranolol in the concentration range 0.02 to 1 ng/ml or 10 pg to 10 ng/ml (see Tables 12, 13 and 14).

REFERENCES

1. **Nieder, M., Strösser, W., and Kappler, J.,** *Arzneim. Forsch.*, 37, 549, 1987.
2. **Schoenwald, R. D. and Hong-Shian, H.,** *J. Pharm. Sci.*, 72, 1266, 1983.
3. **Woods, P. E. and Robinson, M. L.,** *J. Pharm. Pharmacol.*, 33, 172, 1981.
4. **Arendt, R. M. and Greenblatt, D. J.,** *J. Pharm. Pharmacol.*, 36, 400, 1984.
5. **Vila, J. I., Obach, R., Prieto, R., and Moreno, J.,** *Chromatographia,* 22, 48, 1986.
6. **Rekker, R. F. and de Kort, H. M.,** *Eur. J. Med. Chem.*, 14, 479, 1979.
7. **Klopman, G., Namboodin, K., and Schochet, M.,** *J. Comput. Chem.*, 6, 28, 1985.
8. **Mayer, J. M., van de Waterbeemd, H., and Testa, B.,** *Eur. J. Med. Chem.*, 17, 17, 1982.
9. **Higuchi, T. and Kato, K.,** *J. Pharm. Sci.*, 55, 1080, 1966.
10. **Irvin, G. M. and Kostenbauer, M. B.,** *J. Pharm. Sci.*, 58, 313, 1969.
11. **Murthy, L. S. and Lopgraphy, G.,** *J. Pharm. Sci.*, 59, 1281, 1970.
12. **Lee, H. K., Chien, Y. W., Lin, T. K., and Lambert, M. J.,** *J. Pharm. Sci.*, 67, 647, 1978.
13. **Gasco, M. R., Trotta, M., and Carlotti, M. E.,** *J. Pharm. Sci.*, 71, 239, 1982.
14. **Gasco, M. R., Trotta, M., Carlotti, M. E., and Carpignano, R.,** *Int. J. Pharm.*, 18, 235, 1984.
15. **Gasco, M. R., Trotta, M., and Carlotti, M. E.,** *Pharm. Acta Helv.*, 60, 334, 1985.
16. **Szász, Gy.,** *Pharmaceutical Chemistry of adrenergic and cholinergic drugs,* CRC Press, Boca Raton, 1985, 65.
17. **Laxer, M., Capomacchia, A. C., and Hardee, G. E.,** *Talanta,* 28, 976, 1981.
18. **Jen, T. and Kaiser, C.,** *J. Med. Chem.*, 20, 693, 1977.
19. **Balsamo, A., Ceccarelli, G., Grotti, P., Macchia, B., Macchia, F., and Tognetti, P.,** *Eur. J. Med Chem.-Chem. Ther.*, 17, 471, 1982.
20. **Zaagsma, J.,** *J. Med. Chem.*, 22, 441, 1979.
21. **Aboul-Enein, H. Y., Hassan, M. M. A., and Jado, A. I.,** *Spectrosc. Lett.*, 16, 151, 1983.
22. **Carpy, A., Gadret, M., Hickel, D., and Leger, J. M.,** *Acta Crystallogr. Sect. B.*, 35, 185, 1979.
23. **Aunnon, H. L., Howe, D. B., Erhardt, W. D., Balsamo, A., Macchia, B., Macchia, F., and Keefe, V. E.,** *Acta Crystallogr. Sect. B.*, B33, 21, 1977.
24. **Kalenikova, E, I., and Arzamastsev, A. P.,** *Farmatsiya (Moscow),* 32, 67, 1983.
25. **Schunack, W.,** *D. Apoth. Ztg.*, 121, 2879, 1981.
26. **Petterson, C. and Schill, G.,** *J. Chromatogr.*, 204, 179, 1981.
27. **Ahlquist, R. P.,** *Am. J. Physiol.*, 153, 586, 1948.
28. **Powell, C. E. and Slater, I. H.,** *J. Pharmacol. Exp. Ther.*, 122, 480, 1958.
29. **Mills, J.,** U.S. Patent 2, 921, 938, 1960.
30. **Lands, A. M., Arnold, A., McAuliff, J. P., Luduena, F. P., and Brown, T. G., Jr.,** *Nature,* 214, 597, 1967.
31. **Lands, A. M., Luduena, F. P., and Buzzo, H. J.,** *Life Sci.*, 6, 2241, 1967.
32. **Black, J. and Stephenson, J. S.,** *Lancet,* 2, 311, 1962.
33. **Black, J., Crowther, A. F., Shanks, R. G., Smith, L. H., and Dornhorst, A. C.,** *Lancet,* 1, 1080, 1960.

34. **Black, J., Stephenson, J. S., and Shanks, R. G.,** *Br. J. Pharmacol. Chemother.*, 25, 577, 1965.
35. **Crowther, A. F. and Smith, L. H.,** *J. Med. Chem.*, 11, 1009, 1968.
36. **Prichard, B. N. C.,** *Br. J. Clin. Pharmacol.*, 5, 379, 1978.
37. **Karow, A. M., Jr., Riley, M. W., and Ahlquist, R. P.,** *Fortschr. Arzneimittelforsch.*, 15, 103, 1971.
38. **Nickerson, M.,** *Pharmacol. Rev.*, 1, 27, 1949.
39. **Smiths, J. F. M., Coleman, T. G., Smith, T. L., Kasbergen, C. M., van Essen, H., and Struyker-Boudier, H. A. J.,** *J. Cardiovasc. Pharmacol.*, 4, 903, 1982.
40. **Numao, Y., Terui, N., Kumada, M., and Imai, S.,** *Clin. Exp. Hypertens.*, 4, 87, 1982.
41. **Hinderling, P. H., Schmidlin, O., and Seydel, J. K.,** *J. Pharmacokinet. Biopharm.*, 12, 263, 1984.
42. **Shand, D. G.,** *N. Engl. J. Med.*, 293, 280, 1975.
43. **Prichard, B. N. C.,** *Br. J. Clin. Pharmacol.*, 13 (Suppl. 1), 41, 1982.
44. **Barrett, A. M.,** *J. Pharmacol.*, 16 (Suppl. 2), 95, 1985.
45. **McAinsh, J.,** *Methodol. Anal. Toxicol.*, 3, 41, 1985.
46. **Caplar, V., Mikotic-Mihun, Z., Hofman, H., Kuftinec, J., Skreblin, M., Kajfez, F., Nagl, A., and Blazevic, N.,** *Acta Pharm. Jugosl.*, 33, 71, 1983.
47. **Lazowski, J.,** *Farm. Pol.*, 43, 96, 1987.
48. **Staroukine, M., Giot, J. M., Jacobs, W., and Verniory, A.,** *Bull. Mem. Acad. R. Med. Belg.*, 137, 723, 1982.
49. **Bercher, H. and Grisk, A.,** *Wiss. Z. Ernst Moritz Arndt Univ. Greifsw. Math. Naturwiss. Reihe*, 31, 31, 1982.
50. **Bree, F., El Tayar, N., van de Waterbeemed, H., Testa, B., and Tillement, J.,** *J. Recept. Res.*, 6, 381, 1986.
51. **Giudicelli, J. F., Schmitt, H., and Boissier, J. R.,** *J. Pharmacol. Exp. Ther.*, 168, 116, 1969.
52. **Meier, J., Baranowski, E., and Wagner, E.,** *Fortschr. Med.*, 91, 1363, 1973.
53. **Bartsch. W., Dietmann, K., Leinert, H., and Sponer, G.,** *Arzneim. Forsch.*, 27, 1022, 1977.
54. **Chodnekar, M. S., Crowther, A. F., Hepworth, W., Howe, R., McLoughlin, B. J., Mitchell, A., Rao, B. S., Slatcher, R. P., Smith, L. H., and Stevens, M. A.,** *J. Med. Chem.*, 15, 49, 1972.
55. **Saameli, K.,** *Indian Heart. J.*, 24 (Suppl. 1), 146, 1972.
56. **Ablad, B., Brogard, M., Ek, L.,** *Acta Pharmacol. Toxicol.*, 25 (Suppl. 2), 9, 1967.
57. **Ablad, B., Carlsson, E., Ek, L.,** *Life Sci.*, 12, 107, 1973.
58. **Brunner, H., Hedwall, P. R., and Meier, M.,** *Arzneim, Forsch.*, 18, 164, 1968.
59. **Waal, H. J.,** *New Zealand Med. J.*, 67, 291, 1968.
60. **Andersson, O., Berglund, G., Bergman, H., Cramer, K., Fagerberg, S. E., Forsberg, S. Å., Johnsen, V., Lundvist, L., Rutle, O., and Sjølyst, R.,** *Curr. Ther. Res.*, 19, 43, 1976.
61. **Howe, R. and Rao, S. B.,** *J. Med. Chem.*, 11, 1128, 1968.
62. **Somani, P.,** *Am. Heart J.*, 77, 63, 1969.
63. **Baum, T., Rowles, G., Shropshire, A. T., and Gluckman, M. I.,** *J. Pharmacol. Exp. Ther.*, 166, 339, 1971.
64. **Hellerbrecht, D., Müller, K. F., and Grobecker, K.,** *Eur. J. Pharmacol.*, 23, 96, 1972.
65. **Sebestyén, K., Fenyvesi, T., and Hadházi, P.,** *Acta Physiol. Acad. Sci.*, 48, 260, 1976.
66. **Tényi, I., Németh, H. M., Jávor, T., Nemes, J., Bódis, L., Borvendég, J., and Eggenhoffer, J.,** *Curr. Ther. Res.*, 21, 823, 1977.
67. **Härtfelder, J., Lessenich, H., and Schmitt, K.,** *Arzneim. Forsch.*, 22, 930, 1972.
68. **Hansson, B. G. and Hokfelt, B.,** *Eur. J. Clin. Pharmacol.*, 9, 9, 1975.
69. **Crowther, A. F. and Smith, L. H.,** *J. Med. Chem.*, 11, 1009, 1968.
70. **Carissimi, M., Gentili, K., Grumelli, E., Milla, E., Picciola, G., and Ravenna, F.,** *Arzneim. Forsch.*, 26, 506, 1976.
71. **Frishman, W.,** *Am. Heart J.*, 99, 124, 1980.
72. **Jackson, D. A.,** *Br. J. Clin. Pract.*, 34, 211, 1980.
73. **Symposium on Atenolol,** *Postgrad Med. J.*, 53, (Suppl. 3), 1, 1977.
74. **Heel, R. C., Brogden, R. N., Speight, T. M., and Avery, G. S.,** *Drugs*, 17, 425, 1979.
75. **Heel, R. C., Brogden, R. N., Speight, T. M., and Avery, G. S.,** *Med. Progr.*, 2, 13, 1980.
76. **Brogden, R. N., Heel, R. C., Speight, T. M., and Avery, G. S.,** *Drugs*, 14, 321, 1977.
77. **Brunner, H., Gelzer, J., and Stepanrek, J.,** *Clin. Wochenschr.*, 56, (Suppl. 1), 107, 1978.
78. **Bianchetti, G., Blatrix, C., Gomeni, R., Kilborn, J. R., Larribaud, J., Lücker, P. W., Morselli, P. L., Thebault, J. J., and Trocherie, S.,** *Brit. J. Clin. Pharmacol.*, 8, 408P, 1979.
79. **Crowther, A. F., Howe, R., and Smith, L. H.,** *J. Med. Chem.*, 14, 511, 1971.
80. **Gibson, D. G.,** *Drugs*, 7, 12, 1974.
81. **Augstein, J., Cox, D. A., Ham, A. L., Leeming, P. R., and Snarey, P. M.,** *J. Med. Chem.*, 16, 1245, 1973.
82. **Adam, K. R., Baird, J. R. C., Burges, R. A., and Linnel, J.,** *Eur. J. Pharmacol.*, 25, 170, 1974.

83. **Kreighbaum, W. E., Matier, W. L., Dennis, R. D., Minielli, J. L., Deitchman, D., Perhach, J. L., Jr., and Comer, W. T.,** *J. Med. Chem.*, 23, 285, 1980.
84. **Ikezono, K., Uno, T., Watanabe, K., Hoshino, Y., and Ishihara, T.,** *Jpn. J. Pharmacol.*, 28, (Suppl), 139P, 1978.
85. **Shtacher, G., Erez, M., and Cohen, S.,** *J. Med. Chem.*, 16, 516, 1973.
86. **Smith, L. H.,** *J. Med. Chem.*, 20, 705, 1977.
87. **Bodor, N.,** *Advances in Drug Research,* Vol. 13, Testa, B., Ed., Academic Press, New York, 1984.
88. **Bodor, N., Oshiro, Y., Loftsson, T., Katovich, M., and Caldwell, W.,** *Pharm. Res.*, 3, 120, 1984.
89. **Bodor, N., El-Koussi, A. A., Kano, M., and Khalifa, M. M.,** *J. Med. Chem.*, 31, 1651, 1988.
90. **Baldwin, J. J., Denny, G. H., Hirschmann, R., Freedman, M. B., Ponticello, G. S., Gross, D. M., and Sweet, Ch. S.,** *J. Med. Chem.*, 26, 950, 1983.
91. **Hoefle, M. L., Hastings, J. G., Meyer, R. F., Corey, R. M., Holmes, A., and Stratton, Ch. D.,** *J. Med. Chem.*, 18, 148, 1975.
92. **Machin, P. J., Hurst, D. N., Bradshaw, R. M., Blaber, L. C., Burden, D. T., and Melarange, R. A.,** *J. Med. Chem.*, 27, 503, 1984.
93. **Taylor, E. M., Cameron, D., Eden, R. J., Fielden, R., and Owen, D. A. A.,** *J. Cardiovasc. Pharmacol.*, 3, 337, 1981.
94. **Taylor, E. M., Roe, A. M., and Slater, R. A.,** *Clin. Sci.*, 57, 433S, 1979.
95. **Bell, A., Boyce, M. C., Burland, W. L., and Underwood, D. D.,** *Br. J. Clin. Pharmacol.*, 9, 299P, 1980.
96. **Slater, R. A., Howson, W., Swayne, G. T. G., Taylor, E. M., and Reavill, D. R.,** *J. Med. Chem.*, 31, 345, 1988.
97. **Howson, W., Kitteringham, J., Mistry, J., Mitchell, M. B., Novelli, R., Slater, R. A., and Swayne, G. T. G.,** *J. Med. Chem.*, (in Press).
98. **Erhardt, P. W., Woo, C. M., Gorczynski, R. J., and Anderson, W. G.,** *J. Med. Chem.*, 25, 1402, 1982.
99. **Erhardt, P. W., Woo, C. M., Anderson, W. G., and Gorczynski, R. J.,** *J. Med. Chem.*, 25, 1408, 1982.
100. **Zarolinski, J., Borgman, R. J., O'Donnell, J. P., Anderson, W. G., Erhardt, P. W., Kam, S. T., Reynolds, R. D., Lee, R. J., and Gorczynski, R. J.,** *Life Sci.*, 31, 899, 1982.
101. **Kam, S. T., Matier, L. W., Mai, K. X., Barcelon-Yang, C., Borgman, R. J., O'Donnell, J. P., Stampfli, H. F., Sum, Ch. Y., Anderson, W. G., Gorczynski, R. J., and Lee, R. J.,** *J. Med. Chem.*, 27, 1007, 1984.
102. **Lee, R. J.,** *Life Sci.*, 23, 2539, 1978.
103. **Leclerc, G., Bieth, N., and Schwartz, J.,** *J. Med. Chem.*, 23, 620, 1980.
104. **Bouzoubaa, M., Leclerc, G., Decker, N., Schwartz, J., and Andermann, G.,** *J. Med. Chem.*, 27, 1291, 1984.
105. **Macchia, B., Balsamo, A., Lapucci, A., Martinelli, A., Macchia, F., Breschi, M. C., Fantoni, B., and Martinotti, E.,** *J. Med. Chem.*, 28, 153, 1985.
106. **Petrongolo, C.,** *Gazz. Chim. Ital.*, 108, 445, 1978.
107. **Goldblum, A.,** *Mol. Pharmacol.*, 24, 436, 1983.
108. **Loew, G. H., Nienow, J. R., and Poulsen, M.,** *Mol. Pharmacol.*, 26, 19, 1984.
109. **Kocjan, D., Hodoscek, M., and Hadzi, D.,** *J. Med. Chem.*, 29, 1418, 1986.
110. **Van de Waterbeemd, H., Carrupt, P. A., and Testa, B.,** *J. Med. Chem.*, 29, 600, 1986.
111. **Solmajer, T., Lukovits, I. M., and Hadzi, D.,** *J. Med. Chem.*, 25, 1413, 1982.
112. **Solmajer, T., Hodoscek, M., and Hadzi, D.,** *Quant. Struct. Act. Relat.*, 3, 51, 1984.
113. **El-Tayar, N., Carrupt, T. A., Van de Waterbeemd, H., and Testa, B.,** *J. Med. Chem.*, 31, 2072, 1988.
114. **Belg.** Patent 669,402, 1966.
115. **Pfeiffer, S. and Zimmer, H.,** *Pharmazie,* 30, 625, 1975.
116. **Pritchard, J. F., Schneck, D. W., and Hayes, A. H.,** *J. Chromatogr.*, 162, 47, 1979.
117. **Bond, P. A.,** *Nature (London),* 213, 721, 1967.
118. **Welson, W. L. and Burke, T. R., Jr.,** *J. Med. Chem.*, 22, 1082, 1979.
119. **Chen, Ch. H. and Nelson, W. L.,** *J. Pharm. Sci.*, 72, 863, 1983.
120. **Bargar, E. M., Walle, U. K., Bai, S. A. and Walle, T.,** *Drug Metab. Dispos.*, 11, 266, 1983.
121. **Lennard, M. S., Tucker, G. T., and Woods, H. F.,** *Clin. Pharmacokinet.*, 11, 1, 1986.
122. **Walle, T., Walle, U. K., Wilson, M. J., Fagan, T. C., and Gaffney, T. E.,** *Br. J. Clin. Pharmacol.*, 18, 741, 1984.
123. **Nelson, W. L., Bartels, M. J., Bodnarski, P. I., Zhang, S., Messick, K., Horng, J. S., and Ruffolo, R. R., Jr.,** *J. Med. Chem.*, 27, 857, 1984.
124. **Brandys, J. and Piekoszewski, W.,** *Bromatol. Chem. Toksykol.*, 17, 207, 1984.

125. **Brambillo, G., Cajelli, E., Finollo, R., Maura, A., Pino, A., and Robbiano, L.,** *J. Toxicol. Environ.*, 15, 1, 1985.
126. **Abdel-Hamid, M. E., Bedair, M., and Korny, M. A.,** *Pharmazie*, 40, 494, 1985.
127. **Yoshida, H., Morita, I., Masujima, T., and Imai, H.,** *Chem. Pharm. Bull.*, 30, 2287, 1982.
128. **Marko, V. and Soltes, L.,** *Farm. Obz.*, 53, 169, 1984.
129. **Cooper, J. K. and Midha, K. K.,** *Can. J. Pharm. Sci.*, 16, 46, 1981.
130. **Piotrovskii, V. K., Veiko, N. N., Rumjantsev, D. O., Zhirkov, Y. A., Elman, A. R., and Metelitsa, V. I.,** *Farmakol. Toksikol. (Moscow)*, 48, 62, 1985.
131. **Gregg, M. R.,** *Chromatographia*, 20, 129, 1985.
132. **Abernethy, D. R., Todd, E. L., Egan, I. L., and Carrum, G.,** *J. Liq. Chromatogr.*, 9, 2153, 1986.
133. **Isaksson, R. and Lamm, B.,** *J. Chromatogr.*, 362, 436, 1986.
134. **Yuan, Y., Xing, X., Zeng, P., and Zhou, X.,** *Yao Hsueh Hsueh Pao*, 22, 238, 1987.
135. **Ray, K., Trawick, W. G., and Mullins, R. E.,** *Clin. Chem.*, 31, 131, 1985.
136. **Greenblatt, D. J. and Arendt, R. M.,** *Int. J. Clin. Pharmacol. Ther. Toxicol.*, 22, 457, 1984.
137. **Koshakji, R. P. and Wood, A. J. J.,** *J. Pharm. Sci.*, 75, 87, 1986.
138. **El-Yazigi, A. and Martin, C. R.,** *Clin. Chem.*, 31, 1196, 1985.
139. **Bhamsa, R. K., Planagan, R. J., and Holt, D. W.,** *Methodol. Anal. Toxicol.*, 3, 165, 1985.
140. **Yamamura, Y., Uchino, K., Kotaki, H., Isozaki, S., and Saitoh, Y.,** *J. Chromatogr. Biomed. Appl.*, 47, 311, 1986.
141. **Pritchard, F. J., Schneck, D. W., and Hayes, A. H., Jr.,** *J. Chromatogr.*, 162, 47, 1979.
142. **Harrison, P. M., Tonkin, A. M., Cahill, C. M., and McLean, A. J.,** *J. Chromatogr.*, 343, 349, 1985.
143. **Harrison, P. M., Tonkin, A. M., Dixon, S. T., and McLean, A. J.,** *J. Chromatogr. Biomed. Appl.*, 47, 223, 1986.
144. **Harrison, P. M., Tonkin, A. N., Cahill, C. M., McLean, A. J.,** *J. Chromatogr. Biomed. Appl.*, 44, 349, 1985.
145. **Zorz, M., Milivojevic, D., Jankovec, A., and Bano, M.,** *Spectrosc. Int. J.*, 3, 355, 1984.
146. **Tamai, G., Morita, I., Masujima, T., Yoshida, H., and Imai, H.,** *J. Pharm. Sci.*, 73, 1825, 1984.
147. **Koshakji, R. P. and Wood, A. J. J.,** *J. Chromatogr. Biomed. Appl.*, 66, 294, 1987.
148. **Wingstrand, K. H. and Walle, T.,** *J. Chromatogr. Biomed. Appl.*, 30, 250, 1984.
149. **Kwong, E. C. and Shen, D. D.,** *J. Chromatogr.*, 414, 365, 1987.
150. **Smith, M. T., Livingstone, I., Hooper, W. D., Eadie, M. D., and Triggs, E. J.,** *Ther. Drug Monit.*, 5, 87, 1953.
151. **Li, W. and Lanman, R. C.,** *Anal. Lett.*, 20, 603, 1987.
152. **Smith, K. A., Wood, S., and Crous, M.,** *Analyst (London)*, 112, 407, 1987.
153. **Sood, S. P., Green, W. I., and Mason, R. P.,** *Ther. Drug Monit.*, 10, 224, 1988.
154. **Lo, M. W., Silber, B., and Riegelman, S.,** *J. Chromatogr. Sci.*, 20, 126, 1982.
155. **Albany, F., Riva, R., and Baruzzi, A.,** *J. Chromatogr.*, 228, 362, 1982.
156. **Katrurha, S. P. and Kukes, V. G.,** *J. Chromatogr.*, 365, 105, 1986.
157. **Takei, H., Ogata, H., and Ejima, A.,** *Chem. Pharm. Bull.*, 31, 1392, 1983.
158. **Winkler, H., Ried, W., and Lemmer, B.,** *J. Chromatogr.*, 228, 223, 1982.
159. **Li, Y. and Zhang, X.,** *Anal. Chim. Acta*, 196, 255, 1987.
160. **Buehring, K. U. and Garbe, A.,** *J. Chromatogr.*, 382, 215, 1986.
161. **Winkler, H. and Lemmer, B.,** Practical Aspects Modern High Performance Liquid Chromatographic Processes, Molnár, I., Ed., Walter de Gruyter, Berlin, 1981, 293.
162. **Piotrovskii, V. K., Belolipetskaya, V. G., El'man, A. R., and Metelista, V. I.,** *J. Chromatogr. Biomed. Appl.*, 29, 469, 1983.
163. **Bühring, K. U. and Garbe, A.,** *J. Chromatogr.*, 382, 215, 1986.
164. **Mulder, A., Conemans, J. M. H., Pijnenburg, C. C., and Duchateau, A. M. J. A.,** *Ziekenhuisfarmacie*, 3, 1, 1987.
165. **Wilson, M. J. and Walle, T.,** *J. Chromatogr. Biomed. Appl.*, 35, 424, 1984.
166. **Walle, T., Christ, D. D., Walle, U. K., and Wilson, M. J.,** *J. Chromatogr. Biomed. Appl.*, 42, 213, 1985.
167. **Sedman, A. J. and Gal, J.,** *J. Chromatogr. Biomed. Appl.*, 29, 199, 1983.
168. **Gübitz, G. and Mihellyes, S.,** *J. Chromatogr.*, 314, 462, 1984.
169. **Straka, R. J., Lalonde, R. L., and Wainer, I. W.,** *Pharm. Res.*, 5, 187, 1988.
170. **Wainer, I. W., Doyle, T. D., Donn, K. H., and Powell, J. R.,** *J. Chromatogr. Biomed. Appl.*, 31, 405, 1984.
171. **Day, N. H. and Parr, G. D.,** *Anal. Proc. (London)*, 21, 235, 1984.
172. **Kabra, P. M., Chen, S. H., and Marton, L. J.,** *Ther. Drug Monit.*, 3, 91, 1981.
173. **Das Gupta, V.,** *Drug Dev. Ind. Pharm.*, 11, 1931, 1985.
174. **Rao, G. R., Raghuveer, S., Pullarao, Y., and Mohan, K. R.,** *Indian Drugs*, 20, 285, 1983.

175. **Jansson, S. O. and Johansson, M. L.,** *J. Chromatogr.*, 395, 495, 1987.
176. **Lindner, W., Leitner, C., and Uray, G.,** *J. Chromatogr.*, 316, 605, 1984.
177. **Hermansson, J.,** *J. Chromatogr.*, 325, 379, 1985.
178. **Gupta, M. B., Hubbard, J. W., and Midka, K. K.,** *J. Chromatogr.*, 424, 189, 1988.
179. **Pflugmann, G., Spahn, H., and Mutschler, E.,** *J. Chromatogr.*, 421, 161, 1987.
180. **Wang, J., Li, X., Feng, W., and Song, Z.,** *Sepu*, 6, 103, 1988.
181. **Rossel, M. T., Belpaire, F. M., Bekaert, I., and Bogaert, M. G.,** *J. Pharm. Sci.*, 71, 114, 1982.
182. **Johnston, G. D., Nies, A. S., and Gal, J.,** *J. Chromatogr. Biomed. Appl.*, 29, 204, 1983.
183. **Lecaillon, J. B., Souppart, C., and Abadio, F.,** *Chromatographia*, 16, 158, 1982.
184. **Lennard, M. S. and Silas, J. H.,** *J. Chromatogr. Biomed. Appl.*, 23, 205, 1983.
185. **Lennard, M. S.,** *J. Chromatogr. Biomed. Appl.*, 43, 199, 1985.
186. **Gengo, F. M., Ziemniak, M. A., Kinkel, W. R., and McHugh, W. B.,** *J. Pharm. Sci.*, 73, 961, 1984.
187. **Lemer, B. G., Hellenbrecht, D., Bak, I. J., and Grobecker, H.,** *Naunyn-Schmiedeberg's Arch. Pharmacol.*, 275, 299, 1972.
188. **Harrison, P. M., Tonkin, A. M., and McLean, A. J.,** *J. Chromatogr. Biomed. Appl.*, 40, 429, 1985.
189. **Balmer, K., Zhang, Y., Lagerstrom, P. O., and Persson, B. A.,** *J. Chromatogr. Biomed. Appl.*, 61, 357, 1987.
190. **Lecaillon, J. B., Godbillon, J., Abadie, F., and Gosset, G.,** *J. Chromatogr Biomed. Appl.*, 30, 411, 1984.
191. **Godbillon, J. and Duval, M.,** *J. Chromatogr. Biomed. Appl.*, 34, 198, 1984.
192. **Rao, G. R., Raghuveer, S., and Khadgapathi, P.,** *Indian Drugs*, 23, 39, 1985.
193. **Pautler, D. B., and Jusko, W. J.,** *J. Chromatogr.* 228, 215, 1982.
194. **Plavsic, F.,** *Acta Pharm. Jugosl.*, 32, 137, 1982.
195. **Plavsic, F.,** *Acta Pharm. Jugosl.* 32, 67, 1982.
196. **Devi, K. P., Rao, K. V. R., Baveja, S. K., Fathi, M., and Roth, M.,** *J. Chromatogr. Biomed. Appl.*, 70, 229, 1988.
197. **Persson, B. A., Jansson, S. O., Johansson, M. L., and Lagerström, P. O.,** *J. Chromatogr.*, 316, 291, 1984.
198. **Godbillon, J., Duval, M., and Gosset, G.,** *J. Chromatogr. Biomed. Appl.*, 46, 365, 1985.
199. **Tsuei, S. E., Thomas, J., and Moore, R. G.,** *J. Chromatogr.*, 181, 135, 1980.
200. **El-Yazigi, A.,** *J. Pharm. Sci.*, 73, 751, 1084.
201. **Dieterle, W. and Faigle, J. W.,** *J. Chromatogr.*, 259, 301, 1983.
202. **Dieterle, W. and Faigle, J. W.,** *J. Chromatogr.*, 259, 311, 1983.
203. **Wang, J., Bonakdar, M., and Deshmukh, B. K.,** *J. Chromatogr.*, 344, 412, 1985.
204. **Woodman, T. F. and Johnson, B.,** *Ther. Drug Monit.*, 3, 371, 1981.
205. **Ostrovska, V., Svobodova, X., Pechova, A., Kusala, S., and Svoboda, M.,** *J. Chromatogr.*, 446, 323, 1988.
206. **Hidalgo, I. J. and Muir, K. T.,** *J. Chromatogr. Biomed. Appl.*, 30, 222, 1984.
207. **Oosterhuis, B., Van den Berg, M., and Van Boxtel, C. J.,** *J. Chromatogr. Biomed. Appl.*, 226, 259, 1981.
208. **Alton, K. B., Leitz, F., Bariletto, S., Jaworsky, L., Desrivieres, D., and Patrick, J.,** *J. Chromatogr. Biomed. Appl.*, 36, 319, 1984.
209. **Luke, D. R., Matzke, G. R., Clarkson, J. T., and Awni, W. M.,** *Clin. Chem. (Winston-Salem, N.C.)*, 33, 1450, 1987.
210. **Castellani, G., Marai, A., and Vacchi, P.,** *Boll. Chim. Farm.*, 121, 587, 1982.
211. **Rao, E. V., Raghuveer, S., and Rao, G. R.,** *Indian J. Pharm. Sci.*, 47, 134, 1985.
212. **Liu, L. K. and Robinson, M. L.,** *J. Pharm. Biomed. Anal.*, 3, 351, 1985.
213. **Moncrieff, J.,** *J. Chromatogr. Biomed. Appl.*, 43, 206, 1985.
214. **Kinney, C. D.,** *J. Chromatogr. Biomed. Appl.*, 30, 489, 1984.
215. **Gupta, R. N., Haynes, R. B., Logan, A. G., Macdonald, L. A., Pickersgill, R., and Achber, C.,** *Clin. Chem. (Winston-Salem, N.C.)*, 29, 1085, 1983.
216. **Lennard, M. S. and Parkin, S.,** *J. Chromatogr. Biomed. Appl.*, 39, 249, 1985.
217. **Gregg, M. R. and Jack, D. B.,** *J. Chromatogr. Biomed. Appl.*, 30, 244, 1984.
218. **Mazzo, D. J.,** *J. Chromatogr.*, 299, 503, 1984.
219. **Mazzo, D. J. and Snyder, P. A.,** *J. Chromatogr.*, 438, 85, 1988.
220. **Gluth, W. P., Soergel, F., Gluth, B., Braun, J., Geldmacher-v. Mallinckrodt, M.,** *Arzneim. Forsch.*, 38, 408, 1988.
221. **Lemmer, B., Ohm, T., and Winkler, H.,** *J. Chromatogr. Biomed. Appl.*, 34, 187, 1984.
222. **Poirier, J. M., Jaillon, P., and Cheymol, G.,** *Ther. Drug Monit.*, 8, 474, 1986.
223. **Hoyer, G. L.,** *J. Chromatogr. Biomed. Appl.*, 71, 181, 1988.
224. **Karkkainen, S.,** *J. Chromatogr.*, 336, 313, 1984.

225. **Bartek, M. J., Vekhsteyn, M., Boarman, M. P., and Gallo, D. G.,** *J. Chromatogr. Biomed. Appl.*, 65, 309, 1987.
226. **Diquet, B., Nguyen Huu, J. J., and Boutron, H.,** *J. Chromatogr. Biomed. Appl.*, 36, 430, 1984.
227. **Smith, H. T.,** *J. Chromatogr. Biomed. Appl.*, 59, 95, 1987.
228. **Tse, F. L. S. and Caubet, W. L.,** *Biopharm. Drug Dispos.*, 8, 577, 1987.
229. **Shields, B. J., Lima, J. J., Binkley, P. F., Leier, C. V., and MacKiehan, J. J.,** *J. Chromatogr.*, 378, 163, 1986.
230. **Hsyu, P. H. and Giacomini, K. M.,** *J. Pharm. Sci.*, 75, 601, 1986.
231. **Kuesters, E. and Giron, D.,** *J. High Resolut. Chromatogr. Chromatogr. Commun.*, 9, 531, 1986.
232. **Weddle, O. H., Amick, E. N., and Mason, W. D.,** *J. Pharm. Sci.*, 67, 1033, 1978.
233. **Gillilan, R. B. and Mason, W. D.,** *Anal. Lett.*, 16, 941, 1983.
234. **Miller, L. G. and Greenblatt, D. J.,** *J. Chromatogr. Biomed. Appl.*, 54, 201, 1986.
235. **Verghese, C., McLeod, A., and Shand, D.,** *J. Chromatogr.*, 275, 367, 1983.
236. **Keech, A. C., Harrison, P. M., and McLean, A. J.,** *J. Chromatogr. Biomed. Appl.*, 70, 234, 1988.
237. **Bhamra, R. K., Thorley, K. J., Vale, J. A., and Holt, D. W.,** *Ther. Drug Monit.*, 5, 313, 1983.
238. **Sa'sa, S. I.,** *J. Liq. Chromatogr.*, 11, 929, 1988.
239. **Ficarra, R., Ficarra, P., Tommasini, A., Calabro, M. L., and Guarniera, F. C.,** *Farmaco Ed. Prat.*, 40, 307, 1985.
240. **Kushiya, M., Hasegawa, R., Komuro, T., Kanoh, S., and Isaka, H.,** *Eisei Shikensho Hokoku,* Lo 4, 103, 1986.
241. **Okamoto, Y., Kawashima, M., Aburatani, R., Hatada, K., Nishiyama, T., and Masuda, M.,** *Chem. Lett.*, 1237, 1986.
242. **Hermansson, J. and von Bahr, C.,** *J. Chromatogr.*, 227, 113, 1982.
243. **Hermansson, J. and von Bahr, C.,** *J. Chromatogr.*, 221, 109, 1980.
244. **Yamaji, A., Kataoka, K., Kanamori, N., Oishi, M., and Hiraoka, E.,** *Yakugaku Zasshi,* 105, 1179, 1985.
245. **Koenig, W. A. and Ernst, K.,** *J. Chromatogr.*, 280, 135, 1983.
246. **Koenig, W. A., Ernst, K., and Vessman, J.,** *J. Chromatogr.*, 294, 423, 1984.
247. **Koenig, W. A. and Benecke, I.,** *J. Chromatogr.*, 209, 91, 1981.
248. **Holm, G., Kylberg-Hanssen, K., and Svensson, I.,** *Clin. Chem. (Winston-Salem, N.C.),* 31, 868, 1985.
249. **Hata, M., Takahashi, S., Matsubara, K., and Fuki, Y.,** *Nippon Hoigaku Zasshi,* 34, 645, 1980.
250. **Zhdanov, Yu. A., Kalenikova, E. I., Arzamastsev, A. P., Semenov, V. A., and Akalev, A. N.,** *Khim. Farm. Zh.* 17, 1269, 1983.
251. **Stenzel, W. R., Michael, G., and Lyhs, L.,** *Pharmazie,* 40, 360, 1985.
252. **Ballard, K. D., Knapp, D. R., Oatis, J. E., Jr., and Walle, T.** *J. Chromatogr. Biomed. Appl.*, 28, 333, 1983.
253. **Cartoni, G. P., Ciardi, M., Giarrusso, A., and Rosati, F.,** *J. High. Resolut. Chromatogr. Chromatogr. Commun.*, 11, 528, 1988.
254. **Delbeke, F. T., Debackere, M., Desmet, N., and Maestens, F.,** *J. Chromatogr. Biomed. Appl.*, 70, 194, 1988.
255. **Yamaguchi, T., Morimoto, Y., Sekine, Y., and Hashimoto, M.,** *J. Chromatogr.*, 239, 609, 1982.
256. **Maurer, H. and Pfleger, K.,** *J. Chromatogr.*, 382, 147, 1986.
257. **Yamaji, A., Kataoka, K., Kanamori, N., Oishi, M., and Hiraoka, F.,** *Yakugaku Zasshi,* 105, 1179, 1985.
258. **Christophersen, A. S. and Rasmussen, K. E.,** *J. Chromatogr.*, 246, 57, 1982.
259. **Susanto, F. and Reinauer, H.,** *Z. Anal. Chem.*, 318, 425, 1984.
260. **Gyllenhaal, O. and Vessman, J.,** *J. Chromatogr. Biomed. Appl.*, 24, 129, 1983.
261. **Gyllenhaal, O., König, W. A., and Vessman, J.,** *Z. Anal. Chem.*, 325, 526, 1986.
262. **Ervik, M., Kylberg-Hanssen, K., and Johansson, L.,** *J. Chromatogr. Biomed. Appl.*, 54, 168, 1986.
263. **Rohrig, T. P., Rundle, D. A., and Leifer, W. N.,** *J. Anal. Toxicol.*, 11, 231, 1987.
264. **Hoffmann, K. J., Gyllenhaal, O., and Vessman, J.,** *Biomed. Environ. Mass Spectrom.*, 14, 543, 1987.
265. **Gyllenhaal, O., König, W. A., and Vessman, J.,** *J. Chromatogr.*, 350, 328, 1985.
266. **Gaudry, D., Wantiez, D., Richard, J., and Metayer, J. P.,** *J. Chromatogr. Biomed. Appl.*, 40, 404, 1985.
267. **Gyllenhaal, O. and Hoffmann, K. J.,** *J. Chromatogr. Biomed. Appl.*, 34, 317, 1984.
268. **Sioufi, A., Leroux, F., and Sandrenan, N.,** *J. Chromatogr. Biomed. Appl.*, 23, 103, 1083.
269. **Ervic, M., Hoffmann, K. J., and Kylberg-Hanssen, K.,** *Biomed. Mass Spectrom.*, 8, 322, 1981.
270. **Jack, D. B. and Laegher, S. J.,** *Methodol. Anal. Toxicol.*, 3, 117, 1985.
271. **DeBruyne, D., Kinsun, H., Moulin, M. A., and Bigot, M. C.,** *J. Pharm. Sci.*, 68, 511, 1979.
272. **Sioufi, A., Colussi, D., and Mangoni, P.,** *J. Chromatogr. Biomed. Appl.*, 29, 185, 1983.
273. **Gaudry, D., Wantiez, D., and Metayer, J. P.,** *Biomed. Mass Spectrom.*, 12, 269, 1985.

274. **Delbeke, F. T. and Debackere, M.,** *J. Chromatogr. Biomed. Appl.*, 60, 443, 1987.
275. **Cohen, A. I., Jemal, M., Ivaskhiv, E., and Ribick, M.,** *J. Chromatogr. Biomed. Appl.*, 60, 445, 1987.
276. **Cohen, A. I., Devlin, R. G., Ivaskhiv, E., Funke, P. T., and McCormick, T.,** *J. Pharm. Sci.*, 73, 1571, 1984.
277. **Gawrych, Z., Pomazanska, T., and Szyszko, E.,** *Farm. Pol.*, 38, 667, 1982.
278. **Ribick, M., Ivashkiv, E., Jemal, M., and Cohen, A. I.,** *J. Chromatogr.*, 381, 419, 1986.
279. **Cholerton, T. J., Hunt, J. H., and Martin-Smith, M.,** *J. Chromatogr.*, 333, 178, 1985.
280. **Goromaru, T., Matsuki, Y., Matsuura, H., and Baba, S.,** *Yakugaku Zasshi*, 103, 974, 1983.
281. **Tawakkol, M. S., Mohamed, M. E., and Aboul-Enein, H. Y.,** *Chromatographia*, 14, 587, 1981.
282. **Constantinescu, T. and Popa, E.,** *Rev. Chim. (Bucharest)*, 34, 548, 1983.
283. **Beikin, S. G. and Gaponenko, Ya. S.,** *Farm. Zh. (Kiev)*, 6, 63, 1987.
284. **Pflugmann, G., Spahn, H., and Mutschler, E.,** *J. Chromatogr. Biomed. Appl.*, 60, 331, 1987.
285. **Bonicamp, J. M. and Pryor, L.,** *J. Anal. Toxicol.*, 9, 180, 1985.
286. **Cheng, M. L. and Poole, C. F.,** *J. Chromatogr.*, 257, 140, 1983.
287. **Schaefer-Korting, M. and Mutschler, E.,** *J. Chromatogr.*, 230, 461, 1982.
288. **Spahn, H., Prinoth, M., and Mutschler, E.,** *J. Chromatogr. Biomed. Appl.*, 43, 458, 1985.
289. **Zuo, D. and Zhang, S.,** *Yaowy Fenxi Zasshi*, 7, 225, 1987.
290. **Bernhard, W., Fuhrer, A. D., Jeger, A. N., and Rippstein, S. R.,** *Z. Anal. Chem.*, 330, 458, 1988.
291. **Marko, V.,** *Farm. Obz.*, 54, 213, 1985.
292. **Zhou, M., Sun, Y., and Xu, L.,** *Yao Hsueh T'ung Pao*, 17, 660, 1982.
293. **Mohamed, M. E., Tawakkol, M. S., and Aboul-Enein, H. Y.,** *Spectrosc. Lett.*, 15, 609, 1982.
294. **Xu, L.,** *Yiyao Gongye*, 16, 196, 1985.
295. **Zhu, D.,** *Yaowy Fenxi Zasshi*, 7, 83, 1987.
296. **Vetuschi, C., Ragno, G., Mazzeo, P., and Mazzeo-Farina, A.,** *Farmaco Ed. Prat.*, 40, 215, 1985.
297. **Castleden, S. L., Kirkright, G. F., and Long, S. E.,** *Can. J. Spectrosc.*, 26, 244, 1981.
298. **Mohamed, M. E.,** *Pharmazie*, 38, 784, 1983.
299. **Sane, R. T., Kubal, M. L., Nayak, V. G., Malkar, V. B., and Samant, R. S.,** *Indian J. Pharm. Sci.*, 46, 151, 1984.
300. **Radulovic, D., Jovanovic, M. S., and Zivanovic, L.,** *Pharmazie*, 41, 434, 1986.
301. **Radulovic, D., Jovanovic, M. S., and Milosevic, R.,** *Acta Pharm. Jugosl.*, 34, 169, 1984.
302. **Guvener, B.,** *Acta Pharm. Turc.*, 27, 17, 1985.
303. **Dallet, Ph., Dubost, J. P., Audry, E., and Colleter, M. J. C.,** *Bull. Soc. Pharm. Bordeaux*, 120, 140, 1981.
304. **Shingbal, D. M. and Prabhudesai, J. S.,** *Indian Drugs*, 21, 304, 1984.
305. **Shingbal, D. M. and Bhangle, S. R.,** *Indian Drugs*, 24, 270, 1987.
306. **Shingbal, D. M. and Khandepurkar, A. S.,** *Indian Drugs*, 24, 365, 1987.
307. **Korany, M. A., Abdel-Hay, M. H., Galal, S. M., and Elsayed, M. A.,** *J. Pharm. Belg.*, 40, 178, 1984.
308. **Shingbal, D. M. and Sardesai, G. D.,** *Indian Drugs*, 24, 373, 1987.
309. **Rao, G. D. and Raghuveer, S.,** *Indian Drugs*, 23, 626, 1984.
310. **Sane, R. T., Chandrashekar, T. G., and Nayak, V. G.,** *Indian Drugs*, 23, 565, 1986.
311. **Patel, R. B., Patel, A. A., Patel, S. B., and Manakiwala, S. C.,** *Indian Drugs* 25, 424, 1988.
312. **Sultan, S. M.,** *Analyst*, 113, 149, 1988.
313. **Issa, A. S., Mahrous, M. S., Abdel, S. M., and Soliman, N.,** *Talanta*, 34, 670, 1987.
314. **Korany, M. A. and Abdel-Hay, M. H.,** *Indian J. Pharm. Sci.*, 46, 183, 1984.
315. **Mohamed, M. E., Tawakkol, M. S., and Aboul-Enein, H. Y.,** *J. Assoc. Off. Anal. Chem.*, 66, 273, 1983.
316. **Trivedi, D. M., Gohel, M., and Chavda, H.,** *Indian J. Pharm. Sci.*, 48, 142, 1986.
317. **Bateh, R. P. and Winefordner, J. D.,** *J. Pharm. Sci.*, 72, 559, 1983.
318. **Poctova, M. and Kakac, B.,** *Cesk. Farm.*, 34, 222, 1985.
319. **Chiarelli, S. N., Rossi, M. T., Pizzorno, M. T., and Albonico, S. M.,** *J. Pharm. Sci.*, 71, 1178, 1982.
320. **Iorio, M. A., Mazzeo, F. A., and Doldo, A.,** *J. Pharm. Biomed. Anal.*, 5, 1, 1987.
321. **Bishop, E. and Hussein, W.,** *Analyst (London)*, 109, 65, 1984.
322. **Korany, M. A. and Riedel, H.,** *Z. Anal. Chem.*, 314, 678, 1983.
323. **Selinger, K.,** *Chem. Anal. (Warsaw)*, 27, 51, 1982.
324. **Abbasi, U. M., Chand, F., Bhanger, M. I., and Memon, S. A.,** *Talanta*, 33, 173, 1986.
325. **Ahmed, A. K. S., El-Nasser, O., Abdel, R., El Zahaby, M., and Salama, F.,** *J. Pharm. Belg.*, 37, 214, 1982.
326. **Shukla, S., Pathak, V. N., and Shukla, I. C.,** *Indian J. Pharm. Sci.*, 47, 204, 1985.
327. **Pathak, V. N., Shukla, S. R., and Shukla, I. C.,** *Analyst (London)*, 107, 1086, 1982.
328. **Marko, V.,** *Farm. Obz.*, 55, 37, 1986.

329. **Immer, U., Schmidt, H. E., Schmauder, H. P. Kolb, I., Krueger, K., and Herzmann, H.,** *Zentralbl. Pharm. Pharmakother. Laboratoriumsdiagn.*, 122, 172, 1983.
330. **Eller, T. D., Knapp, D. R., and Walle, T.,** *Anal. Chem.*, 55, 1572, 1983.
331. **Herzmann, E. and Reiser, M.,** *Zentralbl. Pharm. Pharmakother. Laboratoriumsdiagn.*, 122, 174, 1983.

Chapter 6

VASODILATORS

I. SODIUM NITROPRUSSIDE

$$\underbrace{\overbrace{Na_2[Fe(CN)_5NO]}^{a} \cdot 2H_2O}_{b}$$

M_r = 261.92 (a)
297.95 (b)

A. PROPERTIES

Sodium nitroprusside (SNp) is a reddish-brown, odorless crystalline substance. It is freely soluble in water, slightly soluble in alcohol, and insoluble in chloroform or benzene. SNp has absorption bands in the UV and visible regions, with maxima at 238, 265, 330, 396, and 498 nm. The corresponding electron transitions have been assigned and interpreted,[1,2] and are used as criteria of identity by USP. The relationship between the light absorbance and photodegradation of SNp has been the subject of a great number of studies. It was found that irradiation in the region of the 265 nm band or lower results in photooxidation, while irradiation in the visible region of the spectrum causes dissociation and/or hydrolytic changes. The most important studies and findings on the photodegradation of SNp up to 1977 have been reviewed by Van Loenen and Hofs-Kemper,[3] and up to 1983 by Leeuwenkamp.[1] The main routes of photodegradation that are important from various aspects may be summarized as follows:[1,4,5,6]

$$[Fe(CN)_5NO]^{2-} \xrightarrow[>300\ nm]{h\nu} [Fe^{II}(CN)_5]^{3-} + NO^+ \quad (1)$$

$$NO^+ + H_2O \rightleftharpoons NO_2^- + 2H^+ \quad (2)$$

$$[Fe(CN)_5]^{3-} + H_2O \rightarrow [Fe(CN)_5H_2O]^{3-} \quad (3)$$

$$[Fe(CN)_5]^{3-} + NO_2^- \rightarrow [Fe(CN)_5NO_2]^{4-} \quad (4)$$

$$[Fe(CN)_5NO]^{2-} \xrightarrow[300\ nm]{h\nu} [Fe^{III}(CN)_5]^{2-} + NO \quad (5)$$

$$[Fe(CN)_5]^{2-} + H_2O \rightarrow [Fe^{III}(CN)_5H_2O]^{2-} \quad (6)$$

The photolysis of SNp is of importance from the aspect of theoretical and practical analytical chemistry (SNp is a fairly frequently used chemical reagent), and the photostability of SNp is interesting from the viewpoint of pharmaceutical chemistry (SNp being a potent antihypertensive agent). Whereas SNp in the solid state is only moderately light-sensitive, in solution its photodegradation takes place extremely rapidly. The optimum conditions of storage in different solvents have been studied by numerous authors. Mechanistic information on the effects of the medium (H_2O, CH_3CN, CH_3OH, etc.) and high pressure on the photooxidation of SNp was given by Stoechel et al.[7] In their work on the photoredox chemistry of SNp, in methanol, Stoechel and Stasicka[8] established the formation of solvent complexes, e.g., $[Fe(CN)_5MeOH]^{2-}$, and the generation of Prussian blue as a result of prolonged irradiation. Additives such as citric acid and edetate seem to be able to slow down the rate of degradation,[9] possibly via the chelation of Fe^{2+} and Fe^{3+} ions, thereby preventing or delaying iron-catalyzed degradation processes.

B. MECHANISM OF ACTION

Nitroprusside is a powerful vasodilator, relaxing both arteriolar and venous smooth muscles. It acts directly on the walls of the blood vessels. Its half-life is only a few minutes due to the rapid inactivation of the compound (see metabolism).

C. STRUCTURE AND STRUCTURE-ACTIVITY RELATIONSHIP

The X-ray investigation by Swinehart[10] indicated C_4 symmetry for SNp. This means that the coplanar arrangement of four cyanide ligands, the fifth cyanide group, and the nitrosyl is axial. The last is situated at a rather short distance from the central Fe^{2+} ion (1.63 ± 2 pm).[10] This bond seems to have a distinct triple bond character. This assumption is confirmed by the high ν (NO) value of the SNp-nitrosyl (1940 cm^{-1}) compared with that of NO gas (1878 cm^{-1}), indicating the shift of the unpaired electron of nitrosyl towards the central cation. The high reactivity of SNp towards compounds with a nucleophilic centre (e.g., primary and secondary amines) is indicative of a formal $+1$ charge for the nitrosyl group and also supports the above assumptions. The structure-activity relationship studies made so far suggest certain similarities between the mechanism of action of SNp and those of organic nitrities and nitrates. The experimental facts confirm the essential role of the positively charged NO^+ in the hypotensive effect. Compounds containing an uncharged NO substituent, e.g., $[Mn(CN)_5NO]^{3-}$, were inactive.[11] On the other hand, the salt-like nitrosyl compounds, such as nitrosyl perchlorate or nitrous acid itself, induce only a very weak and transient hypotensive effect, indicating the carrier function of SNp.

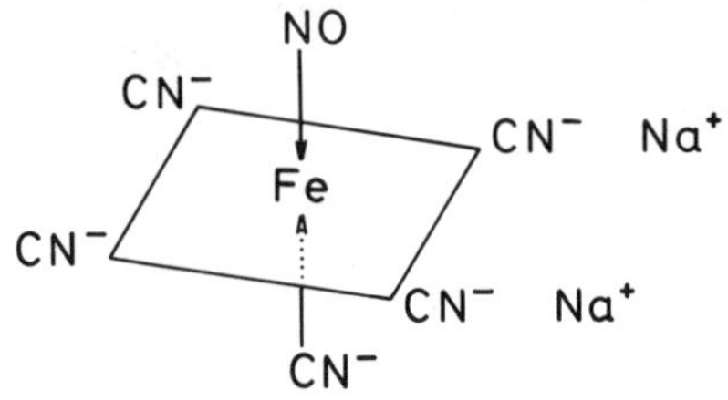

Undoubtedly, SNp is one of the very few inorganic compounds that has an important place among current medicinal compounds, due to its "fascinating physical, chemical, and pharmacological properties".[1] Notwithstanding its strong, and in certain respects unique, antihypertensive efficacy, two potential sources of toxicological danger must be mentioned. One of them is the cyanide ion-releasing property of SNp. Although this probelm has been the subject of numerous investigations,[12-14] the extent to which SNp releases cyanide in the human organism remains an unanswered question, considering the presence of such reaction partners as hemoglobin or -SH compounds such as glutathione. The hypotensive action at a molecular level seems to be linked to its reaction with the -SH group of the glutathione of c-guanylate cyclase.[15,16] Whether this reaction, which leads to activation of the enzyme, is due to NO^+ or $FeNO^{2+}$ intermediate metabolites, is not clear.[15-17] The second point relating to the potential toxicity of SNp is its possible interaction with secondary amines, which may be present either as medicinal compounds administrated together with SNp, or as endogenous compounds. In any event, the formation of nitrosamines (with their potential carcinogenicity) must be taken into consideration.

D. METABOLISM AND STABILITY

The pathway in the metabolism of SNp is the splitting of the complex into ions and radicals containing different amounts of Fe, CN and NO_n species. Part of the released cyanide is transformed to thiocyanate. This transformation can be stimulated by the administration of exogenous sulfur (sodium thiosulfate is a powerful antidote in cases of cyanide intoxication). The mode of action and metabolism of SNp were reviewed recently by Butler et al.[18]

The stability of SNp solutions has been investigated by several authors.

The concentrated injection solution is prepared immediately before clinical use from the commercially available freeze-dried substance in ampoules. This concentrated injection solution (2.5% $SNp \cdot 2H_2O$ in aqueous glucose solution, Nipride, Roche; 2.0% $SNp \cdot 2H_2O$ in aqueous sodium citrate solution, Nipruss, Cedona) proved stable for 2 years when protected from light and stored at room temperature.[19] The improvement of the stability of stored SNp solutions is related to the following conditions: protection from light; pH of the solution between 3 and 5; exclusion of oxygen; avoidance of temperature increase; and addition of iron complexing agents to the solution.[1] In this respect, citric acid, edetate, ascorbate, phosphate, etc. have been applied, with different effects.[20,21]

E. ANALYSIS

Besides the characteristic UV and IR spectra, several chemical reactions are known that can serve for the identification of SNp. From the numerous data on spectroscopic methods of investigating SNp, mention should be made here of the Raman, IR, and NMR data presented by Griffith et al.[22] As concerns the chemical reactions, one of the best known is the reaction with ketones, the Legal reaction. The violet reaction product formed is not characteristic of ketones alone because several other derivatives with nucleophilic moieties react (e.g., thiols, nitriles, uracils, etc.). The essence of the reaction is the nitrosation of the keto group, i.e., the nucleophilic centre of the organic molecule. The color of the reaction product depends on the pH of the solution; it may vary from red-violet to blue-violet.

$$CH_3-C(=O)-CH_3 \xrightarrow{OH^-} [\overset{\ominus}{C}H_2-C(=O)-CH_3]^{1-} + H_2O$$

$$\underset{NP}{[Fe(CN)_5NO]^{2-}} + \overset{\ominus}{C}H_2-C(=O)-CH_3 \longrightarrow [(CN)_5Fe-N(-O)-CH_2-C(=O)-CH_3]^{3-}$$

$$\rightleftharpoons$$

$$\underset{\text{red-violet}}{[(CN)_5Fe-N(-O^{\ominus})=CH-C(=O)-CH_3]^{4-}} \underset{H^+}{\overset{OH^-}{\rightleftharpoons}} \underset{\text{blue-violet}}{[(CN)_5Fe-N(-OH)=CH-C(=O)-CH_3]^{3-}}$$

SNp is polarographically active. At the dropping mercury electrode it is reduced in three steps; the corresponding reduction waves can be observed and interpreted.[23] The different versions of polarography permit a very sensitive determination of SNp. USP XXI applied polarography at a dropping mercury electrode for the assay of Sterile SNp. Leeuvenkamp et al. made a very thorough investigation of the polarographic behavior of SNp.[1] Their pulse polarographic method[24] is suitable for the determination of trace amounts of SNp. The calibration graph was linear for 4 to 1000 ng ml^{-1} of SNp. The limit of detection was 2 ng ml^{-1}. A somewhat modified method of the same research group[25] was used for the determination of therapeutic levels of SNp in serum, plasma, and whole blood samples.

A rapid and convenient method for the determination of SNp is based on the splitting of SNp by the addition of dimethylsulfoxide in alkaline medium. The nitrite formed was determined spectrophotometrically by reaction with sulfanilic acid and l-naphthylamine (azo dye formation).[26] The nonselectivity of the Legal reaction is reflected by the methods developed by Sane et al.,[27] when phenylenediamine or 4-aminophenol were used as color-forming reagents for the colorimetric determination of SNp. A spectrophotometric determination of SNp in tablets was based on the reaction with sodium hydroxide and sodium sulfide.[28] Beer's law was obeyed for 25 to 125 μg ml^{-1} of SNp. An interesting and sensitive determination of SNp in biological fluids was described by Sanchez Perez et al.;[29] this involves the nitroprusside-catalyzed formation of indophenol blue from ammonium ions + phenol + hypochlorite in alkaline medium (λ_{max} 620 nm). The limit of detection was 0.002 μg ml^{-1} (6.7×10^{-9} *M*).

II. HYDRAZINOPHTHALAZINES

A. HISTORY, STRUCTURE-ACTIVITY RELATIONSHIP, AND SYNTHESIS

Phthalazine (**1**) derivatives are condensed heterocycles which were not much investigated, synthetized until a variety of phthalazine derivatives were prepared in the early fifties. Particularly, the 1,4-substituted variants have attracted attention.[30,31]

$$CH_3-\underset{\underset{O}{\|}}{C}-CH_3 \xrightarrow{OH^-} [\overset{\ominus}{CH_2}-\underset{\underset{O}{\|}}{C}-CH_3]^{1-} + H_2O$$

$$\underset{NP}{[Fe(CN)_5NO]^{2-}} + \overset{\ominus}{CH_2}-\underset{\underset{O}{\|}}{C}-CH_3 \longrightarrow [(CN)_5Fe-\underset{\underset{O}{|}}{N}-CH_2-\underset{\underset{O}{\|}}{C}-CH_3]^{3-}$$

$$\underset{\text{red-violet}}{[(CN)_5Fe-\underset{\underset{O^{\ominus}}{|}}{N}=CH-\underset{\underset{O}{\|}}{C}-CH_3]^{4-}} \underset{H^+}{\overset{OH^-}{\rightleftharpoons}} \underset{\text{blue-violet}}{[(CN)_5Fe-\underset{\underset{OH}{|}}{N}=CH-\underset{\underset{O}{\|}}{C}-CH_3]^{3-}}$$

(1-hydrazino-4-R-phthalazine: $NH-NH_2$ at C-1, R at C-4)

R	
H	hydralazine (**2**)
$HNNH_2$	dihydralazine (**3**)

The best-known members of this series are 1-hydrazinophthalazine (hydralazine, **2**) and 1,4-dihydrazinophthalazine (dihydralazine, **3**). Both of them are derivatives with reactive, electron-rich substituents capable of interacting intramolecularly with the aromatic nucleus and/or intermolecularly with biological reaction partners. In spite of the fact that 40 years has passed since their discovery, certain unique features mean that hydralazine and dihydralazine have remained in the armament of antihypertensive therapy, even though they do not belong among the most favored or most frequently used agents. The 1,4-positions are the sites of effective substitution in the phthalazine antihypertensive series. This is indicated by the not too many compounds that have been prepared and introduced into medical practice as late congeners of hydralazine (**4-6**).

R_1	R_2	
$HNNH_2$	$N(CH_2CH_2OH)_2$	oxdralazine[33] (**4**)
$HNNH_2$	$CH_3NCH_2CH(CH_3)OH$	propildiazine[36] (**5**)
$HNNHCOOC_2H_5$	$C_2H_5NCH_2CH(CH_3)OH$	cadralazine (**6**)

Though some of them may be more potent than the parent compounds, the pharmacological drawbacks (side effects) inherent in the phthalazine skeleton hold for these late hydralazine congeners too.

Ring substitution by methyl or pyridinomethyl groups at position 4 resulted in very active antihypertensive compounds, while on the substitution of alkyl groups containing more than one carbon atom the activity was somewhat suppressed.[32] A very reasonable way to gain potent antihypertensive compounds seemed to be to use the pyridazine nucleus which can be regarded as a basic moiety of the phthalazine structure. A characteristic group of the few compounds that have gone through the clinical trials with success comprises the 1-oxyethylamino-4-hydrazino derivatives (**4-6**). Within this type, a linear correlation was found between the π value of the aromatic substituent of the 6-aryl-3-hydrazinopyridazines and antihypertensive activity.[35] It may be interesting from the aspect of further research that the dihydropyridazinones exhibit potent antihypertensive properties without having a hydrazino substituent.[36,37]

A facile synthesis route for the preparation of dihydralazine starts with the reaction of phthalic anhydride (**7**) with hydrazine, followed by the 1,4-chlorination and methanolysis of 1,4-phthalazinone (**8**). A repeated reaction with hydrazine leads to the formation of dihydralazine.

$H_2N—NH_2$ → PCl_5 →

(**7**) (**8**)

$NaOCH_3$ → $H_2N—NH_2$ / $-HCl, CH_3OH$ → dihydralazine (**3**)

B. MECHANISM OF ACTION

Hydralazine, dihydralazine, and most of their congeners produce direct peripheral vasodilation. They have an additional central effect, suppressing the vasoconstrictive activity of the CNS. Their action on the smooth muscles of the arteriolar bed was assumed to involve metal chelation and c-AMP inhibition.[38] The vasodilator activity of hydralazine was reviewed by Kreye.[39]

C. METABOLISM

A scheme (Figure 1) for the metabolism of dihydralazine was given recently by Schneider et al.[40] Acetylated and oxidation products (**2** and **3-4**), hydrazones (**6**), and products of decomposition (**7**) were identified and the possible mechanism of the metabolic processes was also discussed. the metabolites presumed to be formed in the reaction of dihydralazine with nitrite ions could not be detected. The quantitative aspects of the metabolism of dihydralazine were investigated by Schneider et al.[41] After oral administration, dihydralazine is excreted mainly in the form of its metabolites; two thirds of the latter appear in the feces, and one third of the metabolites excrete in the urine. Some methods of metabolism analysis are involved in Table 1.

D. ANALYSIS

The oxidizability served as the basis of a titrimetric method when hydralazine was determined in pharmaceuticals.[42] The titration was performed with a 0.1 *N* solution of $K_3[Fe(CN)_6]$ in alkaline medium.

The UV activity (Figure 2) was also used as a basis for the quantitation of hydralazine. The second-derivative spectrum of hydralazine (hydrochloride) was suitable for determination of the compound in bulk or in combination with hydrochlorothiazide or propranolol hydrochloride.[43]

Hydralazine treated with nitrite forms tetrazolo(5,1,a)-phthalazine, and the latter is measured at 274 nm in the range 4 to 40 ng ml^{-1}. This method[44] has been proposed as a means of avoiding the problems caused by the excipients in the USP XXI direct spectrophotometric procedure for hydralazine hydrochloride tablets. Different color reactions of hydralazine are known, which serve as the basis of its fluorometric or spectrophotometric determination. The method described by Danielson et al.[45] is based on the reduction by hydralazine of 1,2-naphtoquinone-4-sulfonic acid to form the fluorescent product 3,4-dihydroxynaphthalene-l-sulfonic acid (λ_{exc} 320 nm, λ_{emiss} 470 nm). The presence of hydrochlorothiazide interferes with the UV spectrophotometric method, but not with the fluorometric one. The calibration graph was linear from 1 up to 3,5 $\mu g\ ml^{-1}$ hydralazine. Hydralazine in tablets may be determined with good accuracy. A spectrophotometric method involving the same reagent was published by Salman.[46] The tablets were extracted with water and, after the addition of Na 1,2-naphthoquinone-4-sulfonate reagent, the absorbance was measured at 455 nm. The calibration graph covered the range 5.7 to 28.8 $\mu g\ ml^{-1}$. Amines and sulfonamides such as hydrochlorothiazide interfere. The oxidation product of hydralazine formed in the reaction with chloramine and β-alanine was measured in the range 2.5 to 20 $\mu g\ ml^{-1}$ in blood plasma spectrophotometrically by Jendryczko et al.[47] Hydralazine could be determined in tablets or injections by means of spectorphotometry based on the formation of a hydrazone with *p*-dimethylaminocinnamyl aldehyde.[48] Completion of the reaction necessitated 1 h of heating at 60°C. The absorbance of a chloroformic extract was measured at 450 nm. Beer's law was obeyed up to 9.8 $\mu g\ ml^{-1}$ of hydralazine. A simple and sensitive method suitable for the unit dose assay of hydralazine was developed by Issa et al.[49] The method is based on the reaction with 2,3-dichloro-5,6-dicyano-*p*-benzoquinone and the formation of a colored radical anion exhibiting maximum absorption at 460 nm. The redox activity of hydralazine provides a good possibility for its polarographic determination. In a series of papers[50,53] Fijalek and Szyszko reported the polarographic behavior of hydralazine and some other phthalazine antihypertensive agents. The quantitative determination made use of the different versions of polarography (direct current, alternating current, oscillating current, sinusoidal alternating current, and pulse polarography). A sensitive adsorption strip-

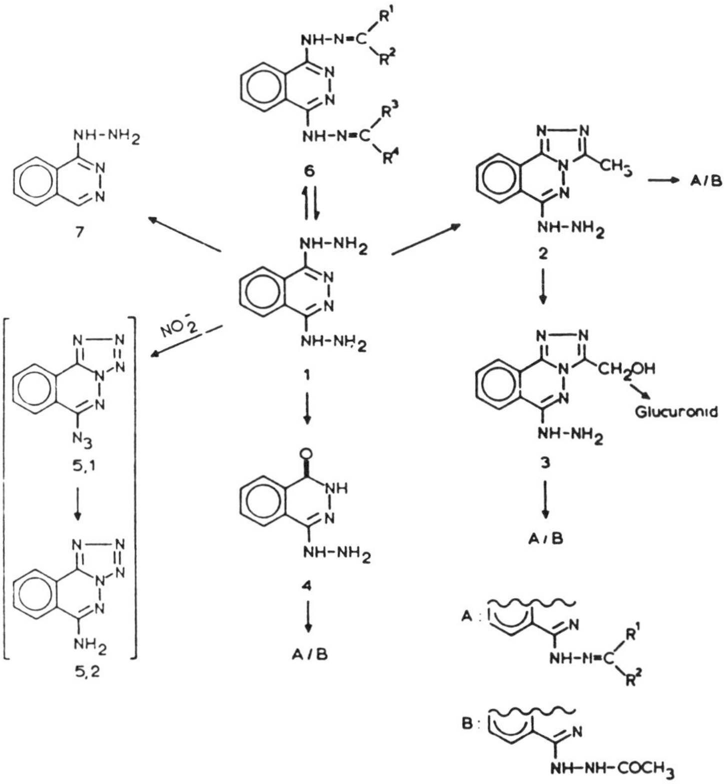

FIGURE 1. Degradations scheme of dihydralazine. (From Schneider, T. et al., *Pharmazie*, 43, 33, 1988. With permission.)

ping voltammetric assay of hydralazine applicable to urine was described by Wang et al.[54] Shah and Stewart[55] published an amperometric determination of hydralazine. The method (vitreous carbon working electrode +650 mV vs. the Ag-AgCl electrode) allows the detection of hydralazine down to 10 ng and its quantitation in the range 1 to 50 μg ml^{-1}. Other antihypertensive drugs and the usual compounds in hydralazine formulations did not interfere.

The potential complex-forming capability of hydralazinophthalazines was verified by several authors. The properties and behavior of complexes with copper,[56] molybdenum,[57] and chromium[58] have been studied.

The most frequently used methods in the pharmaceutical analysis of phthalazines are the chromatographic methods. Representative examples are presented in Table 1.

III. HYDRALAZINE

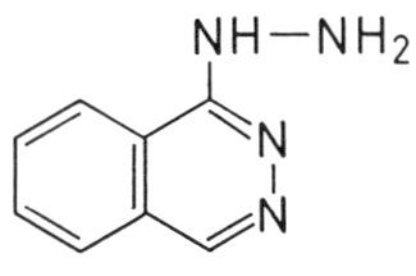

Phthalazine, 1-hydrazino-
1-Hydrazinophthalazine
$C_8H_8N_4$ M_r = 160.2
$C_8H_8N_4 \cdot HCl$ M_r = 196.7

TABLE 1
Chromatographic Methods in the Analysis of Phthalazine Derivatives

Compound	Stationary phase	Mobile phase (carrier gas)	Detection	Sensitivity, linearity	Note	Ref.
HPLC methods						
Hydralazine (H) Phthalazine H	μBondapak phenyl	Methanol-2% acetic acid 3:2	UV, 295 nm	9.7—29.1 ng ml^{-1}	Ph	59
Phenylpropanolamine (I.St.)	μBondapak C_{18}	0.015 *M* KH_2PO_4 (0.1% acetic anhydride, 0.5% MeOH)	UV, 256 nm	—	Ph	60
	μBondapak phenyl	As above, but 2% of MeOH	—	—	Excipients don't interfere	
H		—	F		B	61
Dihydralazine (D)	μBondapak C_{18}		exc: 230 nm emiss: 430 nm	0.5 ng ml^{-1}	Derivatization: with nitrous acid, Na methylate	
H	—	—	—	160 ng ml^{-1}	B Derivatization: *p* nitrobenzaldehyde	62
H					B	63
CH_3-H	μBondapak cyano	Acetonitrile-0.15 *M* Na-acetate (pH = 3) 7:3	UV, 365 nm	160 ng ml^{-1}	Derivatization: with *p*-anisaldehyde	
H CH_3-H	Supelcosil LC-18-DB	0.055 *M* citric acid-0.02 *M* Na_2HPO_4 (pH = 2.3 (66% MeOH)	E	—	B Derivatization: salicylaldehyde	64

Note: F: fluorimetry; Ph: assays from pharmaceutical preparations; B: assays from biological medium; E: electrochemical detector.

FIGURE 2. UV spectrum of hydralazine. Solvents: —— MeOh; - - - 0,1 *N* HCl.

Hydralazine hydrochloride (USP) appears as white or off-white crystals or a crystalline powder. Melting point: at about 275°C (with decompostion). It is soluble in water (4 g in 100 ml), slightly soluble in alcohol (0.2 g in 100 ml), and very slightly soluble in ether. Hydralazine is a moderately strong base with a pK_a value of 7.0. The characteristic UV spectrum and the absorbance at higher wavelengths demonstrate the highly conjugated nature of the hydralazine structure (Figure 2).

REFERENCES

1. **Leeuwenkamp, O. R., Van Bennekom, W. P., Van der Mark, E. J., and Bult, A.,** *Pharm. Weekbl.*, 6, 129, 1984.
2. **Manoharan, P. T. and Gray, H. B.,** *J. Am. Chem. Soc.*, 87, 3340, 1965.
3. **Van Loenen, A. C. and Hofs-Kemper, W.,** *Pharm. Weekbl.*, 113, 1080, 1978.
4. **Mitra, I. P., Sharma, B. K., and Mittal, S. P.,** *J. Inorg. Nucl. Chem.*, 34, 3419, 1972.
5. **Jarzynowski, T., Senkowski, T., and Stasicka, Z.,** *Rocz. Chem.*, 51, 2299, 1977.
6. **Jarzynowski, T., Senkowski, T., and Stasicka, Z.,** *Pol. J. Chem.*, 55, 3, 1981.
7. **Stoechel, G. and Stasicka, Z.,** *Inorg. Chem.*, 25, 3663, 1986.
8. **Stoechel, G. and Stasicka, Z.,** *Polyhedron*, 4, 1887, 1985.
9. **Van Loenen, A. C. and Hofs-Kemper, W.,** *Pharm. Weekbl.*, 114, 424, 1979.
10. **Swinehart, J. H.,** *Coord. Chem. Rev.*, 2, 385, 1967.
11. **Marckwardt, F., Glusa, E., Stürzebecher, J., John, W., and Kaiser, B.,** *Acta Biol. Med. Germ.*, 37, 469, 1978.
12. **Smith, R. P. and Kruszyna, H.,** *J. Pharmacol. Exper. Ther.*, 191, 557, 1974.
13. **Smith, R. P. and Kruszyna, H.,** *Br. J. Anaesth.*, 48, 396, 1976.
14. **Vesey, C. J., Linnell, J. C., and Wilson, J.,** *Br. Med. J.*, 11, 140, 1974.
15. **Napoli, S. A., Gruetter, C. A., Ignarro, L. J., and Kadowitz, P. J.,** *J. Pharmacol. Exper. Ther.*, 212, 469, 1980.
16. **Aliev, D. J. and Vanin, A. F.,** *Zh. Fiz. Khim.*, 56, 2362, 1982.
17. **Aliev, D. J. and Vanin, A. F.,** *56, 2365, 1982.*
18. **Butler, A. R. and Glidwell, Ch.,** *Chem. Soc. Rev.*, 16, 361, 1987.
19. **Vesey, C. J. and Batistoni, G. A.,** *J. Clin. Pharmacol.*, 2, 105, 1977.
20. **Van Loenen, A. C. and Hofs-Kemper, W.,** *Pharm. Weekbl.*, 114, 424, 1979.
21. **Chabrel, B., Mollet, M., and Puisieux, F.,** *Ann. Pharm. Fr.*, 38, 307, 1980.
22. **Griffith, W. P., Mockford, M. J., and Skapski, A. C.,** *Inorg. Chim. Acta*, 126, 179, 1987.
23. **Masek, J. and Maslova, E.,** *Collect. Czeh. Chem. Commun.*, 39, 2141, 1974.
24. **Leeuwenkamp, O. R., Jousma, H., Van der Mark, E. J., Van Bennekom, W. P., and Bult, A.,** *Anal. Chim. Acta*, 166, 51, 1984.
25. **Leeuwenkamp, O. R., Van der Mark, E. J., Jousma, H., Van Bennekom, W. P., and Bult, A.,** *Anal. Chim. Acta*, 166, 129, 1984.
26. **Malbosc, R., Roger, C. R., and Borga, A.,** *Anal. Lett.*, 17 A11, 1317, 1984.
27. **Sane, R. T., Doshi, V. J., and Jukar, S. R.,** *Acta Cienc. Indica*, Ser. Chem., 10c, 264, 1984.
28. **Li, D. and Zhu, B.,** *Yiyao Gongye*, 18, 365, 1987; A.A., 50, 433, 1988.
29. **Sanchez-Perez, A., Hernandez-Mendez, J., and Montero-Garcia, J.,** *Cienc. Ind. Farm.*, 1, 114, 1982.
30. **Druey, J. and Ringier, B. H.,** *Helv. Chim. Acta*, 34, 195, 1951.
31. **Gross, F. J., Druey, J., and Meier, R.,** *Experientia*, 6, 19, 1950.
32. **Schröder, E., Rufer, C., and Smiechen, R.,** *Arzneimittelchemie*, Vol. II., Georg Thieme Verlag, Stuttgart, 1976.
33. **De Ponti, C., Bardi, U., and Marchetti, M.,** *Arzneim. Forsch.*, 26, 2089, 1976.
34. **Pifferi, G., Parravicini, F., Carpi, C., and Dorigotti, L.,** *J. Med. Chem.*, 18, 741, 1975.
35. **LeClerc, G., Wermuth, C-G., Miesch, F., and Schwartz, J.,** *Eur. J. Med. Chem.*, 11, 107, 1976.
36. **Curran, W. C. and Ross, A.,** *J. Med. chem.*, 17, 273, 1974.
37. **McEvoy, F. J. and Allen, G. R., Jr.,** *J. Med. Chem.*, 17, 281, 1974.
38. **Amer, M. S. and Kreighbaum, W. E.,** *J. Pharm. Sci.*, 64, 1, 1975.
39. **Kreye, V. A. W.,** *J. Cardiovasc. Pharmacol.*, 6, 646, 1984.
40. **Schneider, T., Siegmund, W., Zschiesche, M., Kallwellis, R., and Scherber, A.,** *Pharmazie*, 43, 33, 1988.
41. **Schneider, T., Siegmund, W., Zschiesche, M., and Kallwellis, R.,** *Pharmazie*, 43, 704, 1988.
42. **Gaidukevich, O. M., Zhukova, T. V., Zarechenskii, M. A., and Sim, G.,** *Farm. Zh. (Kiev)* 49 (2), 1988.
43. **Bedair, M., Korany, M. A., and El-Yazbi, F. A.,** *Sci. Pharm.*, 54, 31, 1986.
44. **Mopper, B.,** *J. Assoc. Off. Anal. Chem.*, 70, 42, 1987.
45. **Danielson, N. D. and Bartolo, R. G.,** *Anal. Lett.*, 16 (B5) 343, 1983.
46. **Salman, A.,** *Sci. Pharm.*, 55, 255, 1987.
47. **Jendryczko, V., Drozdz, M., and Magner, K.,** *Rev. Roum. Biochim.*, 21, 299, 1984.
48. **Nakashima, K., Shimada, K., and Akiyama, S.,** *Chem. Pharm. Bull.*, 33, 1515, 1985.
49. **Issa, A. S., Mahrous, M. S., Abdel Salam, M., and Soliman, N.,** *Talanta*, 34, 670, 1987.
50. **Fijalek, Z. and Szyszko, E.,** *Acta Pol. Pharm.*, 38, 439, 1981.
51. **Fijalek, Z. and Szyszko, E.,** *Acta Pol. Pharm.*, 39, 395, 1982.

52. **Fijalek, Z. and Szyszko, E.,** *Acta Pol. Pharm.*, 39, 403, 1982.
53. **Fijalek, Z. and Szyszko, E.,** *Acta Pol. Pharm.*, 40, 343, 1983.
54. **Wang, J., Tapia, T., and Bonakdar, M.,** *Analyst (Lonodon)*, 111, 1245, 1986.
55. **Shah, M. H. and Stewart, J. T.,** *J. Pharm. Sci.*, 73, 989, 1984.
56. **Thompson, L. K. and Woon, T. C.,** *Inorg. Chim. Acta*, 111, 45, 1986.
57. **Attanasio, D., Dessy, G., and Fares, V.,** *Inorg. Chim. Acta*, 104, 99, 1985.
58. **Ganescu, I., Várhelyi, Cs., and Brinzan, G.,** *Arch. Pharm.*, 318, 30, 1985.
59. **Molles, R. J., Jr. and Garceau, Y.,** *J. Chromatogr.*, 347, 414, 1985.
60. **Gupta, V. D.,** *J. Liq. Chromatogr.*, 8, 2497, 1985.
61. **Rouan, M. C. and Campestrini, J.,** *J. Pharm. Sci.*, 74, 1270, 1985.
62. **Semple, H. A., Tam, Y. K., Tin, S., and Coutts, R. T.,** *Pharm. Res.*, 5, 383, 1988.
63. **Ludden, T. M., Ludden, L. K., Wade, K. E., and Allerheiligen, S. R. B.,** *J. Pharm. Sci.*, 72, 693, 1983.
64. **Wong, J. K., Joyce, T. H., and Morrow, D. H.,** *J. Chromatogr.*, 385, 261, 1987.
65. **Lacagnin, L. B., Colby, H. D., and O'Donnell, J. P.,** *J. Chromatogr. Biomed. Appl.*, 50, 319, 1986.
66. **Ravichandran, K. and Baldwin, R. P.,** *J. Chromatogr. Biomed. Appl.*, 44, 99, 1985.
67. **Degen, P. H., Brechbuehler, S., Schneider, W., and Zbinden, P.,** *J. Chromatogr. Biomed. Appl.*, 22, 375, 1982.
68. **Jedryczko, A. and Drozdz, M.,** *Chem. Anal. (Warsaw)*, 30, 319, 1983.

INDEX

A

D

E

F

G

H

I

K

L

M

Q

R

S